Lecture Notes in Computer Science 12176

More information about this series at http://www.springer.com/series/7407

Mourad Baïou · Bernard Gendron ·
Oktay Günlük · A. Ridha Mahjoub (Eds.)

Combinatorial Optimization

6th International Symposium, ISCO 2020
Montreal, QC, Canada, May 4–6, 2020
Revised Selected Papers

 Springer

Editors
Mourad Baïou ⓘ
CNRS and Université Clermont Auvergne
Aubière, France

Oktay Günlük ⓘ
School of Operations Research
and Information Engineering
Cornell University
Ithaca, NY, USA

Bernard Gendron ⓘ
Département d'informatique et de recherche
opérationnelle & CIRRELT
University of Montreal
Montréal, QC, Canada

A. Ridha Mahjoub ⓘ
Laboratoire LAMSADE
Université Paris-Dauphine
Paris, France

ISSN 0302-9743   ISSN 1611-3349 (electronic)
Lecture Notes in Computer Science
ISBN 978-3-030-53261-1   ISBN 978-3-030-53262-8 (eBook)
https://doi.org/10.1007/978-3-030-53262-8

LNCS Sublibrary: SL1 – Theoretical Computer Science and General Issues

This Springer imprint is published by the registered company Springer Nature Switzerland AG
The registered company address is: Gewerbestrasse 11, 6330 Cham, Switzerland

Preface

This volume contains the regular papers presented at ISCO 2020, the 6th International Symposium on Combinatorial Optimization, May 4–6, 2020. Originally, the conference was scheduled to take place in Montreal, Canada, but due to the COVID-19 pandemic, the conference was held online, attracting more than 250 registered participants. Past editions of ISCO (Hammamet, Tunisia, March 2010; Athens, Greece, April 2012; Lisboa, Portugal, March 2014; Vietri Sul Mare, Italy, May 2016; Marrakesh, Morocco, April 2018) all included invited talks, short papers, and a doctoral school. Unfortunately, due to the COVID-19 pandemic, these activities, originally scheduled to take place at HEC Montreal and at the University of Montreal, had to be canceled. Nonetheless, the online edition of the conference was a success, with 24 talks of 30 minutes each, grouped into 8 sessions, and every session attracting close to 100 participants from all around the world.

The ISCO series aims to bring together researchers from all communities related to combinatorial optimization, including algorithms and complexity, mathematical programming, operations research, stochastic optimization, graphs, and polyhedral combinatorics. It is intended to be a forum for presenting original research on all aspects of combinatorial optimization, ranging from mathematical foundations and theory of algorithms to computational studies and practical applications, and especially their intersections. In response to the call for papers, ISCO 2020 received 66 regular submissions. Each submission was reviewed by at least two Program Committee members. The submissions were judged on their originality and technical quality, and difficult decisions had to be made. As a result, 25 regular papers were selected to be presented at the symposium, giving an acceptance rate of 38%. One selected paper was withdrawn by the authors, but all other 24 papers were presented at the symposium. The revised versions of the 24 accepted regular papers presented at the conference are included in this volume.

We would like to thank all the authors who submitted their work to ISCO 2020, and the Program Committee members for their remarkable work. They all contributed to the quality of the symposium. Finally, we would like to thank the staff members of CIRRELT, the Interuniversity Research Centre on Enterprise Networks, Logistics and Transportation, which hosted the conference website, for their assistance and support.

May 2020

Mourad Baïou
Bernard Gendron
Oktay Günlük
A. Ridha Mahjoub

Organization

Conference Chairs

Mourad Baïou	Blaise Pascal University, France
Bernard Gendron	University of Montreal, Canada
Oktay Günlük	Cornell University, USA
A. Ridha Mahjoub	Paris Dauphine University, France

Organizing Committee

Bernard Gendron	University of Montreal, Canada
Sanjay Dominik Jena	UQAM, Canada
Issmaïl El Hallaoui	Polytechnique Montreal, Canada

Steering Committee

Mourad Baïou	Blaise Pascal University, France
Pierre Fouilhoux	Pierre and Marie Curie University, France
Luis Gouveia	University of Lisbon, Portugal
Nelson Maculan	UFRJ, Brazil
A. Ridha Mahjoub	Paris Dauphine University, France
Vangelis Paschos	Paris Dauphine University, France
Giovanni Rinaldi	IASI-CNR, Italy

Program Committee

Miguel Anjos	University of Edinburgh, UK
Alper Atamtürk	UC Berkeley, USA
Francisco Barahona	IBM Research, USA
Amitabh Basu	Johns Hopkins University, USA
Daniel Bienstock	Columbia University, USA
Francisco Carrabs	University of Salerno, Italy
Margarida Carvalho	University of Montreal, Canada
Raffaelle Cerulli	University of Salerno, Italy
K. Chandrasekaran	University of Illinois at Urbana-Champaign, USA
Sanjeeb Dash	IBM Research, USA
Abdellatif El Afia	Mohammed V University, Morocco
Issmaïl El Hallaoui	Polytechnique Montreal, Canada
Marcia Fampa	UFRJ, Brazil
Satoru Fujishige	Kyoto University, Japan
Ricardo Fukasawa	University of Waterloo, Canada
Bernard Gendron	University of Montreal, Canada

Contents

Heuristics

Polyhedral Combinatorics

Polyhedra Associated with Open Locating-Dominating and Locating Total-Dominating Sets in Graphs

Gabriela Argiroffo[1], Silvia Bianchi[1], Yanina Lucarini[1],
and Annegret Wagler[2(✉)]

[1] Facultad de Ciencias Exactas, Ingeniería y Agrimensura, Universidad Nacional de Rosario, Rosario, Argentina
{garua,sbianchi,lucarini}@fceia.unr.edu.ar
[2] University Clermont Auvergne (LIMOS, UMR 6158 CNRS),
Clermont-Ferrand, France
wagler@isima.fr

Abstract. The problems of determining open locating-dominating or locating total-dominating sets of minimum cardinality in a graph G are variations of the classical minimum dominating set problem in G and are all known to be hard for general graphs. A typical line of attack is therefore to determine the cardinality of minimum such sets in special graphs.

In this work we study the two problems from a polyhedral point of view. We provide the according linear relaxations, discuss their combinatorial structure, and demonstrate how the associated polyhedra can be entirely described or polyhedral arguments can be applied to find minimum such sets for special graphs.

Keywords: Open locating-dominating code problem · Locating total-dominating code problem · Polyhedral approach

1 Introduction

For a graph G that models a facility, detection devices can be placed at its nodes to locate an intruder (like a fire, a thief or a saboteur). Depending on the features of the detection devices (to detect an intruder only if it is present at the node where the detector is installed and/or also at any of its neighbors), different dominating sets can be used to determine the optimal distribution of the detection devices in G. In the following, we study three problems arising in this context which all have been actively studied during the last decade, see e.g. the bibliography maintained by Lobstein [16].

Let $G = (V, E)$ be a graph. The open neighborhood of a node i is the set $N(i)$ of all nodes of G adjacent to i, and $N[i] = \{i\} \cup N(i)$ is the closed neighborhood of i. A subset $C \subseteq V$ is *dominating* (resp. *total-dominating*) if $N[i] \cap C$ (resp. $N(i) \cap C$) are non-empty sets for all $i \in V$.

© Springer Nature Switzerland AG 2020
M. Baïou et al. (Eds.): ISCO 2020, LNCS 12176, pp. 3–14, 2020.
https://doi.org/10.1007/978-3-030-53262-8_1

A subset $C \subseteq V$ is:

- an *identifying code* (ID) if it is a dominating set and $N[i] \cap C \neq N[j] \cap C$, for distinct $i, j \in V$ [15];
- an *open locating-dominating set* (OLD) if it is a total-dominating set and $N(i) \cap C \neq N(j) \cap C$, for distinct $i, j \in V$ [19];
- a *locating total-dominating set* (LTD) if it is a total-dominating set and $N(i) \cap C \neq N(j) \cap C$, for distinct $i, j \in V - C$ [13].

Figure 1 illustrates the three concepts.

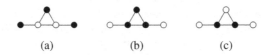

(a)   (b)   (c)

Fig. 1. A graph where the black nodes form a minimum (a) ID-code, (b) OLD-set, (c) LTD-set.

Note that a graph G admits an ID-code (or is *identifiable*) only if there are no true twins in G, i.e., there is no pair of distinct nodes $i, j \in V$ such that $N[i] = N[j]$, see [15]. Analogously, a graph G without isolated nodes admits an OLD-set if there are no false twins in G, i.e., there is no pair of distinct nodes $i, j \in V$ such that $N(i) = N(j)$, see [19].

Given a graph G, for $X \in \{ID, OLD, LTD\}$, the X-problem on G is the problem of finding an X-set of minimum size of G. The size of such a set is called the X-number of G and is denoted by $\gamma_X(G)$. From the definitions, the following relations hold for any graph G (admitting an X-set):

$$\gamma_{LTD}(G) \leq \gamma_{OLD}(G), \tag{1}$$

whereas $\gamma_{ID}(G)$ and $\gamma_{OLD}(G)$ are not comparable in general.

Determining $\gamma_{ID}(G)$ is in general NP-hard [9] and even remains hard for several graph classes where other in general hard problems are easy to solve, including bipartite graphs [9], split graphs and interval graphs [10].

Also determining $\gamma_{OLD}(G)$ is in general NP-hard [19] and remains NP-hard for perfect elimination bipartite graphs and APX-complete for chordal graphs with maximum degree 4 [18]. Concerning the LTD-problem we observe that it is as hard as the OLD-problem by just using the same arguments as in [19].

Typical lines of attack are to determine minimum ID-codes of special graphs or to provide bounds for their size. Closed formulas for the exact value of $\gamma_{ID}(G)$ have been found so far only for restricted graph families (e.g. for paths and cycles by [8], for stars by [12], and for complete multipartite graphs, some suns and split graphs by [2–5]). Closed formulas for the exact value of $\gamma_{OLD}(G)$ have been found so far only for cliques and paths [19], some algorithmic aspects are discussed in [18]. Bounds for the LTD-number of trees are given in [13,14], whereas the LTD-number in special families of graphs, including cubic graphs and grid graphs, is investigated in [14].

As polyhedral methods have been already proved to be successful for several other NP-hard combinatorial optimization problems, it was suggested in [2] to apply such techniques to the ID-problem. For that, the following reformulation as set covering problem has been proposed.

For a 0/1-matrix M with n columns, the set covering polyhedron is $Q^*(M) = $ conv $\left\{\mathbf{x} \in \mathbb{Z}_+^n : M\mathbf{x} \geq \mathbf{1}\right\}$ and $Q(M) = \left\{\mathbf{x} \in \mathbb{R}_+^n : M\mathbf{x} \geq \mathbf{1}\right\}$ is its linear relaxation. A *cover* of M is a 0/1-vector $\mathbf{x}$ such that $M\mathbf{x} \geq \mathbf{1}$, and the *covering number* $\tau(M)$ equals min $\mathbf{1}^T\mathbf{x}, \mathbf{x} \in Q^*(M)$.

We obtain such a constraint system $M\mathbf{x} \geq \mathbf{1}$ for the ID-problem as follows. Consider a graph $G = (V, E)$. Domination clearly requires that any ID-code C intersects the closed neighborhood $N[i]$ of each node $i \in V$; separation means that no two intersections $C \cap N[i]$ and $C \cap N[j]$ are equal. The latter condition can be reformulated that C intersects each symmetric difference $N[i] \triangle N[j]$ for distinct nodes $i, j \in V$. It was shown in [2] that only symmetric differences matter if the nodes $i, j \in V$ have distance $dist(i, j) = 1$ (i.e., are adjacent) or distance $dist(i, j) = 2$ (i.e., are non-adjacent but have a common neighbor).

Hence, determining a minimum ID-code in a graph $G = (V, E)$ can be formulated as set covering problem $\min \mathbf{1}^T\mathbf{x}, M_{ID}(G)\mathbf{x} \geq \mathbf{1}, \mathbf{x} \in \{0, 1\}^{|V|}$ by:

$$\min \mathbf{1}^T \mathbf{x}$$
$$x(N[j]) = \sum_{i \in N[j]} x_i \geq 1 \quad \forall j \in V \qquad \text{(domination)}$$
$$x(N[j] \triangle N[k]) = \sum_{i \in N[j] \triangle N[k]} x_i \geq 1 \quad \forall j, k \in V, j \neq k \text{ (separation)}$$
$$\mathbf{x} \in \{0, 1\}^{|V|}$$

By [2], the matrix $M_{ID}(G)$ encoding row-wise the closed neighborhoods of the nodes and their symmetric differences is called the *identifying code matrix* of G, and the *identifying code polyhedron* of G is defined as

$$P_{ID}(G) = Q^*(M_{ID}) = \text{conv}\{\mathbf{x} \in \mathbb{Z}_+^{|V|} : M_{ID}(G)\ \mathbf{x} \geq \mathbf{1}\}.$$

It is clear by construction that a graph is identifiable if and only if none of the symmetric differences results in a zero-row of $M_{ID}(G)$, and that $\gamma_{ID}(G)$ equals the covering number $\tau(M_{ID}(G))$.

It turned out that studying the ID-problem from a polyhedral point of view can lead to interesting results, see e.g. [2–5]. The aim of this paper is to apply the polyhedral approach to find minimum OLD- or LTD-sets.

In Sect. 2, we give the according definitions of the matrices $M_{OLD}(G)$ and $M_{LTD}(G)$ and of the associated polyhedra, provide some basic properties of the polyhedra $P_{OLD}(G)$ and $P_{LTD}(G)$ and introduce their canonical linear relaxations. Afterwards, we discuss several lines to apply polyhedral techniques.

In Sect. 3, we present cases where $M_{OLD}(G)$ or $M_{LTD}(G)$ are composed of matrices for which the set covering polyhedron is known and we, thus, immediately can obtain a complete description of $P_{OLD}(G)$ or $P_{LTD}(G)$ and the exact value of $\gamma_{OLD}(G)$ or $\gamma_{LTD}(G)$.

This demonstrates how polyhedral techniques can be applied in this context. We close with a discussion on future lines of research, including how the here obtained results can be extended to other classes of graphs.

2 Polyhedra Associated to *OLD-* and *LTD*-Sets

In order to apply the polyhedral approach to the OLD- and the LTD-problem, we first give according reformulations as set covering problem.

Theorem 1. *Let $G = (V, E)$ be a graph.*

(a) Let G have neither isolated nodes nor false twins. $C \subseteq V$ is an OLD-set if and only if C has a non-empty intersection with
> OLD_1 $N(i)$ for all $i \in V$,
> OLD_2 $N(i) \triangle N(j)$ for all distinct $i, j \in V$ with $dist(i,j) = 1$ or $dist(i,j) = 2$;

(b) $C \subseteq V$ is an LTD-set if and only if C has a non-empty intersection with

> LTD_1 $N(i)$ for all $i \in V$,
> LTD_2 $N(i) \triangle N(j)$ for all distinct $i, j \in V$ with $dist(i,j) = 1$,
> LTD_3 $N[i] \triangle N[j]$ for all distinct $i, j \in V$ with $dist(i,j) = 2$.

The matrices $M_{OLD}(G)$ and $M_{LTD}(G)$ encoding row-wise the open neighborhoods and their respective symmetric differences read, therefore, as

$$M_{OLD}(G) = \begin{pmatrix} N(G) \\ \triangle_1(G) \\ \triangle_2(G) \end{pmatrix} \qquad M_{LTD}(G) = \begin{pmatrix} N(G) \\ \triangle_1(G) \\ \triangle_2[G] \end{pmatrix}$$

where every row in $N(G)$ is the characteristic vector of an open neighborhood of a node in G and $\triangle_k(G)$ (resp. $\triangle_k[G]$) is composed of the characteristic vectors of the symmetric difference of open (resp. closed) neighborhoods of nodes at distance k in G. We define by

$$P_X(G) = Q^*(M_X(G)) = \text{conv}\{\mathbf{x} \in \mathbb{Z}_+^{|V|} : M_X(G)\, \mathbf{x} \geq 1\}$$

the *X-polyhedron* for $X \in \{OLD, LTD\}$. We first address the dimension of the two polyhedra. It is known from Balas and Ng [7] that a set covering polyhedron $Q^*(M)$ is full-dimensional if and only if the matrix M has at least two ones per row.

From the submatrix $N(G)$ encoding the open neighborhoods, we see that

$$V_N(G) = \{k \in V : \{k\} = N(i),\ i \in V\}$$

are the cases that result in a row with only one 1-entry. From the submatrix $\triangle_1(G)$, every row has at least two 1-entries (namely i and j for $N(i) \triangle N(j)$). From the submatrix $\triangle_2(G)$, we see that

$$V_2(G) = \{k \in V(G) : \{k\} = N(i) \triangle N(j),\ i, j \in V\}$$

are the cases that result in a row with only one 1-entry, whereas every row from the submatrix $\triangle_2[G]$ has at least two 1-entries (namely i and j for $N[i] \triangle N[j]$). Moreover, if $\{k\} = N(i)$ and $dist(i,j) = 2$, then $k \in N(j)$. Thus $V_2(G) \cap V_N(G) = \emptyset$ follows.

We conclude:

Corollary 1. *Let $G = (V, E)$ be a graph.*

- *Let G have neither isolated nodes nor false twins. We have $dim(P_{OLD}(G)) = |V - V_N(G) - V_2(G)|$.*
- *We have $dim(P_{LTD}(G)) = |V - V_N(G)|$.*

In addition, $M_{OLD}(G)$ and $M_{LTD}(G)$ may contain redundant rows, where we say that $\mathbf{y}$ is *redundant* if $\mathbf{x}$ and $\mathbf{y}$ are two rows of M and $\mathbf{x} \leq \mathbf{y}$. As the covering number of a matrix does not change after removing redundant rows, we define the corresponding clutter matrices $C_{OLD}(G)$ and $C_{LTD}(G)$, obtained by removing redundant rows from $M_{OLD}(G)$ and $M_{LTD}(G)$, respectively. We clearly have

$$P_X(G) = Q^*(C_X(G)) = \mathrm{conv}\{\mathbf{x} \in \mathbb{Z}_+^{|V|} : C_X(G) \, \mathbf{x} \geq 1\}$$

for $X \in \{OLD, LTD\}$. Moreover, also in [7] it is proved that the only facet-defining inequalities of a set covering polyhedron $Q^*(M)$ with integer coefficients and right hand side equal to 1 are those of the system $M\mathbf{x} \geq 1$. Hence we have:

Corollary 2. *All constraints from $C_X(G) \, \mathbf{x} \geq 1$ define facets of $P_X(G)$ for $X \in \{OLD, LTD\}$.*

We obtain the corresponding linear relaxations, the *fractional OLD-polyhedron* $Q_{OLD}(G)$ and the *fractional LTD-polyhedron* $Q_{LTD}(G)$ of G, by considering all vectors satisfying the above inequalities:

$$Q_X(G) = Q(C_X(G)) = \left\{\mathbf{x} \in \mathbb{R}_+^{|V|} : C_X(G) \, \mathbf{x} \geq 1\right\}$$

for $X \in \{OLD, LTD\}$. To study the two problems from a polyhedral point of view, we propose to firstly determine the clutter matrices $C_{OLD}(G)$ and $C_{LTD}(G)$ and then to determine which further constraints have to be added to $Q_{OLD}(G)$ and $Q_{LTD}(G)$ in order to obtain $P_{OLD}(G)$ and $P_{LTD}(G)$, respectively.

3   Complete p-Partite Graphs

In this section, we consider complete p-partite graphs and establish a connection to so-called complete 2-roses of order n. Given $n > q \geq 2$, let $\mathcal{R}_n^q = (V, \mathcal{E})$ be the hypergraph where $V = \{1, \ldots, n\}$ and $\mathcal{E}$ contains all q-element subsets of V. Nobili and Sassano [17] called the incidence matrix of $\mathcal{R}_n^q$ the *complete q-rose of order n* and we denote it by $M(\mathcal{R}_n^q)$. In [6], it was shown:

Theorem 2 ([2,6]). *The covering polyhedron $Q^*(M(\mathcal{R}_n^q))$ is given by the non-negativity constraints and*

$$x(V') \geq |V'| - q + 1$$

for all subsets $V' \subseteq \{1, \ldots, n\}$ with $|V'| \in \{q + 1, \ldots, n\}$.

Complete Bipartite Graphs. First we consider complete bipartite graphs $K_{m,n}$ with bipartition $A = \{1, \ldots, m\}$ and $B = \{m+1, \ldots, m+n\}$. We note that $K_{m,n}$ has false twins (unless $m = 1 = n$) and, thus, no OLD-set, hence we only analyse LTD-sets. We begin with the case of stars $K_{1,n}$, i.e., $A = \{1\}$ and $n \geq 2$. Note that $K_{1,2} = P_3$ and it is easy to see that $\gamma_{LTD}(K_{1,2}) = 2$ holds.

Lemma 1. *For a star $K_{1,n}$ with $n \geq 3$, we have*

$$
C_{LTD}(K_{1,n}) = \begin{pmatrix} 1 & 0 & \cdots & 0 \\ 0 & & & \\ \vdots & & M(\mathcal{R}_n^2) & \\ 0 & & & \end{pmatrix}.
$$

From the above description of the facets of the covering polyhedron associated with complete q-roses by [2], we conclude:

Corollary 3. $P_{LTD}(K_{1,n})$ *with $n \geq 3$ is described by the nonnegativity constraints, the inequalities $x_1 \geq 1$ and $x(B') \geq |B'| - 1$ for all nonempty subsets $B' \subseteq \{2, \ldots, n+1\}$.*

Furthermore, combining $x_1 \geq 1$ and $x(B) \geq |B| - 1$ yields the full rank constraint $x(V) \geq |B|$ which immediately implies $\gamma_{LTD}(K_{1,n}) = |V| - 1 = n$ (and provides an alternative proof for the result given in [14]).

Observe that for $K_{2,2}$, it is easy to see that $\gamma_{LTD}(K_{2,2}) = 2$. For general complete bipartite graphs $K_{m,n}$ with $m \geq 2, n \geq 3$, we obtain:

Lemma 2. *For a complete bipartite graph $K_{m,n}$ with $m \geq 2, n \geq 3$, we have*

$$
C_{LTD}(K_{m,n}) = \begin{pmatrix} M(\mathcal{R}_m^2) & 0 \\ 0 & M(\mathcal{R}_n^2) \end{pmatrix}.
$$

Note that results from [2] show that $C_{ID}(K_{m,n}) = C_{LTD}(K_{m,n})$. Hence, we directly conclude from the facet description of $P_{ID}(K_{m,n})$ by [2]:

Corollary 4. $P_{LTD}(K_{m,n})$ *is given by the inequalities*

1. $x(C) \geq |C| - 1$ *for all nonempty $C \subseteq A$,*
2. $x(C) \geq |C| - 1$ *for all nonempty $C \subseteq B$.*

Moreover, $\gamma_{LTD}(K_{m,n}) = |V| - 2 = m + n - 2$.

This provides an alternative proof for the result given in [14].

Complete p-Partite Graphs. The results above can be further generalized for complete p-partite graphs. Consider $K_{n_1, \ldots, n_p} = (U_1 \cup \cdots \cup U_p, E)$ where each U_i induces a nonempty stable set and all edges between U_i and U_j, $i \neq j$ are present. We use $|U_i| = n_i$ for $i = 1, \ldots, p$, $|V| = n$ and assume $n_1 \leq n_2 \leq \ldots \leq n_p$ as well as $p \geq 3$. For illustration, complete 3-partite and 4-partite graphs are depicted in Fig. 2.

We note that $K_{n_1, \ldots, n_p}$ has false twins and, thus, no OLD-set, unless $n_1 = \cdots = n_p = 1$ and the graph is a clique.

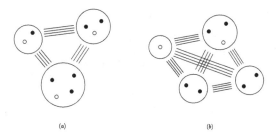

(a)   (b)

Fig. 2. (a) A complete 3-partite graph with $n_1 = 2$, $n_2 = 3$ and $n_3 = 4$, (b) A complete 4-partite graph with $n_1 = 1$, $n_2 = n_3 = 2$ and $n_4 = 3$.

Lemma 3. Let $K_{n_1,n_2,\dots,n_p}$ be a complete p-partite graph.

(a) If $n_1 = \cdots = n_p = 1$, then $K_{n_1,n_2,\dots,n_p}$ equals the clique K_p and

$$C_{OLD}(K_{n_1,n_2,\dots,n_p}) = C_{LTD}(K_{n_1,n_2,\dots,n_p}) = M(\mathcal{R}_p^2).$$

(b) If $n_1 = \cdots = n_r = 1$ with $r \geq 2$ and $n_{r+1} \geq 2$, then

$$C_{LTD}(K_{n_1,n_2,\dots,n_p}) = \begin{pmatrix} M(\mathcal{R}_r^2) & 0 & 0 & \cdots & 0 \\ 0 & M(\mathcal{R}_{n_{r+1}}^2) & 0 & \cdots & 0 \\ \vdots & & \ddots & & \vdots \\ 0 & & & \cdots & M(\mathcal{R}_{n_p}^2) \end{pmatrix}.$$

(c) If $n_1 = 1$ and $n_2 \geq 2$, then

$$C_{LTD}(K_{n_1,n_2,\dots,n_p}) = \begin{pmatrix} 0 & M(\mathcal{R}_{n_2}^2) & 0 & 0 & \cdots & 0 \\ 0 & 0 & M(\mathcal{R}_{n_3}^2) & 0 & \cdots & 0 \\ \vdots & & & \ddots & \ddots & \vdots \\ 0 & 0 & & \cdots & & M(\mathcal{R}_{n_p}^2) \end{pmatrix}.$$

(d) If $n_1 \geq 2$, then

$$C_{LTD}(K_{n_1,n_2,\dots,n_p}) = \begin{pmatrix} M(\mathcal{R}_{n_1}^2) & 0 & 0 & \cdots & 0 \\ 0 & M(\mathcal{R}_{n_2}^2) & 0 & \cdots & 0 \\ \vdots & & \ddots & & \vdots \\ 0 & & & \cdots & M(\mathcal{R}_{n_p}^2) \end{pmatrix}.$$

From the description of the facets of the covering polyhedron associated with complete q-roses by [2] and taking the block structure of the matrices into account, we conclude:

Corollary 5. Let $K_{n_1,n_2,\dots,n_p}$ be a complete p-partite graph.

(a) If $n_1 = \cdots = n_p = 1$, then $P_X(K_{n_1,n_2,\dots,n_p})$ is given by the inequalities

- $x(V') \geq |V'| - 1$ for all nonempty subsets $V' \subseteq V$
and $\gamma_X(K_{n_1,n_2,...,n_p}) = n - 1$ for $X \in \{OLD, LTD\}$.

(b) If $n_1 = \cdots = n_r = 1$ with $r \geq 2$ and $n_{r+1} \geq 2$, then
- $x(V') \geq |V'| - 1$ for all nonempty subsets $V' \subseteq U_1 \cup \cdots \cup U_r$,
- $x(V') \geq |V'| - 1$ for all nonempty subsets $V' \subseteq U_i$ for $i \in \{r+1, \ldots, p\}$
and $\gamma_{LTD}(K_{n_1,n_2,...,n_p}) = n - p + r - 1$.

(c) If $n_1 = 1$ and $n_2 \geq 2$, then $P_{LTD}(K_{n_1,n_2,...,n_p})$ is given by the inequalities
- $x(U_1) \geq 0$,
- $x(V') \geq |V'| - 1$ for all nonempty subsets $V' \subseteq U_i$ for $i \in \{2, \ldots, p\}$,
and $\gamma_{LTD}(K_{n_1,n_2,...,n_p}) = n - p$.

(d) If $n_1 \geq 2$, then $P_{LTD}(K_{n_1,n_2,...,n_p})$ is given by the inequalities
- $x(V') \geq |V'| - 1$ for all nonempty subsets $V' \subseteq U_i$ for $i \in \{1, \ldots, p\}$
and $\gamma_{LTD}(K_{n_1,n_2,...,n_p}) = n - p$.

Corrollary 5(a) provides an alternative proof for the result on OLD-sets in cliques given in [19].

4   Some Families of Split Graphs

A graph $G = (C \cup S, E)$ is a *split graph* if its node set can be partitioned into a clique C and a stable set S. Split graphs are closed under taking complements and form the complementary core of chordal graphs since G is a split graph if and only if G and $\overline{G}$ are chordal or if and only if G is $(C_4, \overline{C}_4, C_5)$-free [11].

Our aim is to study LTD-sets in some families of split graphs having a regular structure from a polyhedral point of view.

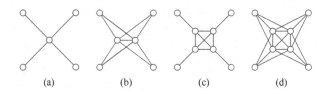

Fig. 3. (a) star, (b) crown, (c) thin headless spider, (d) thick headless spider.

Complete Split Graphs. A *complete split graph* is a split graph where all edges between C and S are present. Complete split graphs can be seen as special case of complete multi-partite graphs studied in Sect. 3. In fact, a complete split graph is a clique if $|S| = 1$, a star if $|C| = 1$, and a crown if $|C| = 2$, see Fig. 3(a), (b). Otherwise, the graph can be seen as a complete multi-partite graph where all parts but one have size 1, i.e. as $K_{n_1,n_2,...,n_p}$ with $n_1 = \cdots = n_{p-1} = 1$ and $n_p \geq 2$ such that $U_1 \cup \cdots \cup U_{p-1}$ induce the clique C and U_p the stable set S. Hence, we directly conclude from Lemma 3 and Corollary 5:

Corollary 6. *Let $G = (C \cup S, E)$ be a complete split graph.*

(a) If $|S| = 1$, then G is a clique,

$$C_X(G) = M(\mathcal{R}^2_{|C|+1})$$

and $\gamma_X(G) = |C|$ for $X \in \{OLD, LTD\}$.

(b) If $|C| = 1$, then G is a star,

$$C_{LTD}(G) = \left(\begin{array}{c|ccc} 1 & 0 & \cdots & 0 \\ \hline 0 & & & \\ \vdots & & M(\mathcal{R}^2_{|S|}) & \\ 0 & & & \end{array} \right).$$

and $\gamma_{LTD}(G) = |S|$.

(c) Otherwise, we have

$$C_{LTD}(G) = \left(\begin{array}{cc} M(\mathcal{R}^2_{|C|}) & 0 \\ 0 & M(\mathcal{R}^2_{|S|}) \end{array} \right)$$

and $\gamma_{LTD}(G) = |S| + |C| - 2$.

Headless Spiders. A *headless spider* is a split graph with $C = \{c_1, \ldots, c_k\}$ and $S = \{s_1, \ldots, s_k\}$; it is *thin* (resp. *thick*) if s_i is adjacent to c_j if and only if $i = j$ (resp. $i \neq j$), see Fig. 3(c), (d) for illustration. Clearly, the complement of a thin spider is a thick spider, and vice-versa. It is easy to see that for $k = 2$, the path P_4 equals the thin and thick headless spider. Moreover, it is easy to check that headless spiders are twin-free.

A thick headless spider with $k = 3$ equals the 3-sun S_3 and it is easy to see that $\gamma_{OLD}(S_3) = 4$ and $\gamma_{LTD}(S_3) = 3$ holds. To describe the clutters for $k \geq 4$, we use the following notations. Let J_n denote the $n \times n$ matrix having 1-entries only and I_n the $n \times n$ identity matrix. Furthermore, let $J_{n-1,n}(i)$ denote a matrix s.t. its i-th column has 0-entries only and removing the i-th column results in J_{n-1}, and $I_{n-1,n}(j)$ denote a matrix s.t. its j-th column has 1-entries only and removing the j-th column results in I_{n-1}.

Lemma 4. *For a thick headless spider $G = (C \cup S, E)$ with $k \geq 4$, we have*

$$C_{OLD}(G) = \left(\begin{array}{cc} M(\mathcal{R}^{|S|-1}_{|S|}) & 0 \\ 0 & M(\mathcal{R}^2_{|C|}) \end{array} \right) \text{ and } C_{LTD}(G) = \left(\begin{array}{cc} 0 & M(\mathcal{R}^{|C|-1}_{|C|}) \\ J_{k-1,k}(k) & I_{k-1,k}(k) \\ \vdots & \vdots \\ J_{k-1,k}(1) & I_{k-1,k}(1) \\ M(\mathcal{R}^2_{|S|}) & M(\mathcal{R}^2_{|C|}) \\ J_{|S|} & I_{|C|} \end{array} \right).$$

The lines associated with $N(c_i) \triangle N(c_j)$ (or equivalently with $N[s_i] \triangle N[s_j]$) in the rows of the matrix $M(R^2_{|C|})$ are ordered according to the rows of $M(R^2_{|S|})$ (that is for the same pairs $(i, j) \in \{1, ..., k\}^2$).

From the description of the polyhedron associated with complete q-roses by [2] and taking the block structure of $C_{OLD}(G)$ into account, we conclude:

Corollary 7. *For a thick headless spider $G = (C \cup S, E)$ with $k \geq 4$, $P_{OLD}(G)$ is given by the inequalities*

- $x_i \geq 0$ *for all $i \in C \cup S$,*
- $x(S') \geq |S'| - k + 2$ *for all $S' \subseteq S$ with $|S'| \geq k - 1$,*
- $x(C') \geq |C'| - 1$ *for all $C' \subseteq C$ with $|C'| \geq 2$,*

and $\gamma_{OLD}(G) = |C| + 1$.

On the other hand, from the clutter matrix $C_{LTD}(G)$, we immediately see that C is an LTD-set. However, C is a minimum LTD-set only if $k = 4$. For thick headless spiders with $k \geq 5$, we can show, using polyhedral arguments, that $k - 1$ is a lower bound for the cardinality of any LTD-set. Exhibiting an LTD-set of size $k - 1$ thus ensures minimality:

Theorem 3. *For a thick headless spider $G = (C \cup S, E)$ with $k \geq 5$, we have $\gamma_{LTD}(G) = k - 1$.*

The situation is different for thin headless spiders:

Lemma 5. *For a thin headless spider $G = (C \cup S, E)$ with $k \geq 3$, we have*

$$C_{OLD}(G) = C_{LTD}(G) = \left(0 \; I_{|C|} \right).$$

We immediately conclude:

Corollary 8. *For a thin headless spider $G = (C \cup S, E)$ with $k \geq 3$, $P_X(G)$ is given by the inequalities*

- $x_i \geq 1$ *for all $i \in C$ and $x_i \geq 0$ for all $i \in S$*

C is the unique X-set of minimum size and $\gamma_X(G) = |C|$ follows for $X \in \{OLD, LTD\}$.

5   Concluding Remarks

In this paper, we proposed to study the OLD- and LTD-problem from a polyhedral point of view, motivated by promising polyhedral results for the ID-problem [2–5]. That way, we were able to provide closed formulas for the LTD-numbers of all kinds of complete p-partite graphs (Sect. 3), and for the studied families of split graphs as well as the OLD-numbers of thin and thick headless spiders (Sect. 4).

In particular, if we have the same clutter matrix for two different X-problems, then we can conclude that every solution of one problem is also a solution for the other problem, and vice versa, such that the two X-polyhedra coincide and the two X-numbers are equal. This turned out to be the case for

– complete bipartite graphs as $C_{ID}(K_{m,n}) = C_{LTD}(K_{m,n})$ holds by Lemma 2 and results from [2],
– thin headless spiders G as $C_{OLD}(G) = C_{LTD}(G)$ holds by Lemma 5.

Furthermore, we were able to provide the complete facet descriptions of

– the LTD-polyhedra for all complete p-partite graphs (including complete split graphs) and for thin headless spiders (see Sect. 3 and Lemma 5),
– the OLD-polyhedra of cliques, thin and thick headless spiders (see Corollary 5 and Sect. 4).

The complete descriptions of some X-polyhedra also provide us with information about the relation between $Q^*(C_X(G))$ and its linear relaxation $Q(C_X(G))$. A matrix M is *ideal* if $Q^*(M) = Q(M)$. For any nonideal matrix, we can evaluate how far M is from being ideal by considering the inequalties that have to be added to $Q(M)$ in order to obtain $Q^*(M)$. With this purpose, in [1], a matrix M is called *rank-ideal* if only 0/1-valued constraints have to be added to $Q(M)$ to obtain $Q^*(M)$. From the complete descriptions obtained in Sect. 3 and Sect. 4, we conclude:

Corollary 9. *The LTD-clutters and OLD-clutters of thin headless spiders are ideal for all $k \geq 3$.*

Corollary 10. *The LTD-clutters of all complete p-partite graphs and the OLD-clutters of cliques and thick headless spiders are rank-ideal.*

Finally, the LTD-clutters of thick headless spiders have a more complex structure such that also a facet description of the LTD-polyhedra is more involved. However, using polyhedral arguments, is was possible to establish that $k-1$ is a lower bound for the cardinality of any LTD-set. Exhibiting an LTD-set of size $k-1$ thus allowed us to deduce the exact value of the LTD-number of thick headless spiders (Theorem 3).

This demonstrates how the polyhedral approach can be applied to find X-sets of minimum size for special graphs G, by determining and analyzing the X-clutters $C_X(G)$, even in cases where no complete description of $P_X(G)$ is known yet.

As future lines of research, we plan to work on a complete description of the LTD-polyhedra of thick headless spiders and to apply similar and more advanced techniques for other graphs in order to obtain either X-sets of minimum size or strong lower bounds stemming from linear relaxations of the X-polyhedra, enhanced by suitable cutting planes.

References

1. Argiroffo, G., Bianchi, S.: On the set covering polyhedron of circulant matrices. Discrete Optim. **6**(2), 162–173 (2009)
2. Argiroffo, G., Bianchi, S., Lucarini, Y., Wagler, A.: Polyhedra associated with identifying codes in graphs. Discrete Appl. Math. **245**, 16–27 (2018)

3. Argiroffo, G., Bianchi, S., Wagler, A.: Study of Identifying code polyhedra for some families of split graphs. In: Fouilhoux, P., Gouveia, L.E.N., Mahjoub, A.R., Paschos, V.T. (eds.) ISCO 2014. LNCS, vol. 8596, pp. 13–25. Springer, Cham (2014). https://doi.org/10.1007/978-3-319-09174-7_2

4. Argiroffo, G., Bianchi, S., Wagler, A.: On identifying code polyhedra of families of suns. In: VIII ALIO/EURO Workshop on Applied Combinatorial Optimization (2014)

5. Argiroffo, G., Bianchi, S., Wagler, A.: Progress on the description of identifying code polyhedra for some families of split graphs. Discrete Optim. **22**, 225–240 (2016)

6. Argiroffo, G., Carr, M.: On the set covering polyhedron of q-roses, In: Proceedings of the VI ALIO/EURO Workshop on Applied Combinatorial Optimization 2008, Buenos Aires, Argentina (2008)

7. Balas, E., Ng, S.M.: On the set covering polytope: I. All the facets with coefficients in $\{0, 1, 2\}$. Math. Program. **43**, 57–69 (1989). https://doi.org/10.1007/BF01582278

8. Bertrand, N., Charon, I., Hudry, O., Lobstein, A.: Identifying and locating dominating codes on chains and cycles. Eur. J. Comb. **25**, 969–987 (2004)

9. Charon, I., Hudry, O., Lobstein, A.: Minimizing the size of an identifying or locating-dominating code in a graph is NP-hard. Theoret. Comput. Sci. **290**, 2109–2120 (2003)

10. Foucaud, F.: The complexity of the identifying code problem in restricted graph classes. In: Lecroq, T., Mouchard, L. (eds.) IWOCA 2013. LNCS, vol. 8288, pp. 150–163. Springer, Heidelberg (2013). https://doi.org/10.1007/978-3-642-45278-9_14

11. Földes, S., Hammer, P.: Split graphs. In: Proceedings of the VIII Southeastern Conference on Combinatorics, Graph Theory and Computing (Baton Rouge, La.), Congressus Numerantium XIX, Winnipeg: Utilitas Math, pp. 311–315 (1977)

12. Gravier, S., Moncel, J.: On graphs having a $V\{x\}$-set as an identifying code. Discrete Math. **307**, 432–434 (2007)

13. Haynes, T.W., Henning, M.A., Howard, J.: Locating and total-dominating sets in trees. Discrete Appl. Math. **154**, 1293–1300 (2006)

14. Henning, M.A., Rad, N.J.: Locating-total domination in graphs. Discrete Appl. Math. **160**, 1986–1993 (2012)

15. Karpovsky, M.G., Chakrabarty, K., Levitin, L.B.: On a new class of codes for identifying vertices in graphs. IEEE Trans. Inf. Theory **44**, 599–611 (1998)

16. Lobstein, A.: Watching systems, identifying, locating-dominating and discriminating codes in graphs. https://www.lri.enst.fr/lobstein/debutBIBidetlocdom.pdf

17. Nobili, P., Sassano, A.: Facets and lifting procedures for the set covering polytope. Math. Program. **45**, 111–137 (1989)

18. Pandey, A.: Open neighborhood locating-dominating set in graphs: complexity and algorithms. In: International Conference on Information Technology (ICIT). IEEE (2015)

19. Seo, S.J., Slater, P.J.: Open neighborhood locating dominating sets. Australas. J. Comb. **46**, 109–119 (2010)

On the p-Median Polytope and the Directed Odd Cycle Inequalities

Mourad Baïou[1] and Francisco Barahona[2(✉)]

[1] CNRS and Université Clermont Auvergne, Campus Universitaire des Cézeaux,
1 rue de la Chebarde, 63178 Aubière Cedex, France
[2] IBM T. J. Watson research Center, Yorktown Heights, NY 10589, USA
`barahon@us.ibm.com`

Abstract. We study the effect of the odd directed cycle inequalities in the description of the polytope associated with the p-median problem. We treat general directed graphs and we characterize all the graphs for which the obvious linear relaxation together with the directed odd cycle inequalities describe the p-median polytope. In a previous work we have shown a similar result for oriented graphs. This result extends the previous work, but its proof depends on the oriented case since it will be the starting point of the proof in this paper.

1 Introduction

This paper follows the study of the classical linear formulation for the p-median problem started in [1–3]. To avoid repetitions, we refer to [1] for a more detailed introduction on the p-median problem.

Let $G = (V, A)$ a directed graph not necessarily connected, where each arc $(u, v) \in A$ has an associated cost $c(u, v)$. Here we make a difference between oriented and directed graphs. In oriented graphs at most one of the the arcs (u, v) or (v, u) exist, while in directed graphs we may have both arcs (u, v) and (v, u). The *p-median problem* (pMP) consists of selecting p nodes, usually called *centers*, and then assign each nonselected node along an arc to a selected node. The goal is to select p nodes that minimize the sum of the costs yielded by the assignment of the nonselected nodes. If the number of centers is not fixed and in stead we have costs associated with nodes, then we get the well known *facility location* problem.

If we associate the variables y to the nodes, and the variables x to the arcs, the following is the classical linear relaxation of the pMP. If we remove equality (1), then we get a linear relaxation of the facility location problem.

© Springer Nature Switzerland AG 2020
M. Baïou et al. (Eds.): ISCO 2020, LNCS 12176, pp. 15–26, 2020.
https://doi.org/10.1007/978-3-030-53262-8_2

$$\sum_{v \in V} y(v) = p, \tag{1}$$

$$y(u) + \sum_{v:(u,v) \in A} x(u,v) = 1 \quad \forall u \in V, \tag{2}$$

$$x(u,v) \le y(v) \quad \forall (u,v) \in A, \tag{3}$$

$$y(v) \ge 0 \quad \forall v \in V, \tag{4}$$

$$x(u,v) \ge 0 \quad \forall (u,v) \in A. \tag{5}$$

Call $pMP(G)$ the p-median polytope, that is the convex hull of integer solutions satisfying (1)–(5).

Now we will introduced a class of valid inequalities based on odd directed cycles. For this we need some additional definitions. A simple *cycle* C is an ordered sequence $v_0, a_0, v_1, a_1, \ldots, a_{t-1}, v_t$, where

- v_i, $0 \le i \le t-1$, are distinct nodes,
- either v_i is the tail of a_i and v_{i+1} is the head of a_i, or v_i is the head of a_i and v_{i+1} is the tail of a_i, for $0 \le i \le t-1$, and
- $v_0 = v_t$.

Let $V(C)$ and $A(C)$ denote the nodes and the arcs of a simple cycle C, respectively. By setting $a_t = a_0$, we partition the vertices of C into three sets: $\hat{C}$, $\dot{C}$ and $\tilde{C}$. Each node v is incident to two arcs a' and a'' of C. If v is the head (resp. tail) of both arcs a' and a'' then v is in $\hat{C}$ (resp. $\dot{C}$) and if v is the head of one of them and a tail of the other, then v is in $\tilde{C}$. Notice that $|\hat{C}| = |\dot{C}|$. A cycle will be called *g-odd* if $|\tilde{C}| + |\hat{C}|$ is odd, that is the number of nodes that are heads of some arcs in C is odd. Otherwise it will be called *g-even*. A cycle C with $V(C) = \tilde{C}$ is a *directed* cycle, otherwise it is called a *non-directed* cycle. Notice that the notion of g-odd (g-even) cycles generalizes the notion of odd (even) directed cycles, that is why we use the letter "g".

Definition 1. *A simple cycle is called a Y-cycle if for every $v \in \hat{C}$ there is an arc $(v, \bar{v})$ in A, where $\bar{v}$ is in $V \setminus \dot{C}$.*

Now consider the following inequalities.

$$\sum_{a \in A(C)} x(a) - \sum_{v \in \hat{C}} y(v) \le \frac{|\tilde{C}| + |\hat{C}| - 1}{2} \quad \text{for each g-odd } Y\text{-cycle } C. \tag{6}$$

The validity of these inequalities is straightforward and comes from the fact that these inequalities are Gomory-Chvátal inequalities of rank one and their separation problem may be reduced to that of the classical odd cycle inequalities in undirected graphs [4]. Hence a natural and important question in polyhedral theory is the characterization of the graphs such that constraints (1)–(5) together with inequalities (6) define an integral polytope. If we remove the equality (1), inequalities (6) remain valid for the facility location problem and the second question is the characterization of the graphs for which inequalities (2)–(5) together with (6) define an integral polytope. As a direct consequence, we get polynomial time algorithms for both the p-median and the facility location problems in this class of graphs. A simplification of both question is to explore first the following subclass of inequalities (6), that we call *odd directed cycle inequalities*:

$$\sum_{a \in A(C)} x(a) \leq \frac{|A(C)| - 1}{2} \quad \text{for each odd directed cycle } C. \tag{7}$$

Now let $PC_p(G)$ to be the polytope defined by (1)–(5) and (7). The main result in this paper is the characterization of the graphs such that $PC_p(G)$ is integral, in other words $PC_p(G) = p\text{MP}(G)$. Still this question is not trivial, since the result in this paper is based on the results of three previous papers [1–3]. In [1] the class of graphs for which the polytope define by inequalities (1)–(5) was characterized. This characterization was useful for the results in [2,3]. In [2], the class of oriented graphs without triangles for which $PC_p(G)$ is integral was characterized and this was used as a starting point to characterize the class of oriented graphs for which $PC_p(G)$ is integral in [3]. Recall that in an oriented graph, for each pair of nodes u and v at most one of the arcs (u, v) or (v, u) exists. So in this paper we conclude the characterization of the graphs for which $PC_p(G)$ is integral in any class of graphs. So here graphs may have both arcs (u, v) and (v, u). The result of this paper although is a generalization of the results in [3], it is not self-contained since it uses [3] as starting point of the induction on the number of pair of nodes u and v where both arcs (u, v) and (v, u) exist.

The paper is organized as follows. In Sect. 2, we state our main theorem with some definitions. In Sect. 3, we give the proof of our main result. This proof is on two parts. The first part is given in the Subsect. 3.1 and the second one in the Subsect. 3.2. We finish this paper with a conclusion that discuss the relationship with the facility location problem and some polyhedral consequences.

2 Preliminaries

The main result of this paper is the following theorem.

Theorem 2. *Let $G = (V, A)$ be a directed graph, then $PC_p(G)$ is integral for any integer p if and only if*

(C1) *it does not contain as a subgraph any of the graphs H_1, H_2, H_3, H_4, H_5, H_6 of Fig. 1, and*

(C2) *it does not contain a non-directed g-odd Y-cycle C with an arc (u, v) with both u and v not in $V(C)$.*

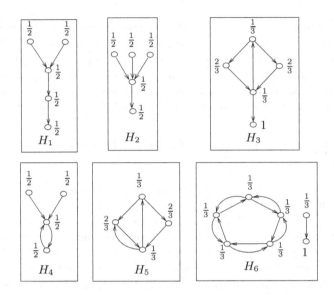

Fig. 1. The six forbidden configurations of condition (C1) of Theorem 2

Given a directed graph $G = (V, A)$, a subgraph induced by the nodes $v_1, \ldots, v_r$ of D is called a *bidirected cycle* if the only arcs in this induced subgraph are (v_i, v_{i+1}) and (v_{i+1}, v_i), for $i = 1, \ldots, r$, with $v_{r+1} = v_1$. We denote it by BIC_r.

For a directed graph $G = (V, A)$ and an arc $(u, v) \in A$, define $G(u, v)$ to be the graph obtained by removing (u, v) from G, and adding a new arc (u, v') (v' is a new pendent node). The rest of this section is devoted to definitions with respect to a feasible point in $PC_p(G)$.

Definition 3. *A vector $(x, y) \in \mathbb{R}^{|A|+|V|}$ will be denoted by z, i.e., $z(u) = y(u)$ for all $u \in V$ and $z(u, v) = x(u, v)$ for all $(u, v) \in A$. Given a vector z and a labeling function $l : V \cup A \to \{-1, 0, 1\}$, we define a new vector z_l from z as follows:*

$$z_l(u) = z(u) + l(u)\epsilon, \text{ for all } u \in V, \text{ and}$$

$$z_l(u, v) = z(u, v) + l(u, v)\epsilon, \text{ for all } (u, v) \in A,$$

where ϵ is a sufficiently small positive scalar. When we assign labels to some nodes and arcs without specifying the labels of the remaining nodes and arcs, it means that they are assigned the label zero.

Definition 4. *When dealing with a vector $z \in PC_p(G)$, we say that the arc (u, v) is tight if $z(u, v) = z(v)$. Also we say that an odd directed cycle C is tight if $z(A(C)) = (|A(C)| - 1)/2$.*

3   The Proof of Theorem 2

We will sketch quickly the necessity part of the proof. With each of the graphs H_1, H_2, H_3 and H_6 in Fig. 1 we show a fractional extreme point of $PC_3(G)$ when G is restricted to these graphs. The graphs H_4 and H_5 show an extreme point of $PC_2(G)$. The numbers near the nodes correspond to y variables. The x variables take the value $\frac{1}{2}$ for H_1, H_2 and H_4. For the graphs H_3, H_5 and H_6 the arcs take the value $\frac{1}{3}$, except the arc in the right in H_6 that takes the value $\frac{2}{3}$.

To prove necessity we just have to notice that the extreme points for the subgraphs H_i, for $i = 1, \ldots, 6$, in Fig. 1 may be extended to extreme points for any graph containing these subgraphs by setting each remaining node variable to one and each remaining arc variable to zero. For condition (C2) when the graph contains a non-directed g-odd Y-cycle C with an arc (u, v) having both nodes u and v not in C, we construct an extreme point of $PC_p(G)$ where $p = \frac{|\tilde{C}| + |\hat{C}| + 1}{2} + |V| - |V(C)| - 1$ as follows. All the nodes in $\dot{C}$ take the value 0, the nodes in $\tilde{C}$ and $\hat{C}$ with the node u take the value $\frac{1}{2}$; all the arcs in C with (u, v) take the value $\frac{1}{2}$. All other nodes take the value 1 and the other arcs take the value 0, except for each unique arc leaving each node in $\hat{C}$ (see the definition of a Y-cycle) they take the value $\frac{1}{2}$. One way to see that these are indeed extreme points is to start adding ϵ to one of the components and try to keep satisfying as equation the same constraints that the original vector satisfies as equation. First we conclude that we have to add or subtract ϵ to other components and this leads to the violation of Eq. (1) or to the impossibility of keeping tight the inequality that are satisfied as equation.

The rest of this section is devoted to the sufficiency part. Denotes by $Pair(G)$ the set of pair of nodes $\{u, v\}$ such that both arcs (u, v) and (v, u) exist. The proof of this theorem will be done by induction on the number of $|Pair(G)|$. This result has been proved in [3] for oriented graphs, that is when $|Pair(G)| = 0$. This case is the starting point of the induction. Assume that Theorem 2 is true for any directed graph H with $|Pair(H)| \leq m$ and let us show that it holds for any directed graph G with $|Pair(G)| = m + 1$. Let $G = (V, A)$ be a directed graph with $|Pair(G)| = m + 1 \geq 1$ satisfying conditions (C1) and (C2) of Theorem 2. Suppose the contrary, that is $PC_p(G)$ is not integral. Let $\bar{z}$ be a fractional extreme point of $PC_p(G)$. Next we will give some useful lemmas and then in Subsect. 3.1 and 3.2 the proof is completed. In Subsect. 3.1 we show the theorem when there is no g-odd Y-cycle and in Subsect. 3.2 we complete the proof when such a cycle exists.

For some arcs (u, v) and (v, u) we construct the graph $G(u, v)$ or $G(v, u)$ to apply the induction hypothesis. In doing so, we have to ensure that conditions (C1) and (C2) of Theorem 2 remain satisfied.

The next lemma states that we may assume that the variables associated with the arcs are positive for each extreme points. Of course we may have variables with zero value, but we may remove the associated arc and we keep working with the resulting fractional point in the resulting graph without altering any hypothesis. We prefer to act that way instead of using induction, this is more simple since we are yet using another induction on the number of pair of nodes.

Lemma 1. *We may assume that* $\bar{z}(u,v) > 0$ *for each extreme point* $\bar{z}$ *of* $PC_p(G)$.

Lemma 2. *For each* $(u,v) \in A$ *such that* (v,u) *is also in* A, *if there is no odd directed cycle tight for* $\bar{z}$ *containing* (u,v), *then* $PC_{p+1}(G(u,v))$ *is not integral.*

Lemma 3. *If* (u,v) *and* (v,u) *are arcs of* G, *then* $G(u,v)$ *and* $G(v,u)$ *satisfies condition (C1).*

From now on and following Lemma 3, when we have two arcs (u,v) and (v,u) and we consider one of the graphs $G(u,v)$ or $G(v,u)$ we implicitly assume that (C1) is satisfied. Then to use the induction we only need to check (C2).

Lemma 4. *The graph* G *does not contain a bidirected cycle.*

Lemma 5. *Let* (u,v) *and* (v,u) *two arcs of* G. *We cannot have a triangle* Δ *containing* (u,v) *and a triangle* Δ' *containing* (v,u).

Lemma 6. *Let* (u,v) *and* (v,u) *two arcs in* G. *If* (u,v) *belongs to a directed cycle of size at least five, then* (v,u) *does not belong to any triangle.*

Proof. Let $C = v_1, \ldots, v_r, v_1$ a directed cycle with $r \geq 5$. Set $u = v_1$ and $v = v_2$. Assume that there is a triangle containing (v_2, v_1), with the arcs (v_1, s) and (s, v_2). We may assume that $s \in V(C)$, otherwise H_1 is present. Let $s = v_k$ with $2 \leq k \leq r$. We must have $k = 3$ or $k = 4$, otherwise the arcs $(v_1, v_2), (s, v_2), (v_2, v_3)$ and (v_3, v_4) induce H_1. Now when $k = 4$ (resp. $k = 3$) we have H_3 (resp. H_1). □

Let us conclude this section by an important lemma concerning the intersection of odd directed cycles.

Lemma 7. *Given a solution* $\bar{z} \in PC_p(G)$, *there is no intersection between two directed odd cycles both of size at least five and both tight for* $\bar{z}$.

Proof. Recall that here we are assuming that the underlying graph satisfies conditions (C1) and (C2). Let C_1 and C_2 be two odd directed cycles both of size at least five tight for $\bar{z}$, and intersecting on some nodes. The fact that (C1) is satisfied implies some interesting properties. Let $C_1 = v_1, \ldots, v_k$. Since C_1 and C_2 intersect, there is at least one node in $V(C_1) \cap V(C_2)$, call it v_1. Let (u, v_1) be an arc of C_2, then we must have $u \in V(C_1)$, otherwise H_1 is present. Moreover, $u \in \{v_2, v_3, v_k\}$, otherwise (C1) is again violated. We continue this reasoning with the next arc in C_2, (v_1, w), if w is not a node of C_1 then to close the cycle C_2 we must have at some stage an arc (s, t) with $t \in V(C_1)$ and $s \notin V(C_1)$, but this is impossible and again we have a violation of (C1). Following the same arguments we must have $w \in \{v_k, v_k - 1, v_2\}$. So each arc in C_2 have its extremities in $V(C_1)$ and vice et versa. Therefore $V(C_1) = V(C_2)$. Moreover, if (u, v) is an arc of C_2, then one of the following hold, where the indices are taken modulo $|V(C_1)| = |V(C_2)|$: (1) (u, v) is also an arc of C_1, that is $u = v_i$ and $v = v_{i+1}$, (2) $u = v_{i+1}$ and $v = v_i$ or (3) $u = v_i$ and $v = v_{i-2}$. Notice that (1) cannot hold for each arc of C_2 since C_1 and C_2 are different. The same (2) cannot hold for each arc of C_2 since from Lemma 4 we cannot have a bidirected cycle. Then we must have at least one arc of type (3). Moreover their number must be even. In fact, If (v_i, v_{i-2}) is an arc of C_2, then one of the following hold

- (v_{i-2}, v_{i-1}) and (v_{i-1}, v_{i-3}) are in C_2, so (v_{i-1}, v_{i-3}) is also an arc of type (3).
- (v_{i-1}, v_i) and (v_{i+1}, v_{i-1}) are in C_2, so (v_{i+1}, v_{i-1}) is also an arc of type (3).

Consequently, the arcs of type (3) come by pair and create two triangles with one common arc of type (1). Let $|V(C_1)| = V(C_2)| = 2k + 1$. Assume that the number of arcs in C_1 of type (2) is r', which is the same as those of C_2. Let Δ contains the set of triangles created by the arcs of type (3). Notice that each triangle contains an arc of C_1 of type (3), an arc of C_2 of type (3) and a common arc of C_1 and C_2 that is of type (1). If we assume that $2r$ is the number of arcs of type (3) that are in C_2, which is the same as those in C_1, then the number of arcs of type (1) is r. Therefore, $2k + 1 = 2r + r + r'$. We have the following

$$\bar{z}(v_i, v_{i+1}) + \bar{z}(v_{i+1}, v_i) \leq 1, \quad \text{for each arc } (v_i, v_{i+1}) \text{ of type (2)}, \quad (8)$$
$$\bar{z}(\delta) \leq 1, \quad \text{for each } \delta \in \Delta. \quad (9)$$

The sum of these inequalities implies $\bar{z}(A(C_1)) + \bar{z}(A(C_2)) \leq r' + 2r = 2k + 1 - r$. Therefore, since C_1 and C_2 are both tight cycles, we must have $r = 1$. But then inequalities (8) and (9) should be tight, which is impossible. In fact, since $r = 1$ and the sizes of C_1 and C_2 are at least five, at least one inequality among (8) should exist, say $z(v_i, v_{i+1}) + z(v_{i+1}, v_i) \leq 1$. We have, $\bar{z}(v_i, v_{i+1}) + \bar{z}(v_{i+1}, v_i) = 1$, $\bar{z}(v_i, v_{i+1}) \leq \bar{z}(v_{i+1})$ and $\bar{z}(v_{i+1}) + \bar{z}(v_{i+1}, v_i) \leq 1$. Combining these constraints, we get $\bar{z}(v_{i+1}) + \bar{z}(v_{i+1}, v_i) = 1$, but since the arc (v_{i+1}, v_{i+2}) exists we must have $\bar{z}(v_{i+1}, v_{i+2}) = 0$, which is impossible from Lemma 1.   □

3.1 G Does Not Contain a Non-directed g-Odd Y-Cycle

When we have two arcs (u, v) and (v, u), from Lemma 3, the graph $G(u, v)$ satisfies condition (C1), which is not the case for condition (C2), even when G does not contain a non-directed g-odd Y-cycle. For example we may have a g-odd cycle C which is not a Y-cycle and an arc (s, t) with both s and t not in $V(C)$. Now if we have an arc (u, v) in $A(C)$ and $(v, u) \in A \setminus A(C)$, and if $v \in \hat{C}$ and $u \in \dot{C}$ this same cycle may become a Y-cycle in $G(v, u)$ and so with the arc (s, t) condition (C2) is violated.

Next we will show that we may always find a pair of arcs (u, v) and (v, u) where at least one of the graphs $G(u, v)$ or $G(v, u)$ satisfies (C2).

Lemma 8. Let $P = v_1, \ldots, v_k$ a maximal bidirected path, different from a bidirected cycle. We have the following

(i) If $G(v_k, v_{k-1})$ contains a non-directed g-odd Y-cycle C, then the unique arc leaving v_k is (v_k, v_{k-1}),

(ii) If $G(v_{k-1}, v_k)$ contains a non-directed g-odd Y-cycle C, then there is at most one another arc leaving v_{k-1} which is (v_{k-1}, v_{k-2}).

Let $P = v_1, \ldots, v_k$ be a maximal bidirected path. From Lemma 4 the extremities of P cannot coincide, that is P is not a bidirected cycle. We will treat two cases (1) none of the arcs (v_{k-1}, v_k) and (v_k, v_{k-1}) belong to an odd directed cycle tight for $\bar{z}$ and of size at least five, (2) at least one of these arcs belong to such a cycle.

Case 1. None of the arcs (v_{k-1}, v_k) and (v_k, v_{k-1}) belong to an odd directed cycle of size at least five. From the lemma above it is easy to see that at least one of the graphs $G(v_k, v_{k-1})$ or $G(v_{k-1}, v_k)$ does not contain a g-odd Y-cycle. In fact, assume that both graphs contain a g-odd Y-cycle. When $G(v_{k-1}, v_k)$ contains a g-odd Y-cycle C, we must have $v_k \in \dot{C}$. But this is impossible since Lemma 8 (i) implies that (v_k, v_{k-1}) is the unique arc leaving v_k. Therefore, we only need to treat the following three cases, ordered as follows.

1. Both $G(v_k, v_{k-1})$ and $G(v_{k-1}, v_k)$ do not contain a g-odd Y-cycle. Then the induction hypothesis applies for both graphs since (C1) and (C2) are satisfied and that $|Pair(G(v_k, v_{k-1}))| = |Pair(G(v_{k-1}, v_k))| < |Pair(G)|$. Hence $PC_{p+1}(G(v_k, v_{k-1}))$ and $PC_{p+1}(G(v_{k-1}, v_k))$ are both integral. Lemma 5 implies that at least one of the arcs (v_k, v_{k-1}) or (v_{k-1}, v_k) does not belong to any triangle, say (v_{k-1}, v_k). Thus Lemma 2 applies and implies that $PC_{p+1}(G(v_{k-1}, v_k))$ is not integral, which provide a contradiction.

2. $G(v_{k-1}, v_k)$ does not contain a g-odd Y-cycle and the arc (v_{k-1}, v_k) belongs to triangle. From the discussion above the graph $G(v_k, v_{k-1})$ must contain a g-odd Y-cycle and so Lemma 8 (i) implies that the unique arc leaving v_k is (v_k, v_{k-1}) and so it must belong to the triangle containing (v_{k-1}, v_k), but this is not possible. Hence the induction hypothesis together with Lemma 2 apply for $G(v_{k-1}, v_k)$ and provide a contradiction.

3. $G(v_k, v_{k-1})$ does not contain a g-odd Y-cycle and the arc (v_k, v_{k-1}) belongs to a triangle, call it Δ. As above, the graph $G(v_{k-1}, v_k)$ must contain a g-odd Y-cycle C. Since there is a triangle containing (v_k, v_{k-1}) we must have another arc than (v_{k-1}, v_k) leaving v_{k-1}. From Lemma 8 (ii) this arc is (v_{k-1}, v_{k-2}) and is unique and so it must be in Δ, therefore $\Delta = \{(v_k, v_{k-1}), (v_{k-1}, v_{k-2}), (v_{k-2}, v_k)\}$. Recall that G does not contain a g-odd Y-cycle but $G(v_{k-1}, v_k)$ contains one, call it C. This means that v_k and v_{k-2} are in $\dot{C}$ and so there is an arc (v_k, s) with $s \neq v_{k-1}, v_{k-2}$ and there is also an arc (v_{k-2}, t) with $t \neq v_{k-1}, v_k$. Notice that $s \neq t$, otherwise the cycle C would be g-even. We must have an arc (s, s'), otherwise C would not be a Y-cycle. We cannot have $s' = v_{k-1}$, otherwise H_4 is present. We cannot have $s' = v_{k-2}$, otherwise there is a g-odd Y-cycle and we can also see that in this case the configurations H_3 and H_5 exist. Therefore the arcs $(v_{k-2}, v_k), (v_{k-1}, v_k), (v_k, s)$ and (s, s') induce H_1 in G, which is not possible. Now since (v_k, v_{k-1}) cannot belong to a triangle, Lemma 2 implies that $PC_{p+1}(G(v_k, v_{k-1}))$ is not integral, but this contradicts the induction hypothesis.

Case 2. One of the arcs (v_k, v_{k-1}) or (v_{k-1}, v_k) belongs to an odd directed cycle of size at least five. Call such a cycle C and w.l.o.g. assume that $(v_k, v_{k-1}) \in A(C)$. Notice that when $(v_{k-1}, v_k) \in A(C)$, we must have $(v_1, v_2) \in A(C)$ which is symmetrically similar to the case $(v_k, v_{k-1}) \in A(C)$.

Lemma 9. *If there is an arc (v_k, s) with $s \neq v_{k-1}$, then s is a pendent node.*

Lemma 9 implies that $G(v_{k-1}, v_k)$ does not contains a g-odd Y-cycle. Hence by induction $PC_{p+1}G(v_{k-1}, v_k)$ is integral. Also this lemma implies that in fact (v_{k-1}, v_k) does not belong to any directed odd cycle and so Lemma 2 implies that $PC_{p+1}G(v_{k-1}, v_k)$ is not integral, a contradiction.

3.2 G Contains a Non-directed g-Odd Y-Cycle

Lemma 10. *Let (u, v) and (v, u) two arcs of G. Then at least one of the graphs $G(u, v)$ or $G(v, u)$ satisfies condition (C2).*

Now from Lemma 3 and Lemma 10, we know that we can always consider one of the graphs $G(u, v)$ or $G(v, u)$, and we get (C1) and (C2) satisfied and so the induction hypothesis applies and implies that at least one of the polytopes $PC_{p+1}(G(u, v))$ or $PC_{p+1}(G(v, u))$ is integral. The idea is to use Lemma 2 to obtain a contradiction.

First let us assume that both graphs $G(u, v)$ and $G(v, u)$ satisfy condition (C2). Then the induction hypothesis implies that both $PC_{p+1}(G(u, v))$ and $PC_{p+1}(G(v, u))$ are integral. From Lemma 7, at least one of the arcs (u, v) or (v, u) does not belong to a directed odd cycle tight for z and of size at least five. Say that this arc is (u, v). If (u, v) belongs to a triangle, then Lemma 6 implies that (v, u) does not belong to a directed cycle of size at least five and Lemma 5 implies that (v, u) does not belong to any triangle. Therefore Lemma 2 applies for $G(v, u)$ and implies that $PC_{p+1}(G(v, u))$ is not integral, which contradicts the induction hypothesis. So we may assume that (u, v) does not belong to any triangle, and in this case Lemma 2 applies for $G(u, v)$ and we obtain again a contradiction. Thus we may assume that (C2) is satisfied by $G(u, v)$ but not by $G(v, u)$. The induction hypothesis applies only for $G(u, v)$ and so $PC_{p+1}(G(u, v))$ is integral. We will also assume that (u, v) belongs to a directed odd cycle D, otherwise we may apply Lemma 2 to obtain the contradiction that $PC_{p+1}(G(u, v))$ is not integral. Since $G(v, u)$ does not satisfies (C2), this means that G contains a g-odd Y-cycle C containing u but not v.

- $u \in \dot{C}$. Call the two nodes of C incident to u, u' and u''. So we have the arcs (u, u') and (u, u'') in $A(C)$.

- Assume that D is of size at least five. Let s and t the neighbors of u and v respectively in D. That is the arcs (s, u) and (v, t) are in D. Since there are two arcs entering u, (s, u) and (v, u), to avoid H_1 any arc leaving u' or u'' must be directed to v or to s which means that both u' and u'' are in D, otherwise the arc leaving u' or u'' to s or t creates H_1, see Fig. 2 (a). From

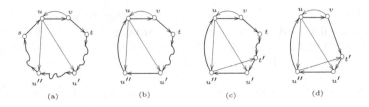

Fig. 2. The odd directed cycle D is in bold.

this figure it is easy to check that in order to avoid H_1, the situation must be as pictured in Fig. 2 (b), that is $s = u''$ and that the path in D directed from u' to u'' consists of one arc (u', u''). Recall that we have a g-odd directed cycle C containing u but not v and it contains the arcs (u, u') and (u, u''). Thus we must have a path from u' to u'' that do not contain nor u nor v and it cannot consist of the arc (u', u''), otherwise C is not g-odd. The only possibility to avoid the forbidden configurations is to have an arc leaving u'' and entering the other neighbor of u' in D, call it t', see Fig. 2 (c). There are no other arcs than the ones pictured in Fig. 2 (c) that are incident to u', u'' or to t', otherwise we have a forbidden configuration. This implies that the g-odd Y-cycle C is composed by the arcs $(u, u'), (u, u''), (u'', t')$ and (t', u'). And since (C2) is satisfied we must have $t = t'$ and hence D is of size five, see Fig. 2 (d).

Now notice that the graph $G(u'', u)$ does not satisfies (C2), since it easy to check that the nodes u, v, t and u' define a g-odd Y-cycle. By Lemma 10, $G(u, u'')$ satisfies (C2) and so the induction hypothesis implies that $PC_{p+1}(G(u, u''))$ is integral. Now it is easy to check that (u, u'') does not belong to any directed odd cycle. In fact let D' such a cycle. The arc (u'', t) must be in D'. D' cannot be of size three, otherwise the arc (t, u) exits with H_2. Then D' is of size at least five. D' must contain the arc (t, u'). But now the only arc leaving u' is (u', u''). Hence Lemma 2 may be applied to $G(u, u'')$ which implies that $PC_{p+1}(G(u, u''))$ is not integral, but this contradicts the induction hypothesis.

- Assume that D is a triangle. Let $(s, u), (u, v), (v, s)$ the arcs of D. We may always assume that $s \neq u''$, otherwise rename u' by u''. To avoid H_1, any arc leaving u'' must be directed to s or to v. The node u'' is not pendent, so there is at least one arc leaving u'' and cannot be directed to v, otherwise (u, v) and (v, u) both belong to triangles, which contradicts Lemma 5. Then we must have an arc (u'', s) which implies that $s \neq u'$, since otherwise the arcs $(v, s), (u, s)$ and (u'', s) represent H_2. But then the arcs $(v, s), (u'', s), (s, u)$ and (u, u') define H_1.

- $u \in \tilde{C}$. Now let (u', u) and (u, u'') the two arcs incident to u in C.

- D is of size at least five. Let (s, u) and (v, t) two arcs of D. We must have $s = u'$, otherwise the arcs $(u, v), (v, u)$ with (s, u) and (u', u) induce H_4. Now

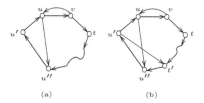

Fig. 3. The odd directed cycle D is in bold.

any arc leaving u'' must be directed to t or to u', and since t and u' are in D, to avoid H_1 we must have (u'', u') an arc of D, see Fig. 3 (a). Now since the g-odd Y-cycle contains u but not v, we must have a path P' between u' and u'' that does not contain nor u nor v and it cannot consist of (u'', u'), otherwise the g-odd Y-cycle C is directed. Moreover, if there is another arc leaving u'' it must be directed to v in order to avoid H_1. But in this case it may be easily seen from Fig. 3 (a) that the arc (u'', v) implies the presence of H_5 . The unique other arc entering u'' is its neighbor in D, call it t'. Hence (t', u'') is in the path P'. Now this path must contain an arc incident to u', by definition it cannot be (u', u) or (u'', u'). This arc cannot enter u', since the only possibility to avoid H_1 is that this arc should be (v, u') or (u, u') which is again impossible by definition. Thus we must have an arc (u', s) in P' and $s = t'$, otherwise $(t', u''), (u, u''), (u'', u')$ and (u', s) induce H_1, see Fig. 3 (b). Hence the path P' consist of two arcs (u', t') and (t', u''). It follows that C is composed by these two arcs with (u', u) and (u, u''). Since the unique arc leaving u'' is (u'', u') and that $u' \in \dot{C}$, the cycle C is not a Y-cycle, a contradiction.

- D is a triangle. This triangle must be composed by the arcs (u, v), (v, u') and (u', u), otherwise H_4 is present. Notice that u'' is not a pendent node, and that any any arc leaving u'' must be directed to v or to u'. If (u'', v) exists, then (u, v) and (v, u) both belong to triangles which contradicts Lemma 5. So we must have the arc (u'', u'). In this case the arcs $(v, u), (v, u'), (u, u''), (u'', u')$ induce a g-odd Y-cycle. This implies that there is only one additional node in C call it t. That is $V(C) = \{u, u', u'', t\}$. Since the unique arc leaving u'' is (u'', u'), we have $(t, u'') \in A(C)$. We cannot have (t, u'), otherwise H_2 is present. Thus C is composed by the arcs $(u, u''), (t, u''), (u', t)$ and (u', u). But then C is not a Y-cycle, a contradiction.

Thus from the discussion above we may conclude that (u, v) does not belong to any odd directed cycle. Lemma 2 applies and implies that $PC_{p+1}(G(u, v))$ is not integral, which contradicts the induction hypothesis that states that $PC_{p+1}(G(u, v))$ is integral.

4   Concluding Remarks

In this paper we characterized all the graphs such that the classical linear relaxation of the p-median problem together with the odd directed cycles define an integral polytope. This result has an interesting computational counterpart. In fact, when optimizing over the polytope $PC_p(G)$ which can be done in polynomial time, the graphs induced by the arcs with positive variables in any optimal fractional solution contain at least one of the configurations H_i, $i = 1, \ldots, 6$, or a non-directed g-odd cycle of condition ($C2$). All these configurations may be found in polynomial time.

Another interesting consequence of this theorem is the relationship with the facility location problem. In fact we have the following result. Call $PC(G)$ the polytope defined by inequalities (2)–(5) and (7).

Corollary 5. *Le $G = (V, A)$ be a directed graph. If G does not contain a nondirected g-odd Y-cycle nor a bicycle of size five and does not contain none of the graphs graphs H_1, H_2 and H_4 of Fig. 1 as a subgraph, then $PC(G)$ is integral.*

Proof. Let $G = (V, A)$ be a directed graph satisfying the conditions of the Corollary. Notice that conditions (C1) and (C2) of Theorem 2 are satisfied as well. Now suppose the contrary, that is $PC(G)$ admits a fractional extreme point $\bar{z}$. Let $k = \sum_{v \in V} \bar{z}(v)$. If k is an integer, then $\bar{z}$ is a fractional extreme point of $PC_k(G)$, but this contradicts Theorem 2. Assume that k is fractional and let $p = \lceil k \rceil$. Add an arc (u, v) to G with both nodes u and v are new. Extend $\bar{z}$ to z' by setting $z' = \bar{z}$ for all nodes and arcs of G and set $z'(u) = p - k$ and $z'(v) = 1$; $z'(u, v) = 1 - p + k$. The resulting graph satisfies conditions (C1) and (C2). Moreover, it is easy to check that z' is an extreme fractional point of $PC_{p+1}(G)$ which again contradicts Theorem 2.

The result above may be easily improved by allowing the bicycles of size five. This can be done by adding a few simple valid inequalities and we get an integral polytope.

References

1. Baïou, M., Barahona, F.: On the linear relaxation of the p-median problem. Discrete Optim. **8**, 344–375 (2011)
2. Baïou, M., Barahona, F.: On the p-median polytope and the directed odd cycle inequalities: triangle-free oriented graphs. Discrete Optim. **22**, 206–224 (2016)
3. Baïou, M., Barahona, F.: On the p-median polytope and the odd directed cycle inequalities: oriented graphs. Networks **71**, 326–345 (2018)
4. Grötschel, M., Lovász, L., Schrijver, A.: Geometric Algorithms and Combinatorial Optimization. AC, vol. 2. Springer, Heidelberg (1993). https://doi.org/10.1007/978-3-642-78240-4

On k-edge-connected Polyhedra: Box-TDIness in Series-Parallel Graphs

Michele Barbato[1], Roland Grappe[2], Mathieu Lacroix[2],
and Emiliano Lancini[2,3(✉)]

[1] Dipartimento di Informatica, Università degli Studi di Milano, Milan, Italy
michele.barbato@unimi.it
[2] Université Sorbonne Paris Nord, LIPN, CNRS, UMR 7030,
93430 Villetaneuse, France
{grappe,lacroix,lancini}@lipn.univ-paris13.fr
[3] Université Paris-Dauphine, PSL, LAMSADE, CNRS, UMR 7243,
75016 Paris, France

Abstract. Given a connected graph $G = (V, E)$ and an integer $k \geq 1$, the connected graph $H = (V, F)$, where F is a family of elements of E, is a k-edge-connected spanning subgraph of G if H remains connected after the removal of any $k - 1$ edges. The convex hull of the k-edge-connected spanning subgraphs of a graph G forms the k-edge-connected spanning subgraph polyhedron of G. We prove that this polyhedron is box-totally dual integral if and only if G is series-parallel. In this case, we also provide an integer box-totally dual integral system describing this polyhedron.

Introduction

Totally dual integral systems—introduced in the late 70's—are strongly connected to min-max relations in combinatorial optimization (Schrijver, 1998). A rational system of linear inequalities $Ax \geq b$ is *totally dual integral (TDI)* if the maximization problem in the linear programming duality:

$$\min\{c^\top x : Ax \geq b\} = \max\{b^\top y : A^\top y = c, y \geq \mathbf{0}\}$$

admits an integer optimal solution for each integer vector c such that the optimum is finite. Every rational polyhedron can be described by a TDI system (Giles and Pulleyblank, 1979). For instance, $\frac{1}{q}Ax \geq \frac{1}{q}b$ is TDI for some positive q. However,

---

M. Barbato—While working on this paper, Michele Barbato was financially supported by Regione Lombardia, grant agreement n. E97F17000000009, project AD-COM.
R. Grappe—Supported by ANR DISTANCIA (ANR-17-CE40-0015).
M. Lacroix, E. Lancini and R. Grappe—This work has been partially supported by the PGMO project *Matrices Totalement Équimodulaires*.

M. Baïou et al. (Eds.): ISCO 2020, LNCS 12176, pp. 27–41, 2020.
https://doi.org/10.1007/978-3-030-53262-8_3

only integer polyhedra can be described by TDI systems with integer right-hand side (Edmonds and Giles, 1984). TDI systems with only integer coefficients yield min-max results that have combinatorial interpretation.

A stronger property is the box-total dual integrality, where a system $Ax \geq b$ is *box-totally dual integral (box-TDI)* if $Ax \geq b, \ell \leq x \leq u$ is TDI for all rational vectors ℓ and u (possibly with infinite components). General properties of such systems can be found in (Cook, 1986) and Chapter 22.4 of (Schrijver, 1998). Note that, although every rational polyhedron $\{x : Ax \geq b\}$ can be described by a TDI system, not every polyhedron can be described by a box-TDI system. A polyhedron which can be described by a box-TDI system is called a *box-TDI polyhedron*. As proved by Cook (1986), every TDI system describing such a polyhedron is actually box-TDI.

Recently, several new box-TDI systems were exhibited. Chen et al. (2008) characterized box-Mengerian matroid ports. Ding et al. (2017) characterized the graphs for which the TDI system of Cunningham and Marsh (1978) describing the matching polytope is actually box-TDI. Ding et al. (2018) introduced new subclasses of box-perfect graphs. Cornaz et al. (2019) provided several box-TDI systems in series-parallel graphs. For these graphs, Barbato et al. (2020) gave the box-TDI system for the flow cone having integer coefficients and the minimum number of constraints. Chen et al. (2009) provided a box-TDI system describing the 2-edge-connected spanning subgraph polyhedron for the same class of graphs.

In this paper, we are interested in integrality properties of systems related to k-edge-connected spanning subgraphs. Given a positive integer k, a *k-edge-connected spanning subgraph* of a connected graph $G = (V, E)$ is a connected graph $H = (V, F)$, with F a family of elements of E, that remains connected after the removal of any $k - 1$ edges.

These objects model a kind of failure resistance of telecommunication networks. More precisely, they represent networks which remain connected when $k - 1$ links fail. The underlying network design problem is the *k-edge-connected spanning subgraph problem (k-ECSSP)*: given a graph G, and positive edge costs, find a k-edge-connected spanning subgraph of G of minimum cost. Special cases of this problem are related to classic combinatorial optimization problems. The 2-ECSSP is a well-studied relaxation of the traveling salesman problem (Erickson et al. 1987) and the 1-ECSSP is nothing but the well-known minimum spanning tree problem. While this latter is polynomial-time solvable, the k-ECSSP is **NP**-hard for every fixed $k \geq 2$ (Garey and Johnson, 1979).

Different algorithms have been devised in order to deal with the k-ECSSP. Notable examples are branch-and-cut procedures (Cornaz et al. 2014), approximation algorithms (Gabow et al. 2009), Cutting plane algorithms Grötschel et al. (1992), and heuristics (Clarke and Anandalingam, 1995). Winter (1986) introduced a linear-time algorithm solving the 2-ECSSP on series-parallel graphs. Most of these algorithms rely on polyhedral considerations.

The *k-edge-connected spanning subgraph polyhedron* of G, hereafter denoted by $P_k(G)$, is the convex hull of all the k-edge-connected spanning subgraphs of G. Cornuéjols et al. (1985) gave a system describing $P_2(G)$ for series-parallel graphs.

Vandenbussche and Nemhauser (2005) characterized in terms of forbidden minors the graphs for which this system describes $P_2(G)$. Chopra (1994) described $P_k(G)$ for outerplanar graphs when k is odd. Didi Biha and Mahjoub (1996) extended these results to series-parallel graphs for all $k \geq 2$. By a result of Baïou et al. (2000), the inequalities in these descriptions can be separated in polynomial time, which implies that the k-ECSSP is solvable in polynomial time for series-parallel graphs.

When studying the k-edge-connected spanning subgraphs of a graph G, we can add the constraint that each edge of G can be taken at most once. We denote the corresponding polyhedron by $Q_k(G)$. Barahona and Mahjoub (1995) described $Q_2(G)$ for Halin graphs. Further polyhedral results for the case $k = 2$ have been obtained by Boyd and Hao (1993), Mahjoub (1994), and Mahjoub (1997). Grötschel and Monma (1990) described several basic facets of $Q_k(G)$. Moreover, Fonlupt and Mahjoub (2006) extensively studied the extremal points of $Q_k(G)$ and characterized the class of graphs for which this polytope is described by cut inequalities and $0 \leq x \leq 1$.

The polyhedron $P_1(G)$ is known to be box-TDI for all graphs (Lancini, 2019). For series-parallel graphs, the system given in (Cornuéjols et al. 1985) describing $P_2(G)$ is not TDI. Chen et al. (2009) showed that dividing each inequality by 2 yields a TDI system for such graphs. Actually, they proved that this system is box-TDI if and only if the graph is series-parallel.

Contribution. Our starting point is the result of Chen et al. (2009). First, their result implies that $P_2(G)$ is a box-TDI polyhedron for series-parallel graphs. However, this leaves open the question of the box-TDIness of $P_2(G)$ for non series-parallel graphs. More generally, for which integers k and graphs G is $P_k(G)$ a box-TDI polyhedron? In this paper, we answer this question and prove that, for $k \geq 2$, $P_k(G)$ is a box-TDI polyhedron if and only if G is series-parallel.

By a result of Giles and Pulleyblank (1979), there exists a TDI system with integer coefficients describing $P_k(G)$. For series-parallel graphs, the system provided by Chen et al. (2009) has noninteger coefficients. Moreover, the system given by Didi Biha and Mahjoub (1996) describing $P_k(G)$ when k is even is not TDI.

We provide an integer TDI system for $P_k(G)$ when G is series-parallel and k is even, and we prove that the description of $P_k(G)$ given by Didi Biha and Mahjoub (1996) when k is odd is TDI if and only if the graph is series-parallel.

The box-totally dual integral characterization of $P_k(G)$ implies that these systems are actually box-TDI if and only if G is series-parallel. By definition of box-TDIness, adding $x \leq 1$ to these systems yields box-TDI systems for $Q_k(G)$ for series-parallel graphs.

We mention that, due to length limitation, some of our proofs are sketched.

1   Definitions and Preliminary Results

This section is devoted to the definitions, notation, and preliminary results used throughout the paper.

1.1   Graphs

Let $G = (V, E)$ be a loopless undirected graph. The graph G is 2-connected if it remains connected whenever a vertex is removed. A 2-connected graph is called *trivial* if it is composed of a single edge. The graph obtained from two disjoint graphs by identifying two vertices, one of each graph, is called a *1-sum*. A subset of edges of G is called a *circuit* if it induces a connected graph in which every vertex has degree 2. Given a subset U of V, the *cut* $\delta(U)$ is the set of edges having exactly one endpoint in U. A *bond* is a minimal nonempty cut. Given a partition $\{V_1, \ldots, V_n\}$ of V, the set of edges having endpoints in two distinct V_i's is called *multicut* and is denoted by $\delta(V_1, \ldots, V_n)$. We denote respectively by $\mathcal{M}_G$ and $\mathcal{B}_G$ the set of multicuts and the set of bonds of G. For every multicut M, there exists a unique partition $\{V_1, \ldots, V_{d_M}\}$ of vertices of V such that $M = \delta(V_1, \ldots, V_{d_M})$, and $G[V_i]$—the graph induced by the vertices of V_i—is connected for all $i = 1, \ldots, d_M$; we say that d_M is the *order* of M.

We denote the symmetric difference of two sets S and T by $S \,\Delta\, T$. It is well-known that the symmetric difference of two cuts is a cut.

We denote by K_n the complete graph on n vertices, that is, the simple graph with n vertices and one edge between each pair of distinct vertices.

A graph is *series-parallel* if its 2-connected components can be constructed from an edge by repeatedly adding edges parallel to an existing one, and subdividing edges, that is, replacing an edge by a path of length two. Duffin (1965) showed that series-parallel graphs are those having no K_4-minor. By construction, simple nontrivial 2-connected series-parallel graphs have at least one vertex of degree 2.

Proposition 1. *For a simple nontrivial 2-connected series-parallel graph, at least one of the following holds:*

(a) two vertices of degree 2 are adjacent,
(b) a vertex of degree 2 belongs to a circuit of length 3,
(c) two vertices of degree 2 belong to a same circuit of length 4.

Proof. We proceed by induction on the number of edges. The base case is K_3 for which (a) holds.

Let G be a simple 2-connected series-parallel graph such that for every simple, 2-connected series-parallel graph with fewer edges at least one among (a), (b), and (c) holds. Since G is simple, it can be built from a graph H by subdividing an edge e into a path f, g. Let v be the vertex of degree 2 added with this operation. By the induction hypothesis, either H is not simple, or one among (a), (b), and (c) holds for H.

Let first suppose that H is not simple, then, by G being simple, e is parallel to exactly one edge e_0. Hence, e_0, f, g is a circuit of G length 3 containing v, hence (b) holds for G.

From now on, suppose that H is simple. If (a) holds for H, then it holds for G.

Suppose that (b) holds for H, that is, in H there exists a circuit C of length 3 containing a vertex w of degree 2. Without loss of generality, we suppose that

$e \in C$, as otherwise (b) holds for G. By subdividing e, we obtain a circuit of length 4 containing v and w, and hence (c) holds for G.

At last, suppose that (c) holds for H, that is, H has a circuit C of length 4 containing two vertices of degree 2. Without loss of generality, we suppose that $e \in C$, as otherwise (c) holds for G. By subdividing e, we obtain a circuit of length 5 containing three vertices of degree 2. Then, at least two of them are adjacent, and so (a) holds for G.   □

1.2   Box-Total Dual Integrality

Let $A \in \mathbb{R}^{m \times n}$ be a full row rank matrix. This matrix is *equimodular* if all its $m \times m$ non-zero determinants have the same absolute value. The matrix A is *face-defining for* a face F of a polyhedron $P \subseteq \mathbb{R}^n$ if $\mathrm{aff}(F) = \{x \in \mathbb{R}^n : Ax = b\}$ for some $b \in \mathbb{R}^m$. Such matrices are the *face-defining matrices of* P.

Theorem 1 (Chervet et al. (2020)). *Let P be a polyhedron, then the following statements are equivalent:*

(i) P is box-TDI.
(ii) Every face-defining matrix of P is equimodular.
(iii) Every face of P has an equimodular face-defining matrix.

The equivalence of conditions (ii) and (iii) stems from the following observation.

Observation 1 (Chervet et al. (2020)). *Let F be a face of a polyhedron. If a face-defining matrix of F is equimodular, then so are all face-defining matrices of F.*

Observation 2. *Let $A \in \mathbb{R}^{I \times J}$ be a full row rank matrix, $j \in J$, $\mathbf{c}$ be a column of A, and $\mathbf{v} \in \mathbb{R}^I$. If A is equimodular, then so are:*

(i) $\begin{bmatrix} A\ \mathbf{c} \end{bmatrix}$, (ii) $\begin{bmatrix} A \\ \pm\chi^j \end{bmatrix}$ *if it is full row rank,* (iii) $\begin{bmatrix} A & \mathbf{v} \\ \mathbf{0}^\top & \pm 1 \end{bmatrix}$, *and* (iv) $\begin{bmatrix} A & \mathbf{0} \\ \pm\chi^j & \pm 1 \end{bmatrix}$.

Observation 3 (Chervet et al. (2020)). *Let $P \subseteq \mathbb{R}^n$ be a polyhedron and let $F = \{x \in P : Bx = b\}$ be a face of P. If B has full row rank and $n - \dim(F)$ rows, then B is face-defining for F.*

1.3 k-edge-connected Spanning Subgraph Polyhedron

The *dominant* of a polyhedron P is $\mathrm{dom}(P) = \{x : x = y + z, \text{for } y \in P \text{ and } z \geq 0\}$. Note that $P_k(G)$ is the dominant of the convex hull of all k-edge-connected spanning subgraphs of G that have each edge taken at most k times. Since the dominant of a polyhedron is a polyhedron, $P_k(G)$ is a polyhedron even though it is the convex hull of an infinite number of points.

From now on, $k \geq 2$. Didi Biha and Mahjoub (1996) gave a complete description of $P_k(G)$ for all k, when G is series-parallel.

Theorem 2. *Let G be a series-parallel graph and k be a positive integer. Then, when k is even, $P_k(G)$ is described by:*

$$
(1) \quad \begin{cases} x(D) \geq k \text{ for all cuts } D \text{ of } G, & (1a) \\ x \geq \mathbf{0}, & (1b) \end{cases}
$$

and when k is odd, $P_k(G)$ is described by:

$$
(2) \quad \begin{cases} x(M) \geq \frac{k+1}{2} d_M - 1 \text{ for all multicuts } M \text{ of } G, & (2a) \\ x \geq \mathbf{0}. & (2b) \end{cases}
$$

The incidence vector of a family F of E is the vector χ^F of $\mathbb{Z}^E$ such that e's coordinate is the multiplicity of e in F for all e in E. Since there is a bijection between families and their incidence vectors, we will often use the same terminology for both. Since the incidence vector of a multicut $\delta(V_1, \ldots, V_{d_M})$ is the half-sum of the incidence vectors of the bonds $\delta(V_1), \ldots, \delta(V_{d_M})$, we can deduce an alternative description of $P_{2h}(G)$.

Corollary 1. *Let G be a series-parallel graph and k be a positive even integer. Then $P_k(G)$ is described by:*

$$
(3) \quad \begin{cases} x(M) \geq \frac{k}{2} d_M \text{ for all multicuts } M \text{ of } G, & (3a) \\ x \geq \mathbf{0}. & (3b) \end{cases}
$$

We call constraints (2a) and (3a) *partition constraints*. A multicut M is *tight for a point* of $P_k(G)$ if this point satisfies with equality the partition constraint (2a) (resp. (3a)) associated with M when k is odd (resp. even). Moreover, M is *tight for a face F* of $P_k(G)$ if it is tight for all the points of F.

The following results give some insight on the structure of tight multicuts.

Theorem 3 (Didi Biha and Mahjoub (1996)). *Let $k > 1$ be odd, let x be a point of $P_k(G)$, and let $M = \delta(V_1, \ldots, V_{d_M})$ be a multicut tight for x. Then, the following hold:*

(i) if $d_M \geq 3$, then $x(\delta(V_i) \cap \delta(V_j)) \leq \frac{k+1}{2}$ for all $i \neq j \in \{1, \ldots, d_M\}$.
(ii) $G \setminus V_i$ is connected for all $i = 1, \ldots, d_M$.

Observation 4. *Let M be a multicut of G strictly containing $\delta(v) = \{f, g\}$. If M is tight for a point of $P_k(G)$, then both $M \setminus f$ and $M \setminus g$ are multicuts of G of order $d_M - 1$.*

Chopra (1994) gave sufficient conditions for an inequality to be facet defining. The following proposition is a direct consequence of (Chopra 1994, Theorem 2.4).

Proposition 2. *Let G be a graph having K_4 as a minor and let $k > 1$ be an odd integer. Then, there exist two disjoint nonempty subsets of edges of G, E' and E'', and a rational b such that*

$$\chi^{E'} + 2\chi^{E''} \geq b, \tag{4}$$

is a facet-defining inequality of $P_k(G)$.

Chen et al. (2009) provided a box-TDI system for $P_2(G)$ for series-parallel graphs.

Theorem 4 (Chen et al. (2009)). *The system:*

$$\begin{cases} \frac{1}{2}x(D) \geq 1 & \text{for all cuts } D \text{ of } G, \\ x \geq 0 \end{cases} \tag{5}$$

is box-TDI if and only if G is a series-parallel graph.

This result proves that $P_2(G)$ is box-TDI for all series-parallel graphs, and gives a TDI system describing this polyhedron in this case. At the same time, Theorem 4 is not sufficient to state that $P_2(G)$ is a box-TDI polyhedron if and only if G is series-parallel.

2   Box-TDIness of $P_k(G)$

In this section we show that, for $k \geq 2$, $P_k(G)$ is a box-TDI polyhedron if and only if G is series-parallel.

When $k \geq 2$, $P_k(G)$ is not box-TDI for all graphs as stated by the following lemma.

Lemma 1. *For $k \geq 2$, if $G = (V, E)$ contains a K_4-minor, then $P_k(G)$ is not box-TDI.*

Proof. When k is odd, Proposition 2 shows that there exists a facet-defining inequality that is described by a non equimodular matrix. Thus, $P_k(G)$ is not box-TDI by Statement (ii) of Theorem 1.

We now prove the case when k is even. Since G is connected and has a K_4-minor, there exists a partition $\{V_1, \ldots, V_4\}$ of V such that $G[V_i]$ is connected and $\delta(V_i, V_j) \neq \emptyset$ for all $i < j \in \{1, \ldots, 4\}$. We prove that the matrix T whose three rows are $\chi^{\delta(V_i)}$ for $i = 1, 2, 3$ is a face-defining matrix for $P_k(G)$ which is not equimodular. This will end the proof by Statement (ii) of Theorem 1.

Let e_{ij} be an edge in $\delta(V_i, V_j)$ for all $i < j \in \{1, \ldots, 4\}$. The submatrix of T formed by the columns associated with edges e_{ij} is the following:

$$\begin{array}{c} \\ \chi^{\delta(V_1)} \\ \chi^{\delta(V_2)} \\ \chi^{\delta(V_3)} \end{array} \begin{array}{c} \begin{matrix} e_{12} & e_{13} & e_{23} & e_{14} & e_{24} & e_{34} \end{matrix} \\ \begin{bmatrix} 1 & 1 & 0 & 1 & 0 & 0 \\ 1 & 0 & 1 & 0 & 1 & 0 \\ 0 & 1 & 1 & 0 & 0 & 1 \end{bmatrix} \end{array}$$

The matrix T is not equimodular as the first three columns form a matrix of determinant -2 whereas the last three ones have determinant 1.

To show that T is face-defining, we exhibit $|E| - 2$ affinely independent points of $P_k(G)$ satisfying the partition constraint (3a) associated with the multicut $\delta(V_i)$, that is, $x(\delta(V_i)) = k$, for $i = 1, 2, 3$.

Let $D_1 = \{e_{12}, e_{14}, e_{23}, e_{34}\}$, $D_2 = \{e_{12}, e_{13}, e_{24}, e_{34}\}$, $D_3 = \{e_{13}, e_{14}, e_{23}, e_{24}\}$ and $D_4 = \{e_{14}, e_{24}, e_{34}\}$. First, we define the points $S_j = \sum_{i=1}^{4} k\chi^{E[V_i]} + \frac{k}{2}\chi^{D_j}$, for $j = 1, 2, 3$, and $S_4 = \sum_{i=1}^{4} k\chi^{E[V_i]} + k\chi^{D_4}$. Note that they are affinely independent.

Now, for each edge $e \notin \{e_{12}, e_{13}, e_{14}, e_{23}, e_{24}, e_{34}\}$, we construct the point S_e as follows. When $e \in E[V_i]$ for some $i = 1, \dots, 4$, we define $S_e = S_4 + \chi^e$. Adding the point S_e maintains affine independence as S_e is the only point not satisfying $x_e = k$. When $e \in \delta(V_i, V_j)$ for some i, j, we define $S_e = S_\ell - \chi^{e_{ij}} + \chi^e$, where S_ℓ is S_1 if $e \in \delta(V_1, V_4) \cup \delta(V_2, V_3)$ and S_2 otherwise. Affine independence comes because S_e is the only point involving e. □

Theorem 5. *For $k \geq 2$, $P_k(G)$ is a box-TDI polyhedron if and only if G is series-parallel.*

Proof. Necessity stems from Lemma 1. Let us now prove sufficiency. When $k = 2$, the box-TDIness of System (5) has been shown by Chen et al. (2009). This implies box-TDIness for all even k: multiplying the right-hand side of a box-TDI system by a positive rational preserves its box-TDIness (Schrijver, 1998, Section 22.5). The system obtained by multiplying the right-hand side of System (5) by $\frac{k}{2}$ describes $P_k(G)$ when k is even. Hence, the latter is a box-TDI polyhedron.

The rest of the proof is dedicated to the case where $k = 2h+1$ for some $h \geq 1$. For this purpose, we prove that every face of $P_{2h+1}(G)$ admits an equimodular face-defining matrix. The characterization of box-TDIness given in Theorem 1 concludes. We proceed by induction on the number of edges of G.

As a base-case of the induction we consider the series-parallel graph G consisting of two vertices connected by a single edge. Then, $P_{2h+1}(G) = \{x \in \mathbb{R}_+ : x \geq 2h + 1\}$ is box-TDI.

(1-sum). Let G be the 1-sum of two series-parallel graphs $G^1 = (W^1, E^1)$ and $G^2 = (W^2, E^2)$. By induction, there exist two box-TDI systems $A^1 y \geq b^1$ and $A^2 z \geq b^2$ describing respectively $P_{2h+1}(G^1)$ and $P_{2h+1}(G^2)$. If v is the vertex of G obtained by the identification, $G \setminus v$ is not connected, hence, by Statement (ii) of Theorem 3, a multicut M of G is tight for a face of $P_{2h+1}(G)$ only if $M \subseteq E^i$ for some $i = 1, 2$. It follows that for every face F of $P_{2h+1}(G)$ there exist two faces F^1 and F^2 of $P_{2h+1}(G^1)$ and $P_{2h+1}(G^2)$ respectively, such that $F = F^1 \times F^2$. Then $P_{2h+1}(G) = \{(y, z) \in \mathbb{R}_+^{E^1} \times \mathbb{R}_+^{E^2} : A^1 y \geq b^1, A^2 z \geq b^2\}$ and so it is box-TDI.

(Parallelization). Let now G be obtained from a series-parallel graph H by adding an edge g parallel to an edge f of H and suppose that $P_{2h+1}(H)$ is

box-TDI. Note that $P_{2h+1}(G)$ is obtained from $P_{2h+1}(H)$ by duplicating f's column and adding $x_g \geq 0$. Hence, by (Chen et al. 2009, Lemma 3.1), $P_{2h+1}(G)$ is a box-TDI polyhedron.

(Subdivision). Let $G = (V, E)$ be obtained by subdividing an edge uw of a series-parallel graph $G' = (V', E')$ into a path of length two uv, vw. By contradiction, suppose there exists a non-empty face $F = \{x \in P_{2h+1}(G) : A_F x = b_F\}$ such that A_F is a face-defining matrix of F which is not equimodular. Take such a face with maximum dimension. Then, every face-defining submatrix of A_F is equimodular. We may assume that A_F is given by the left-hand side of a subset of constraints of System (2). We denote by $\mathcal{M}_F$ the set of multicuts associated with the left-hand sides of constraints (2a) appearing in A_F, and by $\mathcal{E}_F$ the set of edges associated with the nonnegativity constraints (2b) appearing in A_F.

Claim A. $\mathcal{E}_F = \emptyset$.

Proof. Suppose there exists an edge $e \in \mathcal{E}_F$. Let $H = G \setminus e$ and let $A_{F_H} x = b_{F_H}$ be the system obtained from $A_F x = b_F$ by removing the column and the nonnegativity constraint associated with e. The matrix A_F being of full row rank, so is A_{F_H}. Since $M \setminus e$ is a multicut of H for all M in $\mathcal{M}_F$, the set $F_H = \{x \in P_{2h+1}(H) : A_{F_H} x = b_{F_H}\}$ is a face of $P_{2h+1}(H)$. Moreover, deleting e's coordinate of $\mathrm{aff}(F)$ gives $\mathrm{aff}(F_H)$ so A_{F_H} is face-defining for F_H. By the induction hypothesis, A_{F_H} is equimodular, and hence so is A_F by Observation 2-(iii).   □

Claim B. *For all $e \in \{uv, vw\}$, at least one multicut of $\mathcal{M}_F$ different from $\delta(v)$ contains e.*

Proof. Suppose that uv belongs to no multicut of $\mathcal{M}_F$ different from $\delta(v)$.

First, suppose that $\delta(v)$ does not belong to $\mathcal{M}_F$. Then, the column of A_F associated with uv is zero. Let A'_F be the matrix obtained from A_F by removing this column. Every multicut of G not containing uv is a multicut of G' (relabelling vw by uw), so the rows of A'_F are associated with multicuts of G'. Thus, $F' = \{x \in P_k(G') : A'_F x = b_F\}$ is a face of $P_{2h+1}(G')$. Removing uv's coordinate from the points of F gives a set of points of F' of affine dimension at least $\dim(F) - 1$. Since A'_F has the same rank of A_F and one column less than A_F, then A'_F is face-defining for F' by Observation 3. By induction hypothesis, A'_F is equimodular, hence so is A_F.

Suppose now that $\delta(v)$ belongs to $\mathcal{M}_F$. Then, the column of A_F associated with uv has zeros in each row but $\chi^{\delta(v)}$. Let $A^\star_F x = b^\star_F$ be the system obtained from $A_F x = b_F$ by removing the row associated with $\delta(v)$. Then $F^\star = \{x \in P_k(G) : A^\star_F x = b^\star_F\}$ is a face of $P_k(G)$ of dimension $\dim(F) + 1$. Indeed, it contains F and $z + \alpha \chi^{uv}$ for every point z of F and $\alpha > 0$. Hence, $A^\star_F$ is face-defining for $F^\star$. This matrix is equimodular by the maximality assumption on F, and so is A_F by Observation 2-(iv).   □

Claim C. $|M \cap \delta(v)| \neq 1$ *for every multicut $M \in \mathcal{M}_F$.*

Proof. Suppose there exists a multicut M tight for F such that $|M \cap \delta(v)| = 1$. Without loss of generality, suppose that M contains uv and not vw. Then, $F \subseteq \{x \in P_{2h+1}(G) : x_{vw} \geq x_{uv}\}$ because of the partition inequality (2a) associated with the multicut $M \, \Delta \, \delta(v)$. Moreover, the partition inequality associated with $\delta(v)$ and the integrality of $P_{2h+1}(G)$ imply $F \subseteq \{x \in P_{2h+1}(G) : x_{vw} \geq h+1\}$. The proof is divided into two cases.

Case 1. $F \subseteq \{x \in P_{2h+1}(G) : x_{vw} = h+1\}$. We prove this case by exhibiting an equimodular face-defining matrix for F. By Observation 1, this implies that A_F is equimodular, which contradicts the assumption on F.

Equality $x_{vw} = h+1$ can be expressed as a linear combination of rows of $A_F x = b_F$. Let $A'_F x = b'_F$ denote the system obtained by replacing a row of $A_F x = b_F$ by $x_{vw} = h+1$ in such a way that the underlying affine space remains unchanged. Denote by $\mathcal{N}$ the set of multicuts of $\mathcal{M}_F$ containing vw but not uv. If $\mathcal{N} \neq \emptyset$, then let N be in $\mathcal{N}$. We now modify the system $A'_F x = b'_F$ by performing the following operations.

1. Every row associated with a multicut M strictly containing $\delta(v)$ is replaced by the partition constraint (2a) associated with $M \setminus vw$ set to equality.
2. Whenever $\delta(v) \in \mathcal{M}_F$, replace the row associated with $\delta(v)$ by the box constraint $x_{uv} = h$.
3. Replace every row associated with $M \in \mathcal{N} \setminus N$ by the partition constraint (2a) associated with $M \, \Delta \, \delta(v)$ set to equality.
4. Whenever $\mathcal{N} \neq \emptyset$, replace the row associated with N by the box constraint $x_{uv} = h+1$.

These operations do not modify the underlying affine space. Indeed, in Operation 1, $M \setminus vw$ is tight for F because of Observation 4 and $F \subseteq \{x \in P_{2h+1}(G) : x_{vw} = h+1\}$. Operation 2 is applied only if $F \subseteq \{x \in P_{2h+1}(G) : x_{uv} = h\}$. Operations 3 and 4 are applied only if $\mathcal{N} \neq \emptyset$, which implies that $F \subseteq \{x \in P_{2h+1}(G) : x_{uv} = h+1\}$ because of the constraint (2a) associated with $N \, \Delta \, \delta(v)$ and $F \subseteq \{x \in P_{2h+1}(G) : x_{vw} \geq x_{uv}\}$. Note that Operations 2 and 4 cannot be applied both, hence the rank of the matrix remains unchanged.

Let $A''_F x = b''_F$ be the system obtained by removing the row $x_{vw} = h+1$ from $A'_F x = b'_F$. By construction, $A''_F x = b''_F$ is composed of constraints (2a) set to equality and possibly $x_{uv} = h$ or $x_{uv} = h+1$. Moreover, the column of A''_F associated with vw is zero. Let $F'' = \{x \in P_{2h+1}(G) : A''_F x = b''_F\}$. For every point z of F and $\alpha \geq 0$, $z + \alpha \chi^{vw}$ belongs to F'' because the column of A''_F associated with vw is zero, and $z + \alpha \chi^{vw} \in P_{2h+1}(G)$. This implies that $\dim(F'') \geq \dim(F) + 1$.

If F'' is a face of $P_{2h+1}(G)$, then A''_F is face-defining for F'' by Observation 3 and by A'_F being face-defining for F. By the maximality assumption on F, A''_F is equimodular, and hence so is A'_F by Observation 2-(ii).

Otherwise, by construction, $F'' = F^\star \cap \{x \in \mathbb{R}^E : x_{uv} = t\}$ where $F^\star$ is a face of $P_{2h+1}(G)$ strictly containing F and $t \in \{h, h+1\}$. Therefore, there exists a face-defining matrix of F'' given by a face-defining matrix of $F^\star$ and the row χ^{uv}. Such a matrix is equimodular by the maximality assumption of F and

Observation 2-(ii). Hence, A_F'' is equimodular by Observation 1, and so is A_F' by Observation 2-(ii).

Case 2. $F \not\subseteq \{x \in P_{2h+1}(G) : x_{vw} = h + 1\}$. Thus, there exists $z \in F$ such that $z_{vw} > h + 1$. By Claim B, there exists a multicut $N \neq \delta(v)$ containing vw which is tight for F. By Statement (i) of Theorem 3, the existence of z implies that N is a bond. Thus, $uv \notin N$ and $F \subseteq \{x \in P_{2h+1}(G) : x_{vw} = x_{uv}\}$. Consequently, $L = N \triangle \delta(v)$ is also a bond tight for F. Moreover, N is the unique multicut tight for F containing vw. Suppose indeed that there exists a multicut B containing vw tight for F. Then, B is a bond by Statement (i) of Theorem 3 and the existence of z. Moreover, $B \triangle N$ is a multicut not containing vw. This implies that no point x of F satisfies the partition constraint associated with $B \triangle N$ because $x(B \triangle N) = x(B) + x(N) - 2x(B \cap N) = 2(2h+1) - 2x(B \cap N) \leq 4h + 2 - 2x_{vw} \leq 2h$, a contradiction.

Consider the matrix $A_F^\star$ obtained from A_F by removing the row associated with N. Matrix $A_F^\star$ is a face-defining matrix for a face $F^\star \supseteq F$ of $P_{2h+1}(G)$ because $F^\star$ contains F and $z + \alpha\chi^{uv}$ for every point z of F and $\alpha > 0$. By the maximality assumption, the matrix $A_F^\star$ is equimodular. Let B_F be the matrix obtained from A_F by replacing the row χ^N by the row $\chi^N - \chi^L$. Then, B_F is face-defining for F. Moreover, B_F is equimodular by Observation 2-(iv)—a contradiction. $\square$

Let $A_F'x = b_F'$ be the system obtained from $A_Fx = b_F$ by removing uv's column from A_F and subtracting $h + 1$ times this column to b_F. We now show that $\{x \in P_{2h+1}(G') : A_F'x = b_F'\}$ is a face of $P_{2h+1}(G')$ if $\delta(v) \notin \mathcal{M}_F$, and $P_{2h+1}(G') \cap \{x : x_{uw} = h\}$ otherwise. Indeed, consider a multicut M in $\mathcal{M}_F$. If $M = \delta(v)$, then the row of $A_F'x = b_F'$ induced by M is nothing but $x_{uw} = h$. Otherwise, by Observation 4 and Claim C, the set $M \setminus uv$ is a multicut of G' (relabelling vw by uw) of order d_M if $uv \notin M$ and $d_M - 1$ otherwise. Thus, the row of $A_F'x = b_F'$ induced by M is the partition constraint (2a) associated with $M \setminus uv$ set to equality.

By construction, A_F' has full row rank and one column less than A_F. We prove that A_F' is face-defining by exhibiting $\dim(F)$ affinely independent points of $P_{2h+1}(G')$ satisfying $A_F'x = b_F'$. Because of the integrality of $P_{2h+1}(G)$, there exist $n = \dim(F) + 1$ affinely independent integer points $z^1, \ldots, z^n$ of F. By Claim C, every multicut in $\mathcal{M}_F$ contains either both uv and vw or none of them. Then, Claim B and Statement (i) of Theorem 3 imply that $F \subseteq \{x \in \mathbb{R}^E : x_{uv} \leq h + 1, x_{vw} \leq h + 1\}$. Combined with the partition inequality $x_{uv} + x_{vw} \geq 2h + 1$ associated with $\delta(v)$, this implies that at least one of z_{uv}^i and z_{vw}^i is equal to $h + 1$ for $i = 1, \ldots, n$. Since exchanging the uv and vw coordinates of any point of F gives a point of F by Claim C, the hypotheses on $z^1, \ldots, z^n$ are preserved under the assumption that $z_{uv}^i = h + 1$ for $i = 1, \ldots, n - 1$. Let $y^1, \ldots, y^{n-1}$ be the points obtained from $z^1, \ldots, z^{n-1}$ by removing uv's coordinate. Since every multicut of G' is a multicut of G with the same order, $y^1, \ldots, y^{n-1}$ belong to $P_{2h+1}(G')$. By construction, they satisfy $A_F'x = b_F'$ so they belong to a face of

$P_{2h+1}(G')$ or $P_{2h+1}(G') \cap \{x : x_{uw} = h\}$. This implies that A'_F is a face-defining matrix of $P_{2h+1}(G')$ if $\delta(v) \notin \mathcal{M}_F$, and $P_{2h+1}(G') \cap \{x : x_{uw} = h\}$ otherwise.

By induction, $P_{2h+1}(G')$ is a box-TDI polyhedron and hence so is $P_{2h+1}(G') \cap \{x : x_{uw} = h\}$. Hence, A'_F is equimodular by Theorem 1. Since the columns of A_F associated with uv and vw are equal, Observation 2-(i) implies that A_F is equimodular—a contradiction to its assumption of non-equimodularity.   □

3   TDI Systems for $P_k(G)$

Let G be a series-parallel graph. In this section, we study the total dual integrality of systems describing $P_k(G)$. Due to length limitation, some of the proofs of the results below are omitted. They can be found in the appendix.

The following result characterizes series-parallel graphs in terms of TDIness of System (2).

Theorem 6. *For $k > 1$ odd and integer, System (2) is TDI if and only if G is series-parallel.*

Proof (sketch). We first prove that if G is not series-parallel, then System (2) is not TDI. Indeed, every TDI system with integer right-hand side describes an integer polyhedron (Edmonds and Giles, 1977), but, when G has a K_4-minor, System (2) describes a noninteger polyhedron (Chopra, 1994).

Let us sketch the other direction of the proof, that is, when the graph is series-parallel. We proceed by contradiction and consider a minimal counterexample G. First, we show that G is simple and 2-connected. Then, we show that G contains none of the following configurations.

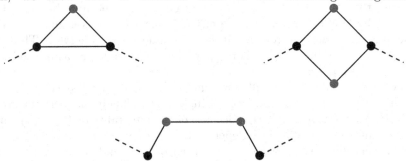

Since the red vertices in the figure above have degree 2 in G, this contradicts Proposition 1.   □

For $k > 1$, by Theorem 5, $P_k(G)$ is a box-TDI polyhedron if and only if G is series-parallel. Together with (Cook 1986, Corollary 2.5), this implies the following.

Corollary 2. *For $k > 1$ odd and integer, System (2) is box-TDI if and only if G is series-parallel.*

The following theorem gives a TDI system for $P_k(G)$ when G is series-parallel and k is even.

Theorem 7. *For a series-parallel graph G and $k > 1$ even, System (3) is TDI.*

The proof of Theorem 7 is based on the characterization of TDIness by means of Hilbert bases that follows. A set of vectors $\{v^1, \ldots, v^k\}$ is a *Hilbert basis* if each integer vector that is a nonnegative combination of $v^1, \ldots, v^k$ can be expressed as a nonnegative integer combination of them. The link between Hilbert basis and TDIness is due to the following result (Schrijver 1998, Theorem 22.5): a system $Ax \geq b$ is TDI if and only if, for each minimal face F of $P = \{x : Ax \geq b\}$, the rows of A associated with constraints tight for F form a Hilbert basis.

Proof (sketch). We only prove the case $h = 1$ since multiplying the right hand side of a system by a positive constant preserves its TDIness (Schrijver 1998, Section 22.5). By the structure of k-edge-connected subgraphs, it is enough to prove the result for 2-connected graphs. We proceed by induction and use the constructive characterization of 2-connected series-parallel graphs. We consider as base case K_2, for which System (3) is TDI.

Let H be a graph for which System (3) is TDI, and let G be obtained by applying a series-parallel operation to H. If this operation is the addition of an edge parallel to an existing one, then the TDIness of System (3) stems from Lemma 3.1 of Chen et al. (2009).

Suppose now that G is obtained from H by subdividing an edge into a path of length 2. For each vertex v of $P_2(G)$, the set of constraints active for v can be built from the set of constraints active for a vertex of $P_2(H)$. The constraints active for such vertex form a Hilbert basis because System (3) is TDI for H. Since the construction preserves this property, the constraints active for v form a Hilbert basis, for all v vertex of $P_2(G)$. This finishes the proof by (Schrijver 1998, Theorem 22.5). $\square$

Again, by Theorem 5 and (Cook 1986, Corollary 2.5), we have the following.

Corollary 3. *For k positive and even, System (3) is box-TDI if and only if G is series-parallel.*

By Corollaries 2 and 3, and by the definition of box-TDI systems, adding $x \leq 1$ to Systems (2) and (3) preserves box-TDIness. For series-parallel graphs, this provides box-TDI systems for the combinatorial version of the k-ECSSP, which is the version where each edge can be taken at most once.

4   Conclusions

In this paper, we studied strong integrality properties of the k-edge-connected spanning subgraph polyhedron $P_k(G)$. We first showed that, for every $k \geq 2$, $P_k(G)$ is a box-TDI polyhedron if and only if G is a series-parallel graph. This result extends and strengthens the work of Chen et al. (2009), who provided a box-TDI system when $k = 2$. When G is series-parallel and k is even, the box-total dual integrality of $P_k(G)$ stems from their result. For k odd, we used a different approach, which relies on the recent characterization of box-TDI polyhedra given by Chervet et al. (2020).

Moreover, we showed the TDIness of the system given by Didi Biha and Mahjoub (1996) describing $P_k(G)$ when G is series-parallel and k odd. When k is even, the only known (box-)TDI system describing $P_k(G)$ has noninteger coefficients—see Theorem 4. We provided a system with integer coefficients describing $P_k(G)$ for all k even, that is TDI if G is series-parallel. By our characterization of the box-TDIness of $P_k(G)$, these systems are box-TDI if and only if G is series-parallel.

Further, we mention that, for series-parallel graphs, our results imply that $Q_k(G)$ is a box-TDI polytope and provide box-TDI systems describing this polytope.

Acknowledgments. The authors wish to express their appreciation to the anonymous referees for their precious comments which helped to improve the presentation of the paper.

References

Baïou, M., Barahona, F., Mahjoub, A.R.: Separation of partition inequalities. Math. Oper. Res. **25**(2), 243–254 (2000)

Barahona, F., Mahjoub, A.R.: On two-connected subgraph polytopes. Discrete Math. **147**(1–3), 19–34 (1995)

Barbato, M., Grappe, R., Lacroix, M., Lancini, E., Wolfler Calvo, R.: The Schrijver system of the flow cone in series-parallel graphs. Discrete Appl. Math. (2020)

Boyd, S.C., Hao, T.: An integer polytope related to the design of survivable communication networks. SIAM J. Discrete Math. **6**(4), 612–630 (1993)

Chen, X., Ding, G., Zang, W.: A characterization of box-mengerian matroid ports. Math. Oper. Res. **33**(2), 497–512 (2008)

Chen, X., Ding, G., Zang, W.: The box-TDI system associated with 2-edge connected spanning subgraphs. Discrete Appl. Math. **157**(1), 118–125 (2009)

Chervet, P., Grappe, R., Robert, L.: Box-total dual integrality, box-integrality, and equimodular matrices. Math. Program. (2020). https://doi.org/10.1007/s10107-020-01514-0

Chopra, S.: The k-edge-connected spanning subgraph polyhedron. SIAM J. Discrete Math. **7**(2), 245–259 (1994)

Clarke, L.W., Anandalingam, G.: A bootstrap heuristic for designing minimum cost survivable networks. Comput. Oper. Res. **22**(9), 921–934 (1995)

Cook, W.: On box totally dual integral polyhedra. Math. Program. **34**(1), 48–61 (1986)

Cornaz, D., Magnouche, Y., Mahjoub, A.R.: On minimal two-edge-connected graphs. In: 2014 International Conference on Control, Decision and Information Technologies (CoDIT), pp. 251–256. IEEE (2014)

Cornaz, D., Grappe, R., Lacroix, M.: Trader multiflow and box-TDI systems in series-parallel graphs. Discrete Optim. **31**, 103–114 (2019)

Cornuéjols, G., Fonlupt, J., Naddef, D.: The traveling salesman problem on a graph and some related integer polyhedra. Math. Program. **33**(1), 1–27 (1985)

Cunningham, W.H., Marsh, A.B.: A primal algorithm for optimum matching. In: Balinski, M.L., Hoffman, A.J. (eds.) Polyhedral Combinatorics, pp. 50–72. Springer, Heidelberg (1978). https://doi.org/10.1007/BFb0121194

Didi Biha, M., Mahjoub, A.R.: k-edge connected polyhedra on series-parallel graphs. Oper. Res. Lett. **19**(2), 71–78 (1996)

Ding, G., Tan, L., Zang, W.: When is the matching polytope box-totally dual integral? Math. Oper. Res. **43**(1), 64–99 (2017)

Ding, G., Zang, W., Zhao, Q.: On box-perfect graphs. J. Comb. Theory Ser. B **128**, 17–46 (2018)

Duffin, R.J.: Topology of series-parallel networks. J. Math. Anal. Appl. **10**(2), 303–318 (1965)

Edmonds, J., Giles, R.: A min-max relation for submodular functions on graphs. In: Annals of Discrete Mathematics, vol. 1, pp. 185–204. Elsevier (1977)

Edmonds, J., Giles, R.: Total dual integrality of linear inequality systems. In: Progress in Combinatorial Optimization, pp. 117–129. Academic Press (1984). ISBN 978-0-12-566780-7

Erickson, R.E., Monma, C.L., Veinott Jr., A.F.: Send-and-split method for minimum-concave-cost network flows. Math. Oper. Res. **12**(4), 634–664 (1987)

Fonlupt, J., Mahjoub, A.R.: Critical extreme points of the 2-edge connected spanning subgraph polytope. Math. Program. **105**(2–3), 289–310 (2006)

Gabow, H.N., Goemans, M.X., Tardos, É., Williamson, D.P.: Approximating the smallest k-edge connected spanning subgraph by LP-rounding. Netw. Int. J. **53**(4), 345–357 (2009)

Garey, M.R., Johnson, D.S.: Computers and Intractability: A Guide to the Theory of NP-Completeness. W. H. Freeman & Co., New York (1979). ISBN 0716710447

Giles, F.R., Pulleyblank, W.R.: Total dual integrality and integer polyhedra. Linear Algebra Appl. **25**, 191–196 (1979)

Grötschel, M., Monma, C.L.: Integer polyhedra arising from certain network design problems with connectivity constraints. SIAM J. Discrete Math. **3**(4), 502–523 (1990)

Grötschel, M., Monma, C.L., Stoer, M.: Computational results with a cutting plane algorithm for designing communication networks with low-connectivity constraints. Oper. Res. **40**(2), 309–330 (1992)

Lancini, E.: TDIness and Multicuts. Ph.D. thesis, Université Sorbonne Paris Nord (2019)

Mahjoub, A.R.: Two-edge connected spanning subgraphs and polyhedra. Math. Program. **64**(1–3), 199–208 (1994)

Mahjoub, A.R.: On perfectly two-edge connected graphs. Discrete Math. **170**(1–3), 153–172 (1997)

Schrijver, A.: Theory of Linear and Integer Programming. Wiley, New York (1998)

Vandenbussche, D., Nemhauser, G.L.: The 2-edge-connected subgraph polyhedron. J. Comb. Optim. **9**(4), 357–379 (2005)

Winter, P.: Generalized steiner problem in series-parallel networks. J. Algorithms **7**(4), 549–566 (1986)

A Polyhedral Study for the Buy-at-Bulk Facility Location Problem

Chaghoub Soraya[1](✉) and Ibrahima Diarrassouba[2]

[1] School of Mathematical Science and Institute of Mathematics, Nanjing Normal University, Nanjing 210023, China
chaghoubsoraya@yahoo.fr

[2] Normandie Univ, UNIHAVRE, LMAH, FR CNRS-3335, 76600 Le Havre, France

Abstract. In this paper, we are interested in Buy-at-Bulk Facility Location problem which arises in network design. The problem has been mainly studied from approximation algorithms perspective, and to the best of our knowledge, only [7] has developed an exact algorithm for the problem. In this work, we address the problem from a polyhedral point of view. First, we give an integer programming formulation which, contrarily to the those in previous works, does not use flow or path variables. Then, we investigate the structure of the polyhedron associated with this formulation and introduce several classes of valid inequalities which defines facets of the polyhedron.

Keywords: Facility location · Buy-at-Bulk · Network design · Polyhedra

1 Introduction

The Buy-at-Bulk facility location problem (BBFLP for short) is defined by an undirected graph $G = (V, E)$, where V denotes the set of nodes and E denotes the set of edges. We are given a set of clients $D \subseteq V$ and a set of facilities $H \subseteq V$. Each client $j \in D$ has a positive demand d_j. Each facility $h \in H$ is associated with an opening cost μ_h. We also have a set K of different cable types. Each cable type $k \in K$ has a capacity u_k and a set-up cost per unit length denoted by γ_k. Finally, for each edge $e \in E$ we consider a length $l_e \in \mathbb{Z}^+$. We assume that the cable types satisfy the so-called economies of scales in that if we let $K = \{1, ..., |K|\}$ such that $u_1 \leq u_2 \leq ... \leq u_K, \gamma_1 \leq \gamma_2 \leq ... \leq \gamma_K$, and $\gamma_1/u_1 \geq \gamma_2/u_2 \geq ... \geq \gamma_K/u_K$. A solution of the BBFLP consists in choosing

- a set of facilities to open,
- an assignment of each client to exactly one opened facility,
- a number of cables of each type to be installed on each edge of the graph in order to route the demands from each client to the facility to which it is assigned.

© Springer Nature Switzerland AG 2020
M. Baïou et al. (Eds.): ISCO 2020, LNCS 12176, pp. 42–53, 2020.
https://doi.org/10.1007/978-3-030-53262-8_4

The BBFLP is closely related to the facility location problem and network loading problem. It has many applications in telecommunications and transportation network design. It is not hard to see that the BBFLP contains the Facility Location problem and hence it is NP-hard.

In the literature, the BBFLP was mainly studied in the perspective of approximation algorithms. It was first studied by Meyrson et al. [17] who considered a cost-distance problem and present the BBFLP as a special case of this latter problem. In their work, they provided an $O(\log |D|)$ approximation algorithm.

Ravi et al. [19] gave an $O(k)$ approximation algorithm for the BBFLP where k is the number of cable types. Later, Friggst et al. [13] considered an integer programming formulation for the BBFLP, and showed that this formulation has an integrality gap of $O(k)$. They also considered the variant of the BBFLP where the opened facilities must be connected and gave an integrality gap of $O(1)$. Recently, Bley et al. [7] presented the first exact algorithm for the problem. They introduced a path-based formulation for the problem and compare it with a compact flow-based formulation. They also design an exact branch-and-price-and-cut algorithm for solving the path-based formulation.

As mentioned before, the BBFLP is related to the Network Loading Problem (NLP) and the Facility Location Problem (FLP). Both problems have received a lot of attention. Concerning, the NLP, Magnanti et al. [14] studied the NLP from a polyhedral point of view. They introduced some classes of valid inequalities and devised a Branch-and-Cut algorithm. In [15], Magnanti et al. considered the NLP with two cable types and some particular graphs and gave a complete description of the associated polyhedron in these cases. Bienstock et al. [6] studied the NLP with two cable types with possible extension to more than three cable types. Barahona [9] addressed the same problem, he used a relaxation without flow variables, this relaxation is based on cut condition for multicommodity flows. Gülnük [11] gave a branch and cut algorithm using spanning tree inequalities and mixed integer rounding inequalities. Agarwal [3] has introduced 4-partition based facets. Agarwal [4] extended his previous work and get a complete description of the 4-node network design problem. Raacker et al. [18] have extended the polyhedral results for cut-based inequalities for network design problem with directed, bidirected and undirected link-capacity models. Agarwal [5] developed the total-capacity inequalities, one-two inequalities and spanning trees inequalities based on a p-partition of the graph and discuss conditions under which these inequalities are facet-defining.

Several authors have solved the problem using benders partitioning based approaches. Avella et al. [2] discussed high rounded metric inequalities for this problem, based on these inequalities they gave a computational results. Mattia [16] used the same approach for designing networks with two layers. Mattia [16] extended the results of Avella et al. [2] by using bi-level programming and reports improved computational results.

Another related problem to BBFLP is facility location problem, the goal of this problem is to open a set of facilities, and assign each client to its nearest open facility, such that the total incurred cost (opening facility cost and assignment cost) is minimized. The BBFLP is a combination of facility location problem and buy-at-bulk in network design problem.

This paper is structured as follows. In Sect. 2, we give an integer programming formulation of the BBFLP and study the associated polyhedron. In Sect. 3, we give some facet defining inequalities from facility location problem and in Sect. 4, we introduce a new family of facet defining inequalities. Finally, in Sect. 5, we give some concluding remarks.

2   Integer Programming Formulation and Polyhedron

Now, we give the so-called cut formulation for the BBFLP. This formulation can be obtained by slightly modifying the flow-based formulation introduced by [7] and projecting out the flow variables. The cut formulation is given below. Variable t_j^h equals 1 if the client j is assigned to facility h, for all $j \in D$ and $h \in H$, and x_e^k is the number of cable of type k installed on edge e, for all $e \in E$ and $k \in K$.

$$\min \ \sum_{h \in H} \mu_h y_h + \sum_{e \in E} \sum_{k \in K} \gamma^k l_e^k x_e^k$$

$$\sum_{h \in H} t_j^h = 1, \qquad\qquad\qquad \text{for all } j \in D \qquad (1)$$

$$t_j^h \leq y_h, \qquad\qquad\qquad \text{for all } h \in H, j \in D$$
$$\qquad\qquad\qquad\qquad\qquad h \in H, j \in D \qquad (2)$$

$$\sum_{e \in \delta(W)} \sum_{k \in K} u_k x_e^k \geq \sum_{j \in W \cap D} \sum_{h \in H \cap \overline{S}} t_j^h d_j + \sum_{j \in W \cap D} \sum_{h \in H \cap S} t_j^h d_j, \text{ for all } W \subseteq D,$$
$$\qquad\qquad\qquad\qquad\qquad S \subseteq V, \overline{S} \subseteq V \backslash S \quad (3)$$

$$t_j^h \geq 0, \qquad\qquad\qquad \text{for all } h \in H, \ j \in D, \quad (4)$$

$$y_h \leq 1, \qquad\qquad\qquad \text{for all } h \in H, \qquad\quad (5)$$

$$x_e^k \geq 0, \qquad\qquad\qquad \text{for all } k \in K, e \in E, \quad (6)$$

$$t_j^h \in \{0, 1\}, \qquad\qquad\qquad \text{for all } h \in H, j \in D, \quad (7)$$

$$y_h \in \{0, 1\}, \qquad\qquad\qquad \text{for all } h \in H, \qquad\quad (8)$$

$$x_e^k \in \mathbb{Z}^+, \qquad\qquad\qquad \text{for all } k \in K, e \in E. \quad (9)$$

The constraints (1) impose that each client must be assigned to exactly one facility. Constraints (2) are the linking constraints and state that the clients can not be assigned to not open facilities. Constraints (3) are the so-called cut-set inequalities ensuring that the capacity on the edges of the graph is enough for routing all the demands. Constraints (4), (5) and (6) are trivial constraints. Constraints (7), (8) and (9) are the integrality constraints.

In the remain of this paper we focus on the cut formulation.

Let $Q = \{(x, y, t) \in \mathbb{R}^{|E||K|} \times \mathbb{R}^{|H|} \times \mathbb{R}^{|H||D|}$ such that (x, y, t) satisfying (1)–(9)$\}$. In the following, we give the dimension of the polyhedron and show that all the trivial inequalities define facets.

Theorem 1. $dim(Q) = |E||K| + |D||H| + |H| - |D|$ *if and only if every connected component of* G *contains at least two facilities.*

Theorem 2. *The following constraints define facets of* Q

1. $t_j^h \geq 0$, for all $j \in D, h \in H$,
2. $t_j^h \leq y_h$, for all $j \in D, h \in H$,
3. $y_h \geq 1$, for all $h \in H$,
4. $x_e^k \geq 0$, for all $k \in K, e \in E$.

3 Facets from Facility Location Problem

One can easily see that the projection of Q on variables y and t corresponds to the solutions of a FLP. Thus every valid inequality (resp. facet) for the FLP polytope, in the space of y and t variables is also valid (resp. facet) for Q. Let l_{jh} be the cost of assigning client j to facility h, recall that the FLP is formulated as follows.

$$\min \sum_{j \in D} \sum_{h \in H} l_{jh} t_j^h + \sum_{h \in H} \mu_h y_h$$

$$\sum_{h \in H} t_j^h = 1, \quad j \in D \tag{10}$$

$$t_j^h \leq y_h, \quad h \in H, j \in D \tag{11}$$

$$t_j^h \in \{0,1\}, \ h \in H, j \in D \tag{12}$$

$$y_i \in \{0,1\} \ h \in H. \tag{13}$$

In what follows, we give some valid inequalities for the FLP which, from the above remarks, are also valid for the BBFLP. Note that under some conditions, these inequalities define facets of Q. For more details on valid inequalities and facets associated with FLP, the reader can refer to [10]. In particularly, the following inequalities are valid for facility location problem.

Circulant and Odd Cycle Inequalities. Cornuejols et al. [8] introduced the following. Let p, q be integers satisfying $2 \leq q < p \leq m$ and $p \leq n$, p is not multiple of q, $s_1, ..., s_p$ be distinct facilities, $m_1, ..., m_p$ be distinct clients, all the indices are modulo p the following inequality is valid for facility location problem.

$$\sum_{i=1}^{p} \sum_{j=i}^{i+q-1} t(s_i, m_j) \leq \sum_{h=1}^{p} y(s_i) + p - \lceil p/q \rceil \tag{14}$$

Where $t(s_i, m_j) = \sum_{i \in s_i} \sum_{i \in m_j} t_j^i, y(Ssi) = \sum_{i \in s_i} y_i$, they called the inequality above circulant inequality. Guignard [12] showed that this inequality defines facet when $p = q+1$, and it is called simple, and when $q = 2$ it is called odd cycle inequality.

(p,q) Inequalities. Ardal et al. [1] addressed a family of valid inequalities (p, q) inequalities. Let p, q be integers, $2 \leq q \leq p \leq n$, p is not multiple of q, $H' \subseteq H$, $|H'| \geq \lceil p/q \rceil$, $D' \subseteq D$, $|D'| = p$, $\tilde{G}$ is a bipartie graph having H' and J' as a node set, and $\forall h \in H'$ degree(h) = q, and the set of edges of $\tilde{G}$ is $\tilde{E}$. The (p, q) inequalities are defined as follows.

$$t(\tilde{E}) \leq y(H') + p - \lceil p/q \rceil \tag{15}$$

where $t(\tilde{E}) = \sum_{\{i,j\} \in \tilde{E}} t^h_j$.

4   New Facets

Contrarily to the FLP, the valid inequalities associated with the network loading problem are not directly valid for the BBFLP. Our idea is to use a lifting procedure to extend these inequalities to the BBFLP. By this procedure, we show that the new inequalities are valid and even define facets under the same conditions.

We particularly consider the BBFLP when there are two cable types. We assume that the larger cable capacity is $C > 1$ and the smaller capacity is 1.

4.1   Network Loading and Valid Inequalities

Several families of valid inequalities were introduced for the problem, in the following we will give the basic valid inequalities that define facets.
The NLP with two cables types is formulated as follows.

Let $G = (V, E)$ be an undirected graph, a set of M commodities $\{(s_1, t_1), ..., (s_{|M|}, t_{|M|})\}$, a commodity $q \in \{1, ..., M\}$ having a demand d_q.

$$\min \sum_{uv \in E} l_{uv} \gamma^1 x^1_{uv} + l_{uv} \gamma^2 x^2_{uv} \tag{16}$$

$$\sum_{v \in V} f^q_{(u,v)} - \sum_{v \in V} f^q_{(v,u)} = \begin{cases} d_q, & \text{if } u = s_q \\ -d_q, & \text{if } u = t_q, \\ 0 & , otherwise. \end{cases} \quad \text{for all } u \in V, \ q \in \{1, ..., M\} \tag{17}$$

$$\sum_{q=1}^{M} (f^k_{(u,v)} + f^k_{(v,u)}) \leq x^1_{uv} + C x^2_{uv} \qquad \text{for all } uv \in E \tag{18}$$

$$x^1_e, x^2_e \geq 0, \qquad \text{for all } e \in E, \tag{19}$$

$$f^k_{(u,v)}, f^k_{(v,u)} \geq 0 \text{ for all } uv \in E, \ k \in K. \tag{20}$$

Rounded Cut-Set Inequalities. The rounded cut-set inequalities are obtained by partitioning the node set into two subsets, namely S and $\overline{S}$. Magnanti et al. [14] introduced the following form for the rounded cut-set inequalities:

$$x^1(\delta(S)) + r x^2(\delta(S)) \geq r \left\lceil \frac{D_{S,\overline{S}}}{C} \right\rceil \tag{21}$$

where $D_{S,\overline{S}}$ is the total demand whose origin lays in S and destination in $\overline{S}$ and vice-versa, and

$$r = \begin{cases} C, & \text{if } D_{S,\overline{S}} \pmod{C} = 0, \\ D_{S,\overline{S}} \pmod{C}, & \text{otherwise.} \end{cases}$$

Theorem 3 *[14]. A rounded cut set inequality defines a facet for the NLP if and only if*

1. *the subgraphs induced by S and by $\overline{S}$ are connected,*
2. *$D_{S,\overline{S}} > 0$.*

Partition Inequalities. Several version of partition inequalities have been developed in the literature, in particular, Magnanti et al. [14] gave a family of three-partition inequalities of the following form:

$$x_{12}^1 + x_{13}^1 + x_{23}^1 + r(x_{12}^2 + x_{13}^2 + x_{23}^2) \geq \left\lceil \frac{r\left(\left\lceil \frac{d_{12}+d_{13}}{C} \right\rceil + \left\lceil \frac{d_{12}+d_{23}}{C} \right\rceil + \left\lceil \frac{d_{13}+d_{23}}{C} \right\rceil\right)}{2} \right\rceil \tag{22}$$

The following metric inequalities are valid for NLP with one cable type:

Metric Inequalities. Metric inequalities can be introduced after projecting out the flow variables from the polyhedron and obtain a "capacity formulation" in the space of capacity variables. In particular Avella et al. [2] gave a family of metric inequalities that completely describe the polyhedron, they take the form

$$\mu x \geq \rho(\mu, G, d), \mu \in Met(G). \tag{23}$$

In the next section we present a lifting procedure that extends the Rounded cut-set inequalities defined for NLP to be valid for BBFLP.

4.2 Lifted Rounded Cut-Set Inequalities

Now, we give the main result of the paper which is a new family of valid inequalities for the BBFLP. These latter inequalities are obtained by lifting the rounded cut-set inequalities.

Consider the situation where all the facilities are opened and all the clients are assigned to a facility. For every $j \in D$, let h_j be the facility to which it is assigned (Fig. 1).

Let $Q_0 = \{(x, y, t) \in Q \text{ such that } t_j^h = 0, \text{for all } h \in H \backslash h_j, j \in D\}$.

Lemma 1. *The solutions of Q_0 are exactly solutions of a NLP where the demand set is composed of the pairs (j, h_j), with j the origin, h_j the destination and d_j is the commodity.*

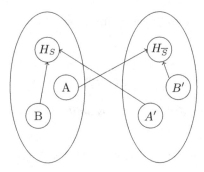

Fig. 1. A rounded cut-set inequality configuration

By Lemma 1, every valid inequality (respectively facet defining) for the corresponding NLP polyhedron are valid for Q_0 (respectively facet defining). This means that rounded cut-set inequalities are valid for Q_0.

Our objective is to extend the rounded cut-set inequalities by a lifting procedure to obtain facet defining inequalities for BBFLP. Let $(x, y, t) \in Q_0$ and $S \subseteq V$ be a node set such that $S \cap V \neq \emptyset \neq \overline{S} \cap V$. Let A (resp. A') be the node subset of $D \cap S$ (resp. $D \cap \overline{S}$) which are assigned to a facility of $H \cap \overline{S}$ (resp. $H \cap S$). Also, let B (resp. B') be the node subset of $D \cap S$ (resp. $D \cap \overline{S}$) which are assigned to a facility of $H \cap S$ (resp. $H \cap \overline{S}$).

The rounded cut-set inequality induced by S is

$$x^1(\delta(S)) + rx^2(\delta(S)) \geq r \left\lceil \frac{d(A \cup A')}{C} \right\rceil, \tag{24}$$

where $r = d(A \cup A') \pmod{C}$. Clearly (24) is valid for Q_0.

For convenience, we let,
$A_1 = \{j \in A \cup A' \text{ such that } d_j \leq r\}$,
$A_2 = \{j \in A \cup A' \text{ such that } d_j > r\}$,
$B_1 = \{j \in B \cup B' \text{ such that } d_j \leq C - r\}$,
$B_2 = \{j \in B \cup B' \text{ such that } d_j > C - r\}$.

Now we give our main result.

Theorem 4. *Let $S \subseteq V$ with $S \cap V \neq \emptyset \neq V \cap \overline{S}$. The inequality*

$$x^1(\delta(S)) + rx^2(\delta(S)) + \sum_{j \in D} \sum_{h \in H \setminus \{h_j\}} C_j^h t_j^h \geq r \left\lceil \frac{d(A \cup A')}{C} \right\rceil \tag{25}$$

where

$$C_j^h = \begin{cases} d_j, \text{for } j \in A_1 \cap A, h \in H \cap S \text{ and for } \in A_1 \cap A', h \in H \cap \overline{S}, \\ r, \text{for } j \in A_2 \cap A, h \in H \cap S \text{ and for } j \in A_2 \cap A', h \in H \cap \overline{S}, \\ C - r - d_j, \text{for } j \in B_1 \cap B, h \in H \cap \overline{S} \text{ and for } j \in B_1 \cap B', h \in H \cap S, \\ 0, \text{otherwise.} \end{cases}$$

is valid for BBFLP. Moreover (25) is facet for Q if (24) is facet for Q_0.

The proof of Theorem 4 is established by lifting inequality (24). The values of C_j^h given in Theorem 4 are the lifting coefficients of all the variables whose values are fixed in Q_0. Recall that the lifting coefficient of a variable depends on the order in which this variable is lifted with respect to the other variables. In particular, if a variable t_j^h, when lifted first, has a coefficient ψ_j^h, then $C_j^h \geq \psi_j^h$ for any order in which t_j^h is lifted. In our proof, the variables are lifted in the following order

1. variables t_j^h, for $j \in A_1 \cap A$, $h \in H \cap S$ and for $j \in A_1 \cap A'$ and $h \in H \cap \overline{S}$,
2. variables t_j^h, for $j \in A_2 \cap A$, $h \in H \cap S$ and for $j \in A_2 \cap A'$ and $h \in H \cap \overline{S}$,
3. variables t_j^h, for $j \in B_2 \cap B$, $h \in H \cap \overline{S}$ and for $j \in B_2 \cap B'$ and $h \in H \cap S$,
4. variables t_j^h, for $j \in B_1 \cap B$, $h \in H \cap \overline{S}$ and for $j \in B_1 \cap B'$ and $h \in H \cap S$.

Since, we do not have enough space, we will give the proof of Theorem 4 for a variable t_j^h with $j \in A_1 \cap A$ and $h \in H \cap S$.

Before, we give a property of the lifting coefficients which is independent of the lifting order.

Lemma 2. *For every $j \in D$ and $h \in H \setminus \{h_j\}$, we have*

1. $C_j^h = 0$, for all $j \in A \cup B'$, $h \in H \cap \overline{S}$, and for all $j \in A' \cup B$, $h \in H \cap S$,
2. $C_j^h = \rho_j$, for all $j \in A \cup B'$, $h \in H \cap S$ and for all $j \in A' \cup B$, $h \in H \cap \overline{S}$,
3. $C_j^h \leq 0$, for all $j \in B$, $h \in H \cap \overline{S}$ and for all $j \in B'$, $h \in H \cap S$.

and this, independently of the order in which the variables are lifted.

Proof. 1) W.l.o.g., we consider a client $j_0 \in A$ and $h_1 \in H \cap \overline{S}$. The proof follows the same lines for the other cases. We denote by $ax + by + ct \geq \theta$ the inequality (25) induced by S. W.l.o.g., we assume that (25) defines a facet of Q, different from those induced by the trivial inequalities. Thus, it exists a solution $(\overline{x}, \overline{y}, \overline{t}) \in Q$ satisfying (25) with equality and such that $\overline{t}_{j_0}^{h_{j_0}} = 1$ (i.e. j_0 is assigned to facility h_0). We can also assume that $\overline{x}_e^k$ is sufficiently large for all $e \in E(S) \cup E(\overline{S})$ and $k \in \{1,2\}$. Now consider the solution $(\overline{x}, \overline{y}, \overline{t}')$ with $\overline{t}_h'^j = \overline{t}_h^j$, for every $j \in D \setminus \{j_0\}$ and $h \in H \setminus \{h_{j_0}, h_1\}$, and $\overline{t}_{j_0}'^{h_0} = 0$ and $\overline{t}_{j_0}'^{h_1} = 1$ (i.e. j_0 is now assigned to facility h_1). It is not hard to see that $(\overline{x}, \overline{y}, \overline{t}')$ induces a feasible solution for the BBFLP. Thus,

$$a\overline{x} + b\overline{y} + c\overline{t}' = a\overline{x} + b\overline{y} + c\overline{t} - C_{j_0}^{h_{j_0}} + C_{j_0}^{h_1} = \theta - C_{j_0}^{h_{j_0}} + C_{j_0}^{h_1} \geq \theta.$$

This implies that $C_{j_0}^{h_1} \geq C_{j_0}^{h_{j_0}}$.

Now, one can consider a feasible solution whose incidence vector satisfies (25) with equality and in which j_0 is assigned to h_1. As before, the solution obtained by modifying this latter solution and now assigning j_0 to h_{j_0} is also feasible for the BBFLP. Using similar arguments as above, we obtain that $C_{j_0}^{h_{j_0}} \geq C_{j_0}^{h_1}$. Therefore, $C_{j_0}^{h_1} = C_{j_0}^{h_{j_0}}$. Finally, since variable $t_{j_0}^{h_{j_0}}$ was not fixed in Q_0, we have that $C_{j_0}^{h_{j_0}} = 0$, which show the result.

2) The proof for this case is very similar to case 1). In fact, for a client $j_0 \in A$ and facility $h_1 \in H \cap S$, we can show using the same arguments that $C_{j_0}^{h_1} = C_{j_0}^{h_0} = \rho_j$, for some $h_0 \in H \cap S$.

Now, one can consider a feasible solution whose incidence vector satisfies (25) with equality and in which j_0 is assigned to h_1. As before, the solution obtained by modifying this latter solution and now assigning j_0 to h_{j_0} is also feasible for the BBFLP. Using similar arguments as above, we obtain that $C_{j_0}^{h_{j_0}} \geq C_{j_0}^{h_1}$. Therefore, $C_{j_0}^{h_1} = C_{j_0}^{h_{j_0}}$. Finally, since variable $t_{j_0}^{h_{j_0}}$ was not fixed in Q_0, we have that $C_{j_0}^{h_{j_0}} = 0$, which show the result.

3) The proof for this case is also similar to case 1). Let $j_0 \in B$ and facility $h_1 \in H \cap \overline{S}$, using the same arguments that we can obtain that $C_{j_0}^{h_1} \leq 0$, for some $h_0 \in H \cap S$.

Considering a feasible solution whose incidence vector satisfies (25) at equality and where j_0 is assigned to h_1. As before, the solution obtained by modifying this latter solution and now assigning j_0 to h_{j_0} is also feasible for the BBFLP. Using similar arguments as above, we obtain that $C_{j_0}^{h_{j_0}} \geq C_{j_0}^{h_1}$. Since variable $t_{j_0}^{h_{j_0}}$ was not fixed in Q_0, we have that $C_{j_0}^{h_1} \leq 0$.   □

Observe that if $D' \subseteq D$ is a set of clients such that the variables t_j^h, $j \in D'$ have been lifted, then the lifting coefficient of a variable $t_{j_0}^{h_0}$, for $j \in D \setminus D'$ and $h \in H \setminus h_{j_0}$, is

$$C_{j_0}^{h_0} = r \left\lceil \frac{d(A \cup A')}{C} \right\rceil - \xi_{j_0}^{h_0}$$

where

$$\xi_{j_0}^{h_0} = \min\{\overline{x}^1(\delta(S)) + r\overline{x}^2(\delta(S)) + \sum_{h \in H} \sum_{j \in D'} C_j^h \overline{t}_j^h, \text{ for all}$$

$$(\overline{x}, \overline{y}, \overline{t}) \in Q \text{ with } \overline{t}_{j_0}^{h_0} = 1 \text{ and } \overline{t}_j^h = 0, \forall j \in D \setminus D', h \in H\}. \tag{26}$$

The above remark means that obtaining the lifting coefficients of Theorem 4 reduces to finding ξ_j^h, for all $j \in D$ and $h \in H \setminus h_j$.

For lack of space, we do not give a complete proof of the following lemmas and Theorem 4. However, we will give the main ideas that lead to the coefficients of Theorem 4.

Before going further, for every $U \subseteq D$, we let

$$r_U = \begin{cases} C, \text{if } d(U) mod(C) = 0, \\ d(U) mod(C), \text{ otherwise.} \end{cases}$$

and

$$\varepsilon_U = \min\{0, r_U - r\}.$$

In the following Lemma we give the form of $\xi_{j_0}^{h_0}$ in the case when $D' = \emptyset$.

Lemma 3. *If $D' = \emptyset$ then, for every $j_0 \in A \cup A'$ and $h_0 \in H \setminus \{h_{j_0}\}$,*

$$\xi_{j_0}^{h_0} = r \left\lceil \frac{d(W)}{C} \right\rceil + \varepsilon_W \tag{27}$$

where $W = (A \cup A') \setminus \{j_0\}$.

Proof. First, let β be the number of cables of type 1 loaded on the edges of $\delta(S)$. Also, w.l.o.g., we assume that the number of cables of type 2 loaded on the edges of $\delta(S)$ is $\lceil \frac{d(W)}{C} \rceil - \lambda$, for some $\lambda \geq 0$. In any feasible solution of the BBFLP, we have that $\beta + C(\lceil \frac{d(W)}{C} \rceil - \lambda) \geq d(W)$.

We have that $x^1(\delta(S)) + rx^2(\delta(S)) = \beta + r(\lceil \frac{d(W)}{C} \rceil - \lambda)$. If $\lambda = 0$, we can show that the minimum value of $x^1(\delta(S)) + rx^2(\delta(S))$ is obtained for $\beta = 0$. Thus, $\xi_{j_0}^{h_0} = r\lceil \frac{d(W)}{C} \rceil$ in this case. Now, if $\lambda \geq 1$, we have that

$$\beta + r(\lceil \frac{d(W)}{C} \rceil - \lambda) \geq (C - r)\lambda - C + r_W + r\lceil \frac{d(W)}{C} \rceil.$$

Here also, we can show that the minimum of $x^1(\delta(S)) + rx^2(\delta(S))$ is obtained for $\lambda = 1$ and $\beta = C(\lambda - 1) + r_W = r_W$, that is $\xi_{j_0}^{h_0} = r_W - r + r\lceil \frac{d(W)}{C} \rceil$.

Thus, $\xi_{j_0}^{h_0} = r\left\lceil \frac{d(W)}{C} \right\rceil + \min\{0, r_W - r\} = r\left\lceil \frac{d(W)}{C} \right\rceil + \varepsilon_W$. $\qquad\square$

Lemma 4. *If $D' = \emptyset$ then, for every $j_0 \in B \cup B'$ and $h_0 \in H \setminus \{h_{j_0}\}$,*

$$\xi_{j_0}^{h_0} = r \left\lceil \frac{d(W)}{C} \right\rceil + \varepsilon_W \tag{28}$$

where $W = A \cup A' \cup \{j_0\}$.

Now, for a node set $U \subseteq D$, we let

$$\alpha_U = \left\lceil \frac{d(A \cup A')}{C} \right\rceil - \left\lceil \frac{d(U)}{C} \right\rceil.$$

From Lemmas 3 and 4, when $D' = \emptyset$, the lifting coefficient of a variable $t_{j_0}^{h_0}$ is of the form

$$C_{j_0}^{h_0} = r\alpha_W - \varepsilon_W$$

where

$$W = \begin{cases} (A \cup A') \setminus \{j_0\}, & \text{for } j_0 \in A \cup A', \\ A \cup A' \cup \{j_0\}, & \text{for } j_0 \in B \cup B'. \end{cases}$$

From this, we can see that the lifting coefficient of variable t_j^h, when $D' = \emptyset$ ({i.e.} when the variable is lifted at first), is

- $C_j^h = r\alpha_W - \varepsilon_W = d_j$, for $j \in A_1$,
- $C_j^h = r\alpha_W - \varepsilon_W = r$, for $j \in A_2$,
- $C_j^h = r\alpha_W - \varepsilon_W = 0$, for $j \in B_1$,
- $C_j^h = r\alpha_W - \varepsilon_W = C - r - d_j$, for $j \in B_2$.

5   Conclusion

In this paper, we have investigated the BBFLP from a polyhedral point of view. We have introduced an integer programming formulation for the problem which does not use flow variables. Then, we have studied the associated polyhedron, and show that several classes of inequalities coming from the FLP are valid for the BBFLP polyhedron. Moreover, we have introduced a new family of facet defining inequalities, coming from the NLP.

As a perspective, one can try to extend other families of valid inequalities coming from the NLP, in particular the partition and the metric inequalities. It is also interesting to devise a Branch-and-Cut algorithm for the BBFLP using these new families of inequalities.

References

1. Ardal, K.: Capacitated facility location: separation algorithms and computational experience. Math. Program. **81**, 149–175 (1998). https://doi.org/10.1007/BF01581103
2. Avella, P., Mattia, S., Sassanoo, A.: Metric inequalities and the network loading problem. Discrete Optim. **4**(1), 103–114 (2007). https://doi.org/10.1016/j.disopt.2006.10.002
3. Agarwal, Y.K.: k-partition-based facets of the network design problem. Networks **47**, 123–136 (2006). https://doi.org/10.1002/net.20098
4. Agarwal, Y.K.: Polyhedral structure of the 4-node network design problem. Networks **54**, 139–149 (2009). https://doi.org/10.1002/net.20317
5. Agarwal, Y.K.: Network loading problem: valid inequalities from 5-and higner partitions. Comput. Oper. Res. **99**, 123–134 (2018)
6. Bienstosk, D., Gülnük, O.: Capacitated network design-Polyhedral structure, and computation. INFORMS J. Comput. **8**(3), 243–259 (1996). https://doi.org/10.1287/ijoc.8.3.243
7. Bley, A., Rezapour, M.: Combinatorial approximation algorithms for buy-at-bulk connected facility location problems. Discrete Appl. Math. **213**, 34–46 (2016). https://doi.org/10.1016/j.dam.2016.05.016
8. Cornuejolis, G., Fischer, M.L., Nemhauser, G.L.: On the uncapacitated location problem. Ann. Discrete Math. **1**, 163–177 (1977). https://doi.org/10.1016/S0167-5060(08)70732-5. 23, 50–74 (1982)
9. Barahona, F.: Network design using cut inequalities. SIAM J. Optim. **6**(3), 823–837. https://doi.org/10.1137/S1052623494279134
10. Galili, L., Letchfrod, A.N., Miller, S.J.: New valid inequalities and facets for the simple plant location problem. Eur. J. Oper. Res. **269**(3), 824–833 (2018). https://doi.org/10.1016/j-ejor2018.03.009
11. Gülnük, D.: A branch-and-cut algorithm for capacitated network design problems. Math. Program. **86**, 17–39 (1999). https://doi.org/10.1007/s1010700500
12. Guinard, M.: Fractional vertices, cuts and facets of simple plant location problem. Math. Program. Study **12**, 150–162 (1980). https://doi.org/10.1007/BFb0120893
13. Friggstad, Z., Rezapour, M., Salavatipour, M.R., Soto, J.A.: LP-based approximation algorithms for facility location in buy-at-bulk network design. In: Dehne, F., Sack, J.-R., Stege, U. (eds.) WADS 2015. LNCS, vol. 9214, pp. 373–385. Springer, Cham (2015). https://doi.org/10.1007/978-3-319-21840-3_31

14. Magnanti, T.L., Mirchandani, P., Vachani, R.: Modeling and solving the two-facility capacitated network loading problem. Oper. Res. 233–250 (1995). https://doi.org/10.1287/opre.43.1.142

15. Magnanti, T.L., Mirchandani, P., Vachani, R.: The convex hull of two core capacitated network design problems. Math. Program. **60**(1–3), 142–157 (1993). https://doi.org/10.1007/BF01580612

16. Matia, S.: Solving survivable two-layer network design problems by metric imequalities. Comput. Optim. Appl. **51**, 809–839 (2011). https://doi.org/10.1007/s10589-010-9364-0

17. Meyerson, A., Munagala, K., Plotkin, S.: Cost distance: two metric network design. In: 41 Annual Symposium Foundations of Computer Science, pp. 624–630 (2000). https://doi.org/10.1109/SFCS.2000.892330

18. Raack, C., Koster, A.M., Orlowski, S., Wessüly, R.: On cut-based inequalities for capacitated network design polyhedra. Network **57**(2), 141–156 (2011). https://doi.org/10.1002/net.20395

19. Ravi, R., Sinha, S.: Approximation algorithms combining facility location and network design. Oper. Res. **54**, 73–81 (2006). https://doi.org/10.1287/opre.1050.0228

Cardinality Constrained Multilinear Sets

Rui Chen[1]([✉]), Sanjeeb Dash[2], and Oktay Günlük[3]

[1] University of Wisconsin-Madison, Madison, WI 53706, USA
rchen234@wisc.edu
[2] IBM T. J. Watson Research Center, Yorktown Heights, NY 10598, USA
sanjeebd@us.ibm.com
[3] School of ORIE, Cornell University, Ithaca, NY 14853, USA
ong5@cornell.edu

Abstract. The problem of minimizing a multilinear function of binary variables is a well-studied NP-hard problem. The set of solutions of the standard linearization of this problem is called the multilinear set, and many valid inequalities for its convex hull are available in the literature. Motivated by a machine learning application, we study a cardinality constrained version of this problem with upper and lower bounds on the number of nonzero variables. We call the set of solutions of the standard linearization of this problem a cardinality constrained multilinear set, and give a complete polyhedral description of its convex hull when the multilinear terms in the problem have a nested structure.

Keywords: Multilinear functions · Valid inequalities · Mixing

1 Introduction

In this paper, we study the convex hull of the set

$$X = \{(x,\delta) \in \{0,1\}^n \times \{0,1\}^m : \delta_i = \prod_{j \in S_i} x_j, \ i = 1,\ldots,m, \ L \le \sum_{j=1}^n x_j \le U\},$$

where m, n are positive integers, $S_i \subseteq N = \{1,\ldots,n\}$ for $i = 1,\ldots,m$ and L, U are integers such that $0 \le L \le U \le n$. We call X a *cardinality constrained multilinear set*. We give a polyhedral characterization of the convex hull of X in the special case that the sets S_i are nested, i.e., $S_1 \subset S_2 \subset \cdots \subset S_m$. Our work for the nested case is closely related to recent work by Fischer, Fischer and McCormick [14].

This paper is motivated by the work in Dash, Günlük and Wei [3], who gave an integer programming (IP) formulation – for the problem of finding boolean rule set based classifiers that have high predictive accuracy – with an exponential number of columns, and tackled it via column generation. Their (column) pricing subproblem can be framed as optimizing a linear function over X.

The set Y obtained from X by dropping the cardinality constraints is called a *multilinear set* and is well-studied in mixed-integer nonlinear optimization.

© Springer Nature Switzerland AG 2020
M. Baïou et al. (Eds.): ISCO 2020, LNCS 12176, pp. 54–65, 2020.
https://doi.org/10.1007/978-3-030-53262-8_5

The *boolean quadric polytope* (Padberg [19]) is equal to $\text{conv}(Y)$ when $|S_i| = 2$ for $i = 1, \ldots, m$, and a cardinality constrained version of this was studied by Mehrotra [18]. Crama and Rodríguez-Heck [2], Del Pia and Khajavirad [4–6], and Del Pia, Khajavirad and Sahinidis [7] recently studied the multilinear set.

Buchheim and Klein [1] studied the QMST-1 problem, a version of the quadratic minimum spanning tree (QMST) problem with a single quadratic term in the objective. They gave a polyhedral description of the standard linearization of QMST-1, and so did Fischer and Fischer [12]. This work was generalized by Fischer, Fischer and McCormick [14] who considered the problem of minimizing a multilinear objective over independent sets of matroids where the nonlinear terms consist of nested monomials. Let $\mathcal{M}$ be the independent set polytope of a matroid on n elements. Fischer, Fischer and McCormick studied the convex hull of $V = \{(x, \delta) \in Y : x \in \mathcal{M}\}$ and characterized its facet-defining inequalities when the sets S_i are nested. When the matroid is a uniform matroid, V is the same as X with $L = 0$. Our results thus generalize their uniform matroid result.

Prior to [3], the maximum monomial agreement problem (MMA) was studied in the context of machine learning by Demirez, Bennett, and Shawe-Taylor [8], Goldberg [16], Eckstein and Goldberg [10,15], and is called the maximum bichromatic discrepancy problem in computer graphics [9]. Given an $m \times n$ binary data matrix A, and weights $w_1, \ldots, w_m$, the MMA problem is

$$\max_{\substack{I,J \subseteq N \\ I \cap J = \emptyset, \ |I|+|J| \geq 1}} \sum_{k=1}^{m} w_k \left(\Pi_{i \in I} A_{ki} \right) \left(\Pi_{j \in J} (1 - A_{kj}) \right). \tag{1}$$

In other words, the goal of the MMA problem is to find disjoint subsets I and J of N and associated *monomial* $p(x) = \Pi_{i \in I} x_i \Pi_{j \in J} (1 - x_j)$ such that the sum of weighted values of the monomial – when evaluated on rows of the data matrix A (by replacing x_j by A_{kj}) – is as large as possible. Goldberg [16], Eckstein and Goldberg [10,15], and Eckstein, Kagawa, and Goldberg [11] present a branch-and-bound method to solve the MMA problem and generate columns within a column generation algorithm for a binary classification problem.

Dash, Günlük and Wei [3] solve a cardinality constrained version of MMA (call it CMMA) by replacing $|I|+|J| \geq 1$ in (1) with $u \geq |I|+|J| \geq 1$ to generate new monomials/clauses to augment an existing list from which a classifier in the form of a disjunction of clauses is created. The CMMA (resp. MMA) problem can be viewed as maximizing a linear function over the set X (resp. Y). Consider the matrix A in (1), and let $A' = [A \quad (\mathbf{1}_{m,n} - A)]$ where $\mathbf{1}_{m,n}$ stands for the $m \times n$ matrix with all components one. Then, for any i we have $1 - A_{ki} = A'_{kj}$ for some $j \in \{1, \ldots, 2n\}$, and therefore CMMA is equivalent to

$$\max_{\substack{I \subseteq \{1,\ldots,2n\}, \\ u \geq |I| \geq 1}} \sum_{k=1}^{m} w_k \left(\prod_{i \in I} A'_{ki} \right) =$$

$$\max \left\{ \sum_{k=1}^{m} w_k \delta_k : \delta_k = \prod_{i : A'_{ki}=0} (1 - z_i), \ 1 \leq \sum_{i=1}^{2n} z_i \leq u, \ \delta \in \{0,1\}^m, z \in \{0,1\}^{2n} \right\}.$$

To see this, note that $\Pi_{i \in I} A'_{ki} = \Pi_{i:A'_{ki}=0}(1 - z_i)$ if $I = \{i : z_i = 1\}$ and $|I| \geq 1$. Dash, Günlük and Wei [3] solve an IP formulation of the problem above using the standard linearization of the expressions $\delta_k = \Pi_{i:A'_{ki}=0}(1 - z_i)$.

2   Preliminaries

Let $I = \{1, \ldots, m\}$, $J = \{1, \ldots, n\}$, $0 \leq l \leq u$ and $u \geq 2$. Let $S_1, \ldots, S_m$ be distinct subsets of J with $2 \leq |S_i| \leq n - l$ for $i = 1, \ldots, m$. Note that the assumptions imply that $n - l \geq 2$. Define $S := \{S_i\}_{i \in I}$. We will study the set

$$X^{l,u} := \{(z, \delta) \in \{0,1\}^n \times \{0,1\}^m : \delta_i = \prod_{j \in S_i}(1 - z_j), \; i \in I, \; l \leq \sum_{j \in J} z_j \leq u\},$$

which is equivalent to the set X in the previous section (let $z_j = 1 - x_j$, $l = n - U$ and $u = n - L$).

Let $\Delta^{l,u} = proj_\delta(X^{l,u})$ denote the orthogonal projection of $X^{l,u}$ onto the space of δ variables. We say that S is closed under nonempty intersection if for each pair $S_i, S_j \in S$ such that $S_i \cap S_j \neq \emptyset$, the set $S_i \cap S_j \in S$.

We first restrict our discussion to the so-called proper families defined as follows.

Definition 1. *A family $S = \{S_i\}_{i \in I}$ of subsets of J is called a proper family if it satisfies the following properties:*

1. *$\Delta^{l,u}$ is a set of exactly $m + 1$ affinely independent vectors in $\mathbb{R}^m$;*
2. *S is closed under nonempty intersection.*

We next present three examples of proper families S.

Example 1. If $S_1 \subset S_2 \subset \ldots \subset S_m$ are nested subsets of J, $l \leq n - |S_m|$ and $u \geq 2$, then $S = \{S_1, S_2, \ldots, S_m\}$ is proper.

Example 2. If S_1, S_2 are two disjoint nonempty subsets of J, $l \leq n - |S_1 \cup S_2|$ and $u \geq 2$, then $S = \{S_1, S_2, S_1 \cup S_2\}$ is proper.

Example 3. If S_1 and S_2 are two subsets of J satisfying $S_1 \cap S_2 \neq \emptyset$, $S_1 \not\subseteq S_2$, $S_2 \not\subseteq S_1$, $l \leq n - |S_1 \cup S_2|$ and $u \geq 2$, then $S = \{S_1 \cap S_2, S_1, S_2, S_1 \cup S_2\}$ is proper.

More generally, given a family S that is closed under union and nonempty intersection, we can show that S is a proper family provided that $\Delta^{l,u} = \Delta^{0,n}$.

For any $\bar{\delta} \in \Delta^{l,u}$, define $X^{l,u}(\bar{\delta}) := \{z \in \{0,1\}^n : (z, \bar{\delta}) \in X^{l,u}\}$. An inequality $\alpha^T z + \beta^T \delta \leq \gamma$ is valid for $X^{l,u}$ if and only if

$$\gamma \geq \max_{(z,\delta) \in X^{l,u}} \{\alpha^T z + \beta^T \delta\} = \max_{\bar{\delta} \in \Delta^{l,u}} \{\beta^T \bar{\delta} + \max_{z \in X^{l,u}(\bar{\delta})} \alpha^T z\}.$$

Therefore, it is valid if and only if for all $\bar{\delta} \in \Delta^{l,u}$ we have

$$\gamma - \beta^T \bar{\delta} \geq \max_{z \in X^{l,u}(\bar{\delta})} \alpha^T z. \tag{2}$$

Theorem 1. *Assume S is a proper family. Then each facet F of $conv(X^{l,u})$ can be defined by an inequality $\alpha^T z + \beta^T \delta \leq \gamma$ where either $\alpha = 0$, or (2) holds as equality for all $\bar{\delta} \in \Delta^{l,u}$. Moreover, we can choose α such that $\alpha \in \{0, \kappa\}^n$ for some $\kappa \in \{-1, +1\}$.*

Proof. Let $\alpha^T z + \beta^T \delta \leq \gamma$ be a facet-defining inequality for $conv(X^{l,u})$ and let F be the associated facet. Assume that the inequality (2) is strict for some $\bar{\delta} \in \Delta^{l,u}$. As S is a proper family, $conv(\Delta^{l,u})$ is a full-dimensional simplex in $\mathbb{R}^m$. Let $(\beta')^T \delta \leq \gamma'$ be the (only one) facet-defining inequality for $conv(\Delta^{l,u})$ such that $\bar{\delta}$ is not contained in the corresponding facet F'. Note that all points in $\Delta^{l,u} \setminus \{\bar{\delta}\}$ satisfy $(\beta')^T \delta = \gamma'$. As inequality (2) is strict, there is no point of the form $(\bar{z}, \bar{\delta})$ in F. Therefore, $\bar{\delta}$ is not contained in $proj_\delta(F)$. We conclude that every point δ' in $proj_\delta(F)$ satisfies $(\beta')^T \delta' = \gamma'$, and so does every point in F. Therefore, F is defined by the inequality $(\beta')^T \delta \leq \gamma'$.

The second part of the proof is omitted due to the page limit. □

3   When S is a Family of Nested Sets

In this section, we consider the special case when $S = \{S_i\}_{i \in I}$ is a family of nested sets. As before, let $I = \{1, \ldots, m\}$, and $J = \{1, \ldots, n\}$. We assume, without loss of generality, that $S_1 \subset S_2 \subset \ldots \subset S_m$, and $S_i = \{1, \ldots, k_i\}$ where $2 \leq k_1 < k_2 < \ldots < k_m$. We assume that $l \leq n - |S_m|$, and $u \geq 2$; therefore $k_m \leq n - l$. Recall, from Example 1, that S is a proper family of sets with respect to $X^{l,u}$. We number the $m + 1$ vectors in $\Delta^{l,u}$ as δ^i, for $i = 0, \ldots, m$, where $\delta^0 = \mathbf{0}$ and δ^i has the first i components equal to 1, and the rest equal to 0, for all $i \geq 1$. Further,

$$\Delta^{l,u} = \{\delta \in \{0,1\}^m : \delta_1 \geq \delta_2 \geq \ldots \geq \delta_m\}, \tag{3}$$

$$conv(\Delta^{l,u}) = \{\delta \in [0,1]^m : \delta_1 \geq \delta_2 \geq \ldots \geq \delta_m\}. \tag{4}$$

The second equation follows from the first because the constraint matrix in the inequality system in (3) is totally unimodular. Note that $conv(\Delta^{l,u})$ is full-dimensional.

For any $(z, \delta) \in X^{l,u}$, we have $\delta_{i+1} \leq \delta_i$ for all $i < m$, as $S_i \subset S_{i+1}$. Moreover, $\delta_{i+1} = \delta_i$ if $z_j = 0$ for all $j \in S_{i+1} \setminus S_i$. Consequently, the following *2-link inequalities* [2] are valid for $conv(X^{l,u})$ for all $i = 1, \ldots, m - 1$:

$$\delta_{i+1} - \delta_i \leq 0, \quad \delta_i - \delta_{i+1} - \sum_{j \in S_{i+1} \setminus S_i} z_j \leq 0. \tag{5}$$

For the nested case the following inequalities together with the 2-link inequalities (5) give a relaxation $conv(X^{l,u})$:

$$l \leq \sum_{j \in J} z_j \leq u, \tag{6}$$

$$z_j + \delta_i \leq 1, \qquad j \in S_i, i \in I, \tag{7}$$

$$1 - \delta_1 - \sum_{j \in S_1} z_j \leq 0, \tag{8}$$

$$\delta_m \geq 0, \quad 1 \geq z_j \geq 0, \qquad\qquad j \in J. \tag{9}$$

Without loss of generality, we can assume that $l < u$ when $\mathcal{S}$ is nested. When $l = u$, the problem can be easily reduced to the case when $l < u$ by considering the substitution $z_n = u - \sum_{j \in J \setminus \{n\}} z_j$. In the case when $l < u$, $\text{conv}(X^{l,u})$ can be shown to be full-dimensional.

We next observe some properties of $\text{conv}(X^{l,u})$.

Lemma 2. *If $\beta^T \delta \leq \gamma$ defines a facet of $\text{conv}(X^{l,u})$ then it is a multiple of an inequality from (5)–(9).*

Proof. If $\beta^T \delta \leq \gamma$ defines a facet of $\text{conv}(X^{l,u})$ then it also defines a facet of $\text{conv}(\Delta^{l,u})$. The only facet-defining inequality for $\text{conv}(\Delta^{l,u})$, see (4), that is not of the form $\delta_{i+1} - \delta_i \leq 0$ or $\delta_m \geq 0$ is $1 \geq \delta_1$. However, $1 \geq \delta_1$ cannot define a facet of $\text{conv}(X^{l,u})$ as it is implied by (7) and (9) for $i = 1$ and any $j \in S_1$. $\quad\square$

3.1   Convex Hull Description of $X^{0,u}$

By Theorem 1, we only need to consider inequalities of the form $\alpha^T z + \beta^T \delta \leq \gamma$ with $\alpha \in \{0,1\}^n$ or $\alpha \in \{0,-1\}^n$. We have already characterized all facets of the form $\beta^T \delta \leq \gamma$ in Lemma 2. We now characterize facet-defining inequalities of the form $\alpha^T z + \beta^T \delta \leq \gamma$ for $\text{conv}(X^{0,u})$ with $\alpha \leq 0$ and $\alpha \neq 0$.

Lemma 3. *Let $\alpha^T z + \beta^T \delta \leq \gamma$ be a facet-defining inequality for $\text{conv}(X^{0,u})$ with $\alpha \in \{0,-1\}^n$, then the inequality is implied by an inequality from (5)–(9).*

Proof. If $\alpha = 0$, then the result follows from Lemma 2. Assume now that $\alpha \neq 0$ and therefore $\alpha \in \{0,-1\}^n \setminus \{0\}$. Then for all $\delta^i \in \Delta^{0,u}$, by Theorem 1,

$$\gamma - \beta^T \delta^i = \max_{z \in X^{l,u}(\delta^i)} \alpha^T z = \max_{j \in S_{i+1} \setminus S_i} \{\alpha_j\}.$$

Let $\theta_i = \max_{j \in S_{i+1} \setminus S_i} \{\alpha_j\} \in \{0,-1\}$. Then

$$\alpha^T z + \beta^T \delta \leq \sum_{i \in I} \theta_{i-1} \sum_{j \in S_i \setminus S_{i-1}} z_j + \sum_{i=1}^{m-1} (\theta_{i-1} - \theta_i)\delta_i + \theta_{m-1}\delta_m$$

$$= \underbrace{\theta_0 (\delta_1 + \sum_{j \in S_1} z_j)}_{\geq 1} + \sum_{i=1}^{m-1} \theta_i \underbrace{(\delta_{i+1} - \delta_i + \sum_{j \in S_{i+1} \setminus S_i} z_j)}_{\geq 0} \leq \theta_0 + 0 = \gamma$$

is implied by inequalities (5), (8) and (9). $\quad\square$

We next derive a family of valid inequalities for $\mathrm{conv}(X^{0,u})$ using the mixing procedure [17]. We start with deriving some valid inequalities from (5)–(9) which we will use as the so-called base inequalities for the mixing procedure. Let $S' \subseteq J$ be fixed and let $M > n$ be a given constant. For any $i \in I$, we can derive the following valid (base) inequality using the fact that for all $j \in S_i$ we have $z_j, \delta_i \le 1$, and $1 - \delta_i - z_j \ge 0$.

$$
\frac{1}{M}\left(u - \sum_{j \in S'} z_j\right) + (1 - \delta_i)
$$
$$
= \frac{1}{M}\left(u - \sum_{j \in S' \setminus S_i} z_j\right) + \frac{1}{M} \sum_{j \in S' \cap S_i} (1 - \delta_i - z_j) + \frac{1}{M}\left(M - |S' \cap S_i|\right)(1 - \delta_i)
$$
$$
\ge \frac{1}{M}\left(u - |S' \setminus S_i|\right).
$$

If $|S' \setminus S_p| \le u-1$ for some $p \in I$, then the right-hand side of $\frac{1}{M}(u - \sum_{j \in S'} z_j) + (1 - \delta_i) \ge \frac{1}{M}(u - |S' \setminus S_i|)$ is strictly between 0 and 1 for all $i = p, p+1, \ldots, m$. Therefore treating the term $\frac{1}{M}(u - \sum_{j \in S'} z_j)$ above as a nonnegative continuous variable, we can apply the type I mixing procedure to these (base) inequalities to obtain the following mixing inequality,

$$
\frac{1}{M}\left(u - \sum_{j \in S'} z_j\right) \ge \frac{1}{M}\left(u - |S' \setminus S_p|\right)\delta_p + \frac{1}{M} \sum_{i=p+1}^{m} \left(|S' \setminus S_{i-1}| - |S' \setminus S_i|\right)\delta_i.
$$

After multiplying the inequality by M and rearranging the terms, we obtain the following valid inequality for $\mathrm{conv}(X^{0,u})$

$$
\sum_{j \in S'} z_j + \left(u - |S' \setminus S_p|\right)\delta_p + \sum_{i=p+1}^{m} \left(|S' \setminus S_{i-1}| - |S' \setminus S_i|\right)\delta_i \le u. \tag{10}
$$

These inequalities are sufficient to describe $\mathrm{conv}(X^{0,u})$.

Theorem 4. *Inequalities (5)–(9) together with inequalities (10), for all $p \in I$ and $S' \subseteq J$ such that $|S' \setminus S_p| \le u-1$, give a complete description of $\mathrm{conv}(X^{0,u})$.*

Proof. Let F be a facet of $\mathrm{conv}(X^{0,u})$ and $\alpha^T z + \beta^T \delta \le \gamma$ be the corresponding facet-defining inequality. By Theorem 1, we can assume that either $\alpha \in \{0,1\}^n$ or $\alpha \in \{0,-1\}^n$. Furthermore, by Lemmas 2 and 3 we have established that if $\alpha \le 0$ (including the case when $\alpha = 0$) the inequality $\alpha^T z + \beta^T \delta \le \gamma$ has to be one of (6)–(9). Therefore, the only remaining case to consider is when $\alpha \in \{0,1\}^n$ and $\alpha \ne 0$.

Let $\bar{S} := \{j \in J : \alpha_j = 1\}$ and therefore $\alpha^T z = \sum_{j \in \bar{S}} z_j$. Also remember that $\Delta^{0,u} = \{\delta^0, \ldots, \delta^m\}$ where the first $p \in I$ components of $\delta^p \in \{0,1\}^m$ is 1, and the rest are 0. Then by Theorem 1, the following equations must hold for all δ^p with $p \in \{0, \ldots, m-1\}$,

$$\gamma - \sum_{i=1}^{p} \beta_i = \max\left\{\alpha^T z \mid z \in X^{0,u}(\delta^p)\right\} = \min\left\{u - \mathbb{1}_{\{\bar{S} \cap S_{p+1} \setminus S_p = \emptyset\}}, |\bar{S} \setminus S_p|\right\},$$
(11)

where we define $\mathbb{1}_A$ to be 1 if condition A is true, and 0, otherwise. Similarly, for δ^m, we have

$$\gamma - \sum_{i=1}^{m} \beta_i = \min\{u, |\bar{S} \setminus S_m|\}.$$
(12)

Let $\bar{S}_i = \bar{S} \cap S_i$ for $i \in I$ and let $\Delta_1 = \bar{S}_1$ and $\Delta_i = \bar{S}_i \setminus \bar{S}_{i-1}$ for $i \in \{2, \ldots, m\}$. Note that $\bar{S} = (\bar{S} \setminus S_m) \cup (\cup_{i=1}^{m} \Delta_i)$. The unique solution to Eqs. (11) and (12) is therefore

$$\gamma = \min\left\{u - \mathbb{1}_{\{\Delta_1 = \emptyset\}}, |\bar{S}|\right\}$$

$$\beta_i = \begin{cases} \min\left\{u - \mathbb{1}_{\{\Delta_i = \emptyset\}}, |\bar{S} \setminus S_{i-1}|\right\} - \min\left\{u - \mathbb{1}_{\{\Delta_{i+1} = \emptyset\}}, |\bar{S} \setminus S_i|\right\}, & i < m, \\ \min\left\{u - \mathbb{1}_{\{\Delta_m = \emptyset\}}, |\bar{S} \setminus S_{m-1}|\right\} - \min\{u, |\bar{S} \setminus S_m|\}, & i = m. \end{cases}$$

We now consider 3 cases:

Case 1: $|\bar{S} \setminus S_m| \geq u$. In this case, $|\bar{S} \setminus S_i| \geq u$ also holds for all $i \in I$ and

$$\gamma = u - \mathbb{1}_{\{\Delta_1 = \emptyset\}}, \quad \beta_i = \begin{cases} \mathbb{1}_{\{\Delta_{i+1} = \emptyset\}} - \mathbb{1}_{\{\Delta_i = \emptyset\}}, & i \in \{1, \ldots, m-1\}, \\ -\mathbb{1}_{\{\Delta_m = \emptyset\}}, & i = m. \end{cases}$$

In this case, using inequalities (5), (6), (8), (9), we can write

$$\alpha^T z + \beta^T \delta = \sum_{j \in \Delta_1} z_j + \mathbb{1}_{\{\Delta_1 = \emptyset\}} \underbrace{(-\delta_1)}_{\leq -1 + \sum_{j \in S_1} z_j}$$

$$+ \sum_{i=1}^{m-1} \left[\sum_{j \in \Delta_{i+1}} z_j + \mathbb{1}_{\{\Delta_{i+1} = \emptyset\}} \underbrace{(\delta_i - \delta_{i+1})}_{\leq \sum_{j \in S_{i+1} \setminus S_i} z_j}\right] + \sum_{j \in \bar{S} \setminus S_m} z_j$$

$$\leq \sum_{j \in J} z_j - \mathbb{1}_{\{\Delta_1 = \emptyset\}} \leq u - \mathbb{1}_{\{\Delta_1 = \emptyset\}} = \gamma.$$

Case 2a: $|\bar{S} \setminus S_m| \leq u - 1$ and $|\bar{S}| \leq u - 1$. In this case, (11) and (12) imply

$$\gamma = |\bar{S}|, \quad \text{and} \quad \beta_i = |\bar{S} \setminus S_{i-1}| - |\bar{S} \setminus S_i| = |\Delta_i|, \quad i \in I.$$

In this case, using inequalities (7) and (9), we can write

$$\alpha^T z + \beta^T \delta = \sum_{i=1}^{m}\left[\sum_{j \in \Delta_i}(z_j + \delta_i)\right] + \sum_{j \in \bar{S} \setminus S_m} z_j \leq \sum_{i=1}^{m} |\Delta_i| + |\bar{S} \setminus S_m| = \gamma.$$

Case 2b: $|\bar{S} \setminus S_m| \leq u - 1$ and $|\bar{S}| \geq u$. Let $h := \min\{i \in I : |\bar{S} \setminus S_i| \leq u - 1\}$. In this case,

$$\gamma = u - \mathbb{1}_{\{\Delta_1 = \emptyset\}}, \quad \beta_i = \begin{cases} \mathbb{1}_{\{\Delta_{i+1} = \emptyset\}} - \mathbb{1}_{\{\Delta_i = \emptyset\}}, & i \in \{1, \ldots, h-1\}, \\ u - \mathbb{1}_{\{\Delta_h = \emptyset\}} - |\bar{S} \setminus S_h|, & i = h \\ |\Delta_i| = |\bar{S} \setminus S_{i-1}| - |\bar{S} \setminus S_i|, & i \in \{h+1, \ldots, m\}. \end{cases}$$

In this case, using inequalities (5), (8), (9) and inequality (10) with $S' = \bar{S} \cup S_h$ and $p = h$, we can write

$$\begin{aligned}
\alpha^T z + \beta^T \delta &= \sum_{j \in \Delta_1} z_j + \mathbb{1}_{\{\Delta_1 = \emptyset\}} \underbrace{(-\delta_1)}_{\leq -1 + \sum_{j \in S_1} z_j} + \sum_{i=1}^{h-1} \left[\sum_{j \in \Delta_{i+1}} z_j \right. \\
&\quad + \mathbb{1}_{\{\Delta_{i+1} = \emptyset\}} \underbrace{(\delta_i - \delta_{i+1})}_{\leq \sum_{j \in S_{i+1} \setminus S_i} z_j} \left. \right] + \left[\sum_{j \in \Delta_{h+1}} z_j + (u - |\bar{S} \setminus S_h|) \delta_h \right] \\
&\quad + \sum_{j=h+1}^{m} \left[\sum_{j \in \Delta_{i+1}} z_j + |\Delta_i| \delta_i \right] + \sum_{j \in \bar{S} \setminus S_m} z_j \\
&\leq -\mathbb{1}_{\{\Delta_1 = \emptyset\}} + \sum_{j \in \bar{S} \cup S_h} z_j + (u - |\bar{S} \setminus S_h|) \delta_h + \sum_{i=h+1}^{m} |\Delta_i| \delta_i \\
&\leq u - \mathbb{1}_{\{\Delta_1 = \emptyset\}} = \gamma.
\end{aligned}$$

In all three cases, $\alpha^T z + \beta^T \delta \leq \gamma$ is implied by inequalities (5)–(9) and (10) for all $p \in I$ and $S' \subseteq J$ such that $|S' \setminus S_p| \leq u - 1$. $\square$

Clearly, there is an exponential number of such inequalities as one can write a mixing inequality for each $S' \subseteq J$ and $p \in I$. However, given a fractional solution $(\hat{z}, \hat{\delta}) \in \mathbb{R}_+^{n+m}$, the separation problem can be solved simply by checking if the inequality (10) is violated for $S' = S_p^*$ for all $p \in I$, where

$$S_p^* := S_p \cup \left(\bigcup_{i=p+1}^{m} \{j \in S_i \setminus S_{i-1} : \hat{z}_j - \hat{\delta}_p + \hat{\delta}_i > 0\} \right) \cup \{j \in J \setminus S_m : \hat{z}_j - \hat{\delta}_p > 0\}.$$

We next present a set of necessary conditions for inequalities (10) to be facet-defining for $\text{conv}(X^{0,u})$.

Theorem 5. *Let S be nested and let $p \in I$ and $S' \subseteq J$ be such that $|S' \setminus S_p| \leq u - 1$. Then, without loss of generality, the following three conditions are necessary for the associated inequality (10) to define a facet of $\text{conv}(X^{0,u})$:*
(U1) $S' \supseteq S_p$,   (U2) $|S' \setminus S_{p-1}| \geq u$ if $p \geq 2$,   (U3) $|S'| \geq u + 1$.

Proof. If condition (U1) is not satisfied, then replacing S' with $S' \cup S_p$ in inequality (10) leads to a stronger inequality as $z_j \geq 0$ for all $j \in J$. Similarly, if condition (U2) is not satisfied, then replacing p with $p - 1$ in inequality (10) leads to a stronger inequality as $\delta_p \leq \delta_{p-1}$.

If condition (U3) is not satisfied, then $|S'| \leq u$ and

$$\sum_{j \in S'} z_j + \left(u - |S' \setminus S_p|\right)\delta_p + \sum_{i=p+1}^{m} \left(|S' \setminus S_{i-1}| - |S' \setminus S_i|\right)\delta_i$$

$$= \sum_{j \in S' \cap S_p} (z_j + \delta_p) + \sum_{i=p+1}^{m} \sum_{j \in S' \cap (S_i \setminus S_{i-1})} (z_j + \delta_i) + \sum_{j \in S' \setminus S_m} z_j + \underbrace{\left(u - |S'|\right)}_{\geq 0}\delta_p$$

$$\leq |S' \cap S_p| + \sum_{i=p+1}^{m} |S' \cap (S_i \setminus S_{i-1})| + |S' \setminus S_m| + (u - |S'|) = u,$$

where the last inequality is implied by the fact that $z_j \leq 1$ for all $j \in J$ and $z_j + \delta_i \leq 1$ for all $j \in J$, $i \in I$. Therefore, if condition (U3) is not satisfied then inequality (10) is implied by other valid inequalities. As $\mathrm{conv}(X^{0,u})$ is full-dimensional, conditions (U1)–(U3) are necessary for inequality (10) to define a facet.   □

3.2   Convex Hull Description of $X^{l,n}$

In [14], the authors study the convex hull description of the following set:

$$\left\{ (x, \delta) \in \{0,1\}^{|J|+|I|} : \delta_i = \prod_{j \in S_i} x_j \quad \text{for } i \in I, \ x \in P_{\mathcal{M}} \right\} \tag{13}$$

where $\{S_i\}_{i \in I}$ is a family of nested subsets of a given set J and $P_{\mathcal{M}}$ is the convex hull of incidence vectors associated with independent sets $\mathcal{U}$ of the matroid $\mathcal{M} = (J, \mathcal{U})$ defined on the ground set J. Note that if we let $\mathcal{U}$ to be the set of all subsets of J with cardinality at most k for some $k \in \mathbb{Z}_+$, the constraint $x \in P_{\mathcal{M}}$ simply becomes $\sum_{j \in J} x_j \leq k$. Consequently, taking $k = n - l$ to define the independent sets and replacing x_j with $(1 - z_j)$ for $j \in J$, precisely gives the set $X^{l,n}$. Due to the complementation of the x variables in (13), the upper bound on the sum of the x variables becomes a lower bound on the sum of the z variables. Also note that $X^{0,u}$ is not of the form (13) due to the difference in the multilinear terms.

Using the particular uniform matroid described above, we next translate the results from [14] to our context. We define $S_0 = \emptyset$ for convenience.

Theorem 6 ([14]). *Inequalities (5)–(9) together with*

$$-\sum_{j \in S'} z_j + (|S' \cup S_p| - n + l)\delta_p + \sum_{i=p+1}^{m} (|S' \cup S_i| - |S' \cup S_{i-1}|)\delta_i \leq 0, \tag{14}$$

for all $p \in I$ and $S' \subset J$ that satisfy $|S' \cup S_{p-1}| \leq n - l < |S' \cup S_p|$ give a complete description of $\mathrm{conv}(X^{l,n})$.

Notice that similar to inequalities (10), inequalities (14) above are also defined for subsets of J and both (10) and (14) have the term $\sum_{j \in S'} z_j$ as well as a telescopic sum involving the δ variables. In fact, we can show that inequalities (14) can also be derived by the mixing procedure using the following base inequalities:

$$\frac{1}{M}\sum_{j\in S'} z_j + (1-\delta_i) \geq \frac{|S'\cup S_i| - n + l}{M} \quad \text{for } i = p, p+1, \ldots, m$$

where $S' \subseteq J$, S_p satisfies $|S'\cup S_p| \geq n - l + 1$ and $M > n$ is a large constant.

In [14], the following two conditions (i) $|S'\cup S_{p-1}| \leq n-l$, and (ii) $n-l+1 \leq |S'\cup S_p|$, are implicitly imposed as necessary conditions for (14) to be facet-defining. We next present a stronger characterization of the necessary conditions.

Theorem 7. *Let S be nested and let $p \in I$ and $S' \subset J$ be such that $|S'\cup S_{p-1}| \leq n-l < |S'\cup S_p|$. Then the following conditions are necessary for inequality (14) to define a facet of $\operatorname{conv}(X^{l,n})$: (L1) $S' \cap S_p = \emptyset$ (L2) $|S'| \leq n-l-1$.*

Proof. If condition (L1) is not satisfied, then replacing S' with $S' \setminus S_p$ in inequality (14) leads to a stronger inequality. If condition (L2) is not satisfied, then $|S'| \geq n - l$. By valid inequalities (8), (9), $\delta_p \leq 1$ and $\sum_{j\in J} z_j \geq l$,

$$-\sum_{j\in S'} z_j + (|S'\cup S_p| - n + l)\delta_p + \sum_{i=p+1}^{m} (|S'\cup S_i| - |S'\cup S_{i-1}|)\delta_i$$

$$= -\sum_{j\in S'} z_j + \underbrace{[|S'| - (n-l)]}_{\geq 0}\delta_p + (|S_p \setminus S'|)\delta_p + \sum_{i=p+1}^{m} (|S_i \setminus S_{i-1} \setminus S'|)\delta_i$$

$$\leq -\sum_{j\in S'} z_j + |S'| - (n-l) + \sum_{j\in S_p\setminus S'} (1 - z_j) + \sum_{i=p+1}^{m} \sum_{j\in S_i\setminus S_{i-1}\setminus S'} (1 - z_j)$$

$$= -\sum_{j\in S'\cup S_m} z_j + |S'\cup S_m| - (n-l)$$

$$= -\sum_{j\in J} z_j + \sum_{j\in J\setminus(S'\cup S_m)} z_j + |S'\cup S_m| - (n-l)$$

$$\leq -l + |J \setminus (S'\cup S_m)| + |S'\cup S_m| - (n-l) = 0$$

where the first inequality is implied by the fact that $\delta_p \leq 1$ and $z_j + \delta_i \leq 1$, for all $j \in J$, $i \in I$ and the second inequality is implied by the fact that $\sum_{j\in J} z_j \geq l$, $z_j \leq 1$ for all $j \in J$. Therefore, if condition (L2) is not satisfied then inequality (14) is implied by other valid inequalities. As $\operatorname{conv}(X^{l,n})$ is full-dimensional, conditions (L1) and (L2) are necessary for inequality (14) to define a facet. $\square$

3.3   Convex Hull Description of $X^{l,u}$

We now consider $\operatorname{conv}(X^{l,u})$ when l is not necessarily equal to 0 and u can be strictly less than n. Note that $X^{l,u} = X^{l,n} \cap X^{0,u}$. We next show that

$$\operatorname{conv}(X^{l,u}) = \operatorname{conv}(X^{0,u}) \cap \operatorname{conv}(X^{l,n}).$$

Remember that by Theorem 1 and Lemma 2 we know that if an inequality $\alpha^T z + \beta^T \delta \leq \gamma$ with $\alpha \neq 0$ is facet-defining for $X^{l,u}$, then we can assume

$$\gamma - \beta^T \delta^i = \max_{z \in X^{l,u}(\delta^i)} \alpha^T z \text{ for } i = 0, \ldots, m \qquad (15)$$

where $\delta^0 = \mathbf{0}$ and for $i \in \{1, \ldots, m\}$, δ^i has the first i components equal to 1, and the rest equal to 0.

For a given $\alpha \in \mathbb{R}^n$ and $S \subseteq I$, let $\alpha_S \in \mathbb{R}^{|S|}$ denote the vector with entries $\{\alpha_j : j \in S\}$ and let α_S^* denote the largest entry of α_S, i.e. $\alpha_S^* = \max_{j \in S} \alpha_j$. Notice that for $i \in \{0, \ldots, m-1\}$, the right-hand side of (15) becomes $\alpha_{S_{i+1} \setminus S_i}^*$ plus the largest sum of at least $(l-1)$ and at most $(u-1)$ remaining entries of $\alpha^{J \setminus S_i}$, where S_0 stands for $\emptyset$. Similarly, for $i = m$, the right-hand side becomes the largest sum of at least l and at most u entries of $\alpha^{J \setminus S_m}$.

We next give a characterization of $\text{conv}(X^{l,u})$.

Theorem 8. *The convex hull of $X^{l,u}$ is equal to $\text{conv}(X^{0,u}) \cap \text{conv}(X^{l,n})$.*

Proof. As $X^{l,u} = X^{0,u} \cap X^{l,n}$, we have $\text{conv}(X^{0,u}) \cap \text{conv}(X^{l,n}) \supseteq \text{conv}(X^{l,u})$. We next argue that $\text{conv}(X^{0,u}) \cap \text{conv}(X^{l,n}) \subseteq \text{conv}(X^{l,u})$.

Let $\alpha^T z + \beta^T \delta \leq \gamma$ be a facet-defining inequality for $\text{conv}(X^{l,u})$.

If $\alpha = 0$, then, by Lemma 2, the inequality is a multiple of an inequality from (5)–(9). On the other hand, if $\alpha \neq 0$, then by Theorem 1, we can assume either $\alpha \geq 0$ or $\alpha \leq 0$.

If $\alpha \geq 0$, then using the observation on the right-hand side of (15) above, (β, γ) satisfies

$$\gamma - \sum_{k=1}^{i} \beta_i \geq \alpha_{S_{i+1} \setminus S_i}^* + \text{ sum of largest } (u-1) \text{ remaining entries of } \alpha^{J \setminus S_i} \qquad (16)$$

for $i \in \{0, \ldots, m-1\}$, where S_0 stands for $\emptyset$, and

$$\gamma - \sum_{k=1}^{m} \beta_i \geq \text{ sum of largest } u \text{ remaining entries of } \alpha^{J \setminus S_m}. \qquad (17)$$

This implies that the inequality $\alpha^T z + \beta^T \delta \leq \gamma$ is also valid for $\text{conv}(X^{0,u})$.

On the other hand, if $\alpha \leq 0$, then we can use an identical argument to observe that (α, β, γ) satisfies (16) and (17) with u replaced with l. This implies that $\alpha^T z + \beta^T \delta \leq \gamma$ is also valid for $\text{conv}(X^{l,n})$.

Combining these two cases we conclude that facet-defining inequalities for $\text{conv}(X^{l,u})$ are valid for $\text{conv}(X^{0,u}) \cap \text{conv}(X^{l,n})$ and therefore $\text{conv}(X^{0,u}) \cap \text{conv}(X^{l,n}) \subseteq \text{conv}(X^{l,u})$. $\square$

4   Conclusions

In this paper, we gave a polyhedral characterization of the convex hull of the cardinality constrained multilinear set when the monomial terms have a nested structure. In a separate study, we have obtained a polyhedral characterization when there are just two non-nested monomial terms, and observe that the polyhedral structure is significantly more complicated in the non-nested case. See also [13] for an extension of their matroid work to non-nested multilinear sets.

References

1. Buchheim, C., Klein, L.: Combinatorial optimization with one quadratic term: spanning trees and forests. Discrete Appl. Math. **177**, 34–52 (2014)
2. Crama, Y., Rodríguez-Heck, E.: A class of valid inequalities for multilinear 0–1 optimization problems. Discrete Optim. **25**, 28–47 (2017)
3. Dash, S., Gunluk, O., Wei, D.: Boolean decision rules via column generation. In: Advances in Neural Information Processing Systems, pp. 4655–4665 (2018)
4. Del Pia, A., Khajavirad, A.: A polyhedral study of binary polynomial programs. Math. Oper. Res. **42**(2), 389–410 (2016)
5. Del Pia, A., Khajavirad, A.: The multilinear polytope for acyclic hypergraphs. SIAM J. Optim. **28**(2), 1049–1076 (2018)
6. Del Pia, A., Khajavirad, A.: On decomposability of multilinear sets. Math. Program. **170**(2), 387–415 (2018)
7. Pia, D., Alberto, K., Aida, S., Nikolaos, V.: On the impact of running intersection inequalities for globally solving polynomial optimization problems. Math. Program. Comput. 1–27 (2019). https://doi.org/10.1007/s12532-019-00169-z
8. Demiriz, A., Bennett, K.P., Shawe-Taylor, J.: Linear programming boosting via column generation. Mach. Learn. **46**, 225–254 (2002)
9. Dobkin, D.P., Gunopulos, D., Maass, W.: Computing the maximum bichromatic discrepancy, with applications to computer graphics and machine learning. J. Comput. Syst. Sci. **52**, 453–470 (1996)
10. Eckstein, J., Goldberg, N.: An improved branch-and-bound method for maximum monomial agreement. INFORMS J. Comput. **24**(2), 328–341 (2012)
11. Eckstein, J., Kagawa, A., Goldberg, N.: REPR: rule-enhanced penalized regression. INFORMS J. Optim. **1**(2), 143–163 (2019)
12. Fischer, A., Fischer, F.: Complete description for the spanning tree problem with one linearised quadratic term. Oper. Res. Lett. **41**, 701–705 (2013)
13. Fischer, A., Fischer, F., McCormick, S.T.: Matroid optimisation problems with monotone monomials in the objective (2017, preprint)
14. Fischer, A., Fischer, F., McCormick, S.T.: Matroid optimisation problems with nested non-linear monomials in the objective function. Math. Program. **169**(2), 417–446 (2017). https://doi.org/10.1007/s10107-017-1140-9
15. Goldberg, N., Eckstein, J.: Boosting classifiers with tightened l_0-relaxation penalties. In: 27th International Conference on Machine Learning, Haifa, Israel (2010)
16. Golderg, N.: Optimization for sparse and accurate classifiers. Ph.D. thesis, Rutgers University, New Brunswick, NJ (2012)
17. Günlük, O., Pochet, Y.: Mixing mixed-integer inequalities. Math. Program. **90**(3), 429–457 (2001)
18. Mehrotra, A.: Cardinality constrained boolean quadratic polytope. Discrete Appl. Math. **79**, 137–154 (1997)
19. Padberg, M.: The boolean quadric polytope: some characteristics, facets and relatives. Math. Program. **45**(1–3), 139–172 (1989)

On the Multiple Steiner Traveling Salesman Problem with Order Constraints

Raouia Taktak[1(✉)] and Eduardo Uchoa[2(✉)]

[1] ISIMS & LT2S/CRNS, Université de Sfax, Sfax, Tunisia
raouia.taktak@isims.usf.tn
[2] Departamento de Engenharia de Produção, Universidade Federal Fluminense, Niterói, RJ, Brazil
eduardo.uchoa@gmail.com

Abstract. The paper deals with a variant of the Traveling Saleman Problem (TSP), called the Multiple Steiner TSP with Order Constraints (MSTSPOC). Consider an undirected weighted graph, and a set of salesmen such that with each salesman is associated a set of ordered terminals. The MSTSPOC consists in finding a minimum-cost subgraph containing for each salesman a tour going in order through its terminals. We propose an ILP formulation for the problem, study the associated polytope and investigate the facial aspects of the basic constraints. We further identify new families of valid inequalities. Some computational results are also presented.

Keywords: Steiner TSP · Order constraints · Polytope · Facet-defining · Valid inequalities · Branch-and-Cut

1 Introduction and Related Works

We consider a variant of the Traveling Salesman Problem (TSP), that is the Multiple Steiner Traveling Salesman Problem with Order Constraints (MSTSPOC). The problem was introduced in [10,11], motivated by reliability issues in multilayer telecommunication networks, and is proved to be NP-hard even for a single salesman [11]. Let $G = (V, E)$ be an undirected weighted graph, and K the set of salesmen. For each salesman $k \in K$, there is a set $T_k \subseteq V$ of *terminals*, and a set $S_k \subset V$ of *Steiner nodes* such that $T_k \cap S_k = \emptyset$. The MSTSPOC consists in finding a set of edges $F \subseteq E$, with minimum total weight, such that for each salesman $k \in K$ there is a tour that visits all terminals in T_k in a predefined cyclic order. Steiner nodes not belonging to T_k are optional. They may be traversed between visits to terminals. Moreover, due to some survivability restrictions explained in [10,11], tours must be *elementary*, that is nodes and edges are not allowed to be visited more than once. Mahjoub et al. [19] propose an ILP compact formulation for the MSTSPOC. The formulation is based on a layered view of the problem. An extensive computational study shows the efficiency of the proposed formulation in solving hard instances.

© Springer Nature Switzerland AG 2020
M. Baïou et al. (Eds.): ISCO 2020, LNCS 12176, pp. 66–77, 2020.
https://doi.org/10.1007/978-3-030-53262-8_6

Several variants of the TSP are in close relationship with the MSTSPOC. The first variant is the so-called *Steiner Traveling Salesman Problem (STSP)* in which only a given subset of nodes, called terminals, must be visited in a minimum-weight cycle. The STSP was first introduced by Cornuéjols et al. [12]. The problem is considered in the graphical case, and the associated polytope is investigated. In [5], Baïou and Mahjoub give a complete polyhedral description of the STSP in series-parallel graphs. In [21], Steinová proposes some approximation results for the STSP. Letchford et al. propose in [16] several compact formulations for the STSP deduced from the ones known for the TSP. In [15], Interian and Ribeiro develop an efficient GRASP heuristic-based algorithm for the problem. The second variant is the *TSP with Precedence Constraints*. This arises when the circuit starts and ends at a given node, and precedence constraints between some pairs of nodes are considered. This variant was also widely studied in the literature. In [6], Balas et al. study the polytope of the Asymmetric TSP with precedence constraints. In [4], Ascheuer et al. propose a Branch-and-Cut algorithm to solve the problem. In [13], Gouveia et al. propose several formulations for the Asymmetric TSP and the related precedence constrained version combining precedence variable based formulations with network flow based formulations. A further interesting variant is the so-called *Multiple TSP* which consists in finding a set of tours for a predefined number of salesmen, each tour starting from and coming back to a depot node while visiting exactly once the other intermediate nodes. Bektas [7] presents a survey of formulations and solution approaches for the problem assuming that all salesmen start from the same depot. In [8], Benavent and Martínez study the problem when salesmen are assumed to start from different depots. The authors propose an ILP formulation for the problem. Several families of valid inequalities are identified and a substantial polyhedral study is presented. The authors devise an efficient Branch-and-Cut algorithm to solve the problem. In [20], Sarin et al. study the multiple asymmetric TSP with precedence constraints and investigate the performances of 32 formulations modeling the problem.

The paper is organized as follows. In the next section, we give necessary notations and propose an ILP formulation for the MSTSPOC. In Sect. 3, we define the associated polytope and study the facial aspect of the basic constraints. In Sect. 4, we describe some valid inequalities and give necessary conditions and sufficient conditions for these inequalities to be facet defining. Due to space limits, all the polyhedral and facet-defining proofs will be skipped in the paper. In Sect. 5, we present a preliminary computational study. Some concluding remarks and indications for future work are given in Sect. 6.

2   Integer Linear Programming Formulation

2.1   Notations

Let $G = (V, E)$ be an edge-weighted undirected graph, such that with each edge $e \in E$ is associated a non-negative weight w_e. An edge e between two nodes u and v in V will be denoted by $e = uv$. We suppose given a set K of salesmen, each having to visit a set of *terminals* $T_k = \{w_1^k, w_2^k, \ldots, w_{|T_k|}^k\}$, $k \in K$. The order of terminals' visitation for each tour (salesman) is assumed to follow the cyclic order of the indices j in w_j^k. Consider a terminal $w_j^k \in T_k$, terminals in $T_k \setminus \{w_{j-1}^k, w_{j+1}^k\}$ are said to be *non-successive* for w_j^k, $j = 1, 2, \ldots |T_k|$, where $w_{|T_k|+1}^k$ refers to w_1^k. Nodes of V that are not terminals for $k \in K$ are called *Steiner nodes* $S_k \subset V$, with $T_k \cap S_k = \emptyset$. Consider $k \in K$ and denote by $q_j^k = (w_j^k, w_{j+1}^k)$ the *section* defined by the two successive terminals w_j^k and w_{j+1}^k, $j \in \{1, 2, \ldots, |T_k|\}$. Each tour for a salesman $k \in K$ can be seen as union of node-disjoint section paths. Let $\mathcal{T}_k$ denote the set of these sections for $k \in K$. For each section $q_j^k = (w_j^k, w_{j+1}^k) \in \mathcal{T}_k$, $k \in K$ and $j \in \{1, 2, \ldots, |T_k|\}$, we associate a reduced graph denoted by $G_j^k = (V_j^k, E_j^k)$, in which a path joining w_j^k to w_{j+1}^k has to be found. The reduced graph G_j^k is obtained from the original graph G by deleting all the terminals of T_k, except w_j^k and w_{j+1}^k, as well as their incident edges. Let $W \subset V$ be a subset of nodes of V. Recall that $\delta_G(W)$ denotes the *cut* calculated in G, that is the set of edges of G having one node in W and the other in $\overline{W} = V \setminus W$. When $W = \{w\}$, $w \in V$, we will write $\delta_G(w)$. In the sequel we simply write $\delta(W)$ and $\delta(w)$, if the cut is calculated in G.

2.2   ILP Formulation

Consider a salesman $k \in K$ and an edge $e \in E$. We define the binary variable x_e^k that is equal to 1 if the tour of salesman k uses edge e, and 0 otherwise. We also define the binary variable y_e for each $e \in E$, that is equal to 1 if edge e is considered in the final solution, 0 otherwise. The MSTSPOC can hence be formulated as follows.

$$min \sum_{e \in E} w_e y_e \tag{1}$$

$$\sum_{e \in \delta_{G_j^k}(W)} x_e^k \geq 1 \qquad \begin{array}{l} \text{for all } k \in K,\ q_j^k = (w_j^k, w_{j+1}^k) \in \mathcal{T}_k, \\ W \subset V_j^k : w_j^k \in W \text{ and } w_{j+1}^k \in \overline{W}, \end{array} \tag{2}$$

$$\sum_{e \in \delta(w)} x_e^k \leq 2 \qquad \text{for all } w \in V,\ k \in K, \tag{3}$$

$$x_e^k \leq y_e \qquad \text{for all } e \in E,\ k \in K, \tag{4}$$

$$0 \leq x_e^k \qquad \text{for all } e \in E,\ k \in K, \tag{5}$$

$$y_e \leq 1 \qquad \text{for all } e \in E, \tag{6}$$

$$x_e^k \in \{0, 1\} \qquad \text{for all } e \in E,\ k \in K, \tag{7}$$

$$y_e \in \{0, 1\} \qquad \text{for all } e \in E. \tag{8}$$

Inequalities (2) are called *section cut inequalities*. They ensure for each section $q_j^k = (w_j^k, w_{j+1}^k)$ corresponding to a salesman $k \in K$ a path in the reduced graph G_j^k. Hence, this guarantees for each salesman a tour going in order through its terminals. Inequalities (3) are called *disjunction inequalities*. They ensure that the different sections for a salesman $k \in K$ are disjoint, and hence that the associated tour is elementary. Inequalities (4) are the *linking inequalities* which express the fact that if an edge $e \in E$ is not considered, that is $y_e = 0$, then it can not be used in any tour for the salesmen K. Finally, inequalities (5) and (6) are the *trivial inequalities*, and (7) and (8) are the *variables' integrality constraints*.

3   Polyhedral Analysis

An instance of the MSTSPOC corresponds to the triplet (G, K, T), where G is a graph, K a set of salesmen, each salesman has a set T_k of terminals and $T = \bigcup_{k \in K} T_k$. We denote by MSTSPOC$(G, K, T)$ the polytope associated with the MSTSPOC, that is the convex hull of the solutions of formulation (2)–(8) corresponding to K and T in G, i.e.

$$\text{MSTSPOC}(G, K, T) = conv\{w^T y \,|\, (x, y) \in \{0, 1\}^{|E|(|K|+1)} : (x, y) \text{ satisfies } (2)\text{–}(6)\}.$$

In what follows, G is assumed to be complete and for each salesman $k \in K$ $T_k \neq V$. These assumptions are not restrictive. First, because if the graph is not complete, one can consider a complete graph by associating very high costs to the non-existent edges. Moreover, if there exists a salesman $k \in K$ such that $S_k = \emptyset$, then the solution is unique for this salesman. In this case, the problem reduces to solving the MSTSPOC for the $K \setminus \{k\}$ remaining salesmen.

The dimension of polytope MSTSPOC(G, K, T) is given in the following result.

Theorem 1 (Polytope Dimension)

$$dim(MSTSPOC(G, K, T)) = (|K| + 1)|E| - \sum_{k \in K} \frac{|T_k|(|T_k| - 1)}{2}.$$

In what follows, we study the facial structure of the polytope MSTSPOC (G, K, T). In particular, we state necessary and sufficient conditions for inequalities of the ILP formulation to be facet defining (as mentioned above, all the proofs will be skipped due to space limitation).

Theorem 2 (x trivial inequalities). *Consider $k \in K$ and $e \in E$. Inequality $x_e^k \geq 0$ defines a facet of MSTSPOC(G, K, T) if and only if e is not between non-successive terminals of T_k.*

Theorem 3 (y trivial inequalities). *Consider $e \in E$. Inequality $y_e \leq 1$ defines a facet of MSTSPOC(G, K, T).*

Theorem 4 (Section cut inequalities). *Consider $k \in K$ and a section $q_j^k = (w_j, w_{j+1}) \in T_k$. Consider W a subset of nodes of V_j^k such that $w_j \in W$ and $w_{j+1} \in \overline{W}$. Inequality (2) defines a facet of MSTSPOC(G, K, T) if and only if $W \cap S_k \neq \emptyset \neq \overline{W} \cap S_k$.*

Theorem 5 (Disjunction inequalities). *Consider $k \in K$ and $w \in V$. Inequality $\sum_{e \in \delta(w)} x_e^k \leq 2$ defines a facet for MSTSPOC(G, K, T) if and only if w is not a terminal of T_k.*

4   Valid Inequalities

4.1   Steiner Cut Inequalities

The first family of valid inequalities is a straight consequence related to the connectivity requirements of the problem. Consider the graph of Fig. 1 which consists of four nodes, three terminals numbered $1, 2$ and 3 and a Steiner node, namely node 4. The instance consists of a tour going in order through terminals $(1, 2, 3, 1)$. Figure 1 shows a fractional solution for this instance. Let $\bar{x}$ be the solution given by $\bar{x}_{e_i} = \frac{1}{2}$ for $i = 1, ..., 6$. Clearly, $\bar{x}$ satisfies all the constraints of the linear relaxation (2)–(6). However, $\bar{x}$ violates the inequality $x_{e_1} + x_{e_2} + x_{e_3} \geq 2$, which is valid for the MSTSPOC(G, K, T) polytope. In the following proposition, we state that this inequality belongs to a more general class of valid inequalities.

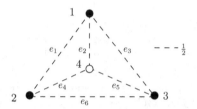

Fig. 1. First fractional solution

Proposition 1. *Consider a salesman $k \in K$ and let $W \subset V$ such that $W \cap T_k \neq \emptyset \neq \overline{W} \cap T_k$. Then*

$$\sum_{e \in \delta(W)} x_e^k \geq 2 \tag{9}$$

is valid for MSTSPOC(G, K, T), and is called Steiner cut inequality.

Theorem 6. *Inequality $\sum_{e \in \delta(W)} x_e^k \geq 2$ defines a facet of MSTSPOC(G, K, T) if and only if the following conditions hold.*

1. *W and $\overline{W}$ do not contain non-successive terminals of T_k,*
2. *If $|W \cap T_k| \geq 3$ (resp. $|\overline{W} \cap T_k| \geq 3$), then $S_k \subset W$ (resp. $S_k \subset \overline{W}$),*
3. *If $|W \cap T_k| = 2$ (resp. $|\overline{W} \cap T_k| = 2$), then $W \cap S_k \neq \emptyset$ (resp. $\overline{W} \cap S_k \neq \emptyset$).*

4.2   Steiner Non-successive Terminals Inequalities

In this section, we introduce a new family of valid inequalities for the MST-SPOC. These come enhancing the constraints related to sections' disjunction and order between the terminals of a given salesman. In Fig. 2, is presented a graph consisting of six nodes, four terminals (nodes $1, 2, 3$ and 4) and two Steiner nodes (nodes 5 and 6). Figure 2 illustrates a solution for an instance looking for a

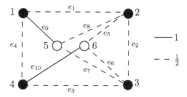

Fig. 2. Second fractional solution

tour going in order through terminals $(1, 2, 3, 4, 1)$. Figure 2 presents a fractional solution for this instance. Let $\bar{x}$ be the solution given by $\bar{x}_{e_i} = \frac{1}{2}$ for $i = 1, ..., 8$ and $\bar{x}_{e_i} = 1$ for $i = 9, 10$. Clearly, $\bar{x}$ satisfies all the constraints of (2)–(6) as well as the Steiner cut inequalities (9) previously introduced. Observe that, in this solution, the Steiner node 5 has three incident edges e_7, e_8 and e_9 such that $\bar{x}_{e_7} = \frac{1}{2}$, $\bar{x}_{e_8} = \frac{1}{2}$ and $\bar{x}_{e_9} = 1$. This implies that the Steiner node 5 is used to route two sections, namely section $(1, 2)$ and section $(2, 3)$, violating hence the sections' disjunction constraint that ensure an elementary tour. In order to cut this fractional point, one can add the inequality $x_{e_8} \geq x_{e_9}$, which is valid for $MSTSPOC(G, K, T)$. This inequality expresses the fact that if edge e_9 is considered in a solution, it should be used to route only one among the sections adjacent to terminal 1, that is to say either section $(1, 2)$ or section $(4, 1)$. As in this case the Steiner node 5 can be used to route section $(1, 2)$, this means that if edge e_9 is considered, then edge e_8 must also be taken in the solution. This can be generalized as follows. Consider a salesman $k \in K$ and let w_j be a terminal of T_k. Consider a Steiner node s of S_k and denote $f = sw_j$. Denote the edges linking the Steiner node s with the terminals of T_k non-successive to w_j by $e_1, e_2, ..., e_p$ (see Fig. 3). Remark that if the edge f is considered in a solution $\mathcal{S}$, it can be used to route only one among the sections (w_{j-1}, w_j) and (w_j, w_{j+1}). Thus, none of the edges $e_1, e_2, ..., e_p$ could be considered in the solution $\mathcal{S}$. This can be expressed by the inequality

$$\sum_{e \in \delta'(s)} x_e^k \geq x_f^k, \tag{10}$$

where $\delta'(s) = \delta(s) \setminus \{f, e_1, e_2, ..., e_p\}$.

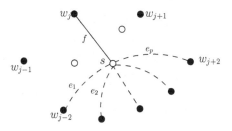

Fig. 3. Steiner non-successive terminals inequalities (configuration 1)

Note that inequality (10) can be viewed otherwise. Indeed, one could say that the flow entering from terminal w_j to the Steiner node s must be conserved when leaving the Steiner node s, and must be used to route only sections (w_{j-1}, w_j) and (w_j, w_{j+1}). In the following, we propose a generalization of inequality (10). Consider a salesman $k \in K$ such that $|T_k| \geq 4$ and let w_j be a terminal of T_k. Consider a subset of Steiner nodes $S \subset S_k$ and let $\Pi = (V_0, V_1, ..., V_p)$, $p \geq 4$ be a partition of V (see Fig. 4) such that:

1. $V_0 = S$,
2. $V_1 \cap T_k = \{w_{j-l}, ..., w_{j-2}, w_{j-1}\}$, that is V_1 contains a sequence of successive terminals ending by w_{j-1},
3. $V_2 = \{w_j\}$,
4. $V_3 \cap T_k = \{w_{j+1}, w_{j+2}, ..., w_{j+l'}\}$, that is V_3 contains a sequence of successive terminals ending by $w_{j+l'}$,
5. $V_4, ..., V_p$ are such that $V_i \cap T_k \neq \emptyset$ and $V_i \cap S_k = \emptyset$, $i = 4, ..., p$.

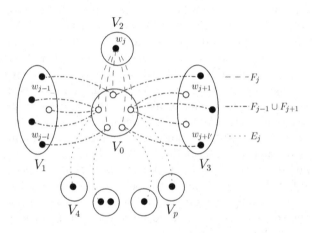

Fig. 4. Steiner non-successive terminals inequalities (configuration 2)

Denote F_{j-1}, F_j, F_{j+1} and E_j the sets of edges of E given by: $F_{j-1} = [V_0, V_1]$, $F_j = [V_0, V_2]$, $F_{j+1} = [V_0, V_3]$, and $E_j = \bigcup_{i=4}^{p} ([V_0, V_i])$. With partition Π and the sets of edges F_{j-1}, F_j, F_{j+1} and E_j, we associate the following inequality

$$\sum_{e \in \delta'(S)} x_e^k \geq \sum_{e \in F_j} x_e^k, \tag{11}$$

where $\delta'(S) = F_{j-1} \cup F_{j+1} = \delta(S) \setminus \{E_j, F_j\}$. Inequality (11) implies the following. The flow going from w_j to a subset of Steiner nodes $S \subseteq S_k$ must be conserved in S and used only to route sections that are adjacent to w_j.

Proposition 2. *Inequality* (11) *is valid for MSTSPOC(G, K, T), and is called Steiner non-successive terminals inequality.*

Theorem 7. *Inequality* (11) *defines a facet of MSTSPOC(G, K, T) if and only if $V_1 \cap T_k = \{w_{j-1}\}$ and $V_3 \cap T_k = \{w_{j+1}\}$.*

4.3   F-partition Inequalities

The F-partition inequalities were first introduced by Mahjoub in 1994 [17]. Further works have shown the efficiency of this class of inequalities to solve different variants of the survivable network design problem (see for instance [9,14,18]). In what follows, we discuss the F-partition inequalities for the MSTSPOC. First, we give a fractional solution which is cut by an F-partition inequality. In Fig. 5 is illustrated a graph that consists of six nodes, three terminals $1, 2$ and 3, and three Steiner nodes $4, 5$ and 6. The instance consists of a tour going in order through terminals $(1, 2, 3, 1)$. Let $\bar{x}$ be the solution given by $\bar{x}_{e_i} = 1$ for $i = 1, 2, 3$ and $\bar{x}_{e_i} = \frac{1}{2}$ for $i = 4, ..., 9$. It is not hard to see that $\bar{x}$ satisfies all the constraints of the linear relaxation (2)–(6) and all the valid inequalities previously introduced, namely the Steiner cut inequalities (9) and the Steiner non-successive terminals inequalities (11). However, the fractional solution of Fig. 5 violates a valid inequality as it will be shown in the following. Consider the partition $\Pi = (V_0, V_1, V_2, V_3)$ of V given by $V_0 = \{4, 5, 6\}$, $V_1 = \{1\}$, $V_2 = \{2\}$ and $V_3 = \{3\}$. Let $F = \{e_1, e_2, e_3\}$ (see Fig. 6). One can easily check that $\bar{x}$ violates the inequality $x_{e_4} + x_{e_5} + x_{e_6} \geq 3 - \lfloor \frac{|F|}{2} \rfloor = 2$, which is valid for MSTSPOC(G, K, T). This inequality is a special case of a more general class of inequalities stated in the following proposition.

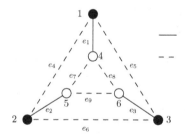

Fig. 5. Third fractional solution

Proposition 3. *Consider a salesman $k \in K$ and let $\Pi = (V_0, ..., V_p)$, $p \geq 2$ be a partition of V such that $|V_i \cap T_k| \geq 1, i = 1, ..., p$. Let $F \subseteq \delta(V_0)$ such that $|F|$ is odd. Then*

$$x^k(\delta(V_0, ..., V_p) \setminus F) \geq p - \lfloor \frac{|F|}{2} \rfloor \tag{12}$$

is valid for MSTSPOC(G, K, T), and is called Steiner F-partition inequality.

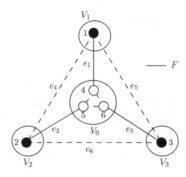

Fig. 6. F-partition inequalities configuration

5   Computational Results

We devise a Branch-and-Cut algorithm for the proposed ILP formulation without and with adding valid inequalities. Due to space limitations, the separation routines of inequalities (2), (9), (11) and (12) are skipped. Note that this order of the inequalities is used when generating them. The Branch-and-Cut algorithm is implemented in C++ with ABACUS 3.2 [1], and makes use of CPLEX 12.6 [2] as linear solver. Experimentations are performed on a Bi-Xeon quad-core E5507 2.27 GHz with 8 Gb of RAM, running under Linux, setting a time limit to 3 h. Tests are computed on small and medium sized instances taken from the TSPLib [3] with some slight modifications in order to adapt the instances to the MSTSPOC. We particularly generate our instances based on data from $a280$, $bier127$, $eil101$, $lin105$ and $tsp225$. Depending on the size of the instance, we choose the $|V|$ first nodes from the original TSPLib instance, and then randomly generate the sets T_k for each $k \in K$. The first series of experimentations aim to show the efficiency of the valid inequalities, and are reported in Table 1. Entries of the table are the following: Instance is the name of the instance, $|V|$ is the number of nodes in the graph, $|K|$ is the number of salesmen, Nodes-i is the number of nodes in the Branch-and-Cut tree, Gap-$i(\%)$ is the relative error between the best upper bound and the lower bound obtained at the root and CPU-i(s) is the total time of execution in seconds ($i = 1, 2$). $i = 1$ for the basic ILP formulation (2)–(8) without valid inequalities, and $i = 2$ is the same formulation for which we added all the valid inequalities (*i.e.*,inequalities (9), (11) and (12)). We present the results obtained for instances having up to 12 nodes in G, and exactly 10 salesmen.

It appears from Table 1 that the formulation with valid inequalities performs better than the basic one for all the instances. In fact, we notice that the number of the Branch-and-Cut tree's nodes for the basic formulation is larger than the one of the formulation with valid inequalities. See, for example, instance $(a, 12, 10)$ for which the Branch-and-Cut tree explored 3899 nodes with the basic formulation and only 153 when adding valid inequalities. For the same instance, the gap and the total time of execution were larger using the basic formulation. And this remark can be generalized for all the tested instances, for which we can clearly

note that the gap and the total time of execution were always better with the
formulation using the valid inequalities. Notice also that, using the basic formulation, some of the instances like $(tsp, 8, 10)$ have not been solved to optimality
within 3 h. However, for the same instances optimality has been reached within
some seconds when adding the valid inequalities. All these observations lead us
to conclude about the importance of the added valid inequalities for a better
resolution of the MSTSPOC.

We also may see the unexpected behaviour when solving instances of size
$|V| = 8$ compared to those of size $|V| = 10$ using the basic formulation. Note that
strangely instances with $|V| = 10$ are solved better than those with $|V| = 8$. This
is due to the fact that instances and mainly terminals are randomly generated,
which implies that difficulty of the instances does not only depend on size, but
also on the generated terminals.

Table 1. Impact of valid inequalities

Instance	$|V|$	$|K|$	Nodes-1	Gap-1(%)	CPU-1(s)	Nodes-2	Gap-2(%)	CPU-2(s)
a	8	10	17	5.18	120.81	3	1.73	1.22
bier	8	10	19	5.39	3063.59	15	2.23	0.96
eil	8	10	7	1.90	2533.11	7	0.72	0.61
lin	8	10	17	5.86	10800.00	5	1.64	1.10
tsp	8	10	21	5.25	10800.00	15	4.20	1.80
a	10	10	79	8.54	7.10	25	4.38	0.88
bier	10	10	83	7.16	6.82	47	6.11	1.87
eil	10	10	43	7.37	6.42	21	5.85	0.92
lin	10	10	43	8.38	3.63	9	2.52	1.16
tsp	10	10	163	11.03	15.88	25	7.78	3.12
a	12	10	3899	12.97	1218.84	153	6.41	3.92
bier	12	10	1479	11.63	398.21	161	8.37	7.23
eil	12	10	489	8.22	128.49	59	6.05	4.58
lin	12	10	1845	13.56	10800.00	181	7.10	1.78
tsp	12	10	457	11.06	10800.00	83	7.64	5.52

A second series of experiments is reported in Table 2 for which we generate
each time 5 different instances of the same size in order to have average results.
For each pair $(|V|, |K|)$, we tested over instances generated from the 5 TSPlib
instances as previously described. Entries of the table are the following: $|V|$ the
number of nodes (ranging from 8 to 16), $|K|$ the number of salesmen (ranging
from 6 to 10), T_k the average number of terminals for each $k \in K$, Nodes the
number of nodes in the Branch-and-Cut tree, Ncut (resp. NScut, NSNST, and
NSFP) the number of generated inequalities of type (2) (resp. (9), (11) and (12)),
Gap(%) is the relative error between the best upper bound (if any) and the lower

bound obtained at the root, Opt the number of instances over 5 that have been solved to optimality and CPU is the total time of execution. All the entries are an average of 5 tested instances of the same size.

Table 2 shows that the difficulty of solving an instance depends mainly on its size (i.e. size of the graph and number of salesmen). In fact, none of the instances with a graph of 16 nodes could be solved to optimality within the time limit. Note also, that for the majority of group of instances we generate a significant number of cuts, Steiner cut and Steiner non-successive terminals inequalities. This means that these inequalities are helpful for the tested instances. However, only small numbers of the Steiner F-partition inequalities have been separated. This can be explained first by the structure of instances, and second by the order of inequalities separation.

Table 2. Branch-and-Cut average results for TSPLib instances

$|V|$	$|K|$	T_k	Nodes	Ncut	NScut	NSNST	NSFP	Gap(%)	Opt	CPU(s)
8	8	4.12	21.8	120.8	227.4	110.8	0.2	3.38	05/05	19.45
8	10	4.10	16.2	130.4	249.4	125.2	1.8	4.72	05/05	1.14
10	8	4.75	37.0	191.2	551.6	144.2	0.4	3.04	05/05	12.65
10	10	5.10	82.2	314.4	999.0	189.0	1.0	4.53	05/05	7.97
12	8	5.62	371.8	1200.2	471.4	277.2	0.6	5.53	05/05	608.45
12	10	5.70	633.8	1650.2	663.8	347.8	0.8	6.49	05/05	476.78
14	8	4.87	1251.2	561.4	673.4	306.0	0.0	–	01/05	13000.47
14	10	4.90	976.7	527.2	864.0	394.2	0.0	–	03/05	10413.94
16	6	4.83	809.6	3632.0	622.6	278.6	0.0	–	0/5	18000.00
16	8	4.87	661.6	863.4	1127.4	386.0	0.0	–	0/5	18000.00

6   Conclusion

In this paper, we study the Multiple Steiner TSP with Order Contraints (MST-SPOC). We propose an ILP formulation for the problem, study the corresponding polytope and the facial aspect of its basic constraints. We also identify new families of valid inequalities, and state necessary conditions and sufficient conditions for these inequalities to be facet-defining. Preliminary experimentations show the efficiency of valid inequalities to tight the linear relaxation of our ILP formulation. In the future, it would be interesting to investigate other valid inequalities and solve larger instances. Another possible extention is to study a close version of the problem which is the Multiple Assymettric Steiner TSP with Precedence Constraints.

References

1. http://www.informatik.uni-koeln.de/abacus/

2. http://www.ilog.com/products/cplex/

3. http://comopt.ifi.uni-heidelberg.de/software/TSPLIB95/

4. Ascheuer, N., Jünger, M., Reinelt, G.: A branch & cut algorithm for the asymmetric traveling salesman problem with precedence constraints. Comput. Optim. Appl. **17**, 61–84 (2000)

5. Baïou, M., Mahjoub, A.R.: Steiner 2-edge connected subgraph polytopes on series-parallel graphs. SIAM J. Discrete Math. **10**, 505–514 (2002)

6. Balas, E., Fischetti, M., Pulleyblank, W.R.: The precedence-constrained asymmetric traveling salesman polytope. Math. Program. **68**, 241–265 (1995)

7. Bektas, T.: The multiple traveling salesman problem: an overview of formulations and solution procedures. Omega **34**(3), 209–219 (2006)

8. Benavent, E., Martínez, A.: Multi-depot multiple TSP: a polyhedral study and computational results. Ann. Oper. Res. **207**(1), 7–25 (2013)

9. Bendali, F., Diarrassouba, I., Mahjoub, A.R., Didi Biha, M., Mailfert, J.: A branch-and-cut algorithm for the k-edge connected subgraph problem. Networks **55**(1), 13–32 (2010)

10. Borne, S., Gabrel, V., Mahjoub, R., Taktak, R.: Multilayer survivable optical network design. In: Pahl, J., Reiners, T., Voß, S. (eds.) INOC 2011. LNCS, vol. 6701, pp. 170–175. Springer, Heidelberg (2011). https://doi.org/10.1007/978-3-642-21527-8_22

11. Borne, S., Mahjoub, A.R., Taktak, R.: A branch-and-cut algorithm for the multiple Steiner TSP with order constraints. Electron. Notes Discrete Math. **41**, 487–494 (2013)

12. Cornuéjols, G., Fonlupt, J., Naddef, D.: The traveling salesman problem on a graph and some related integer polyhedra. Math. Program. **33**, 1–27 (1985)

13. Gouveia, L., Pesneau, P., Ruthmair, M., Santos, D.: Combining and projecting flow models for the (precedence constrained) asymmetric traveling salesman problem. Networks **71**, 451–465 (2017)

14. Huygens, D., Mahjoub, A., Pesneau, P.: Two edge-disjoint hop-constrained paths and polyhedra. SIAM J. Discrete Math. **18**(2), 287–312 (2004)

15. Interian, R., Ribeiro, C.C.: A GRASP heuristic using path-relinking and restarts for the Steiner traveling salesman problem. Int. Trans. Oper. Res. **24**(6), 1307–1323 (2017)

16. Letchford, A.N., Nasiri, S.D., Theis, D.O.: Compact formulations of the Steiner traveling salesman problem and related problems. Eur. J. Oper. Res. **228**(1), 83–92 (2013)

17. Mahjoub, A.R.: Two edge connected spanning subgraphs and polyhedra. Math. Program. **64**(1–3), 199–208 (1994)

18. Mahjoub, A.R., Pesneau, P.: On the steiner 2-edge connected subgraph polytope. RAIRO - Oper. Res. **42**(3), 259–283 (2008)

19. Mahjoub, A.R., Taktak, R., Uchoa, E.: A layered compact formulation for the multiple Steiner TSP with order constraints. In: 2019 6th International Conference on Control, Decision and Information Technologies (CoDIT), pp. 1462–1467 IEEE (2019)

20. Sarin, S.C., Sherali, H.D., Judd, J.D., Tsai, P.-F.J.: Multiple asymmetric traveling salesmen problem with and without precedence constraints: performance comparison of alternative formulations. Comput. Oper. Res. **51**, 64–89 (2014)

21. Steinová, M.: Approximability of the minimum Steiner cycle problem. Comput. Inform. **29**(6+), 1349–1357 (2012)

Integer Programming

On the Linear Relaxation of the $s − t$-cut Problem with Budget Constraints

Hassene Aissi$^{(\boxtimes)}$ and A. Ridha Mahjoub

Paris Dauphine University, Paris, France
{aissi,mahjoub}@lamsade.dauphine.fr

Abstract. We consider in this paper a generalization of the minimum $s − t$ cut problem. Suppose we are given a directed graph $G = (V, A)$ with two distinguished nodes s and t, k non-negative arcs cost functions $c^1, \ldots, c^k : A \to \mathbb{Z}_+$, and $k − 1$ budget bounds $b_1, \ldots, b_{k−1}$ where k is a constant. The goal is to find a $s − t$ cut C satisfying budget constraints $c^h(C) \leqslant b_h$, for $h = 1, \ldots, k − 1$, and whose cost $c^k(C)$ is minimum. We study the linear relaxation of the problem and give necessary and sufficient conditions for which it has an integral optimal basic solution.

Keywords: Integer programming · $s − t$ cut problem · Budget constraints

1 Introduction

We consider in this paper a generalization of the well known minimum $s − t$ cut problem. Consider a directed graph $G = (V, A)$ with two distinguished nodes s and t, k arcs cost functions or *criteria* $c^1, \ldots, c^k : A \to \mathbb{Z}_+$ defined on its arcs, and bounds b_h associated to criteria c^h, for $h = 1, \ldots, k − 1$. We assume that k is a given constant. A *cut* C of G is a subset $C \subseteq V$ such that $\emptyset \neq C \neq V$. For a given cut C, $\delta^+(C)$ is the set of arcs such that the heads are in C and the tails in $V \setminus C$. The *cost* of cut C w.r.t. criterion h is $c^h(C) \equiv c^h(\delta^+(C))$. A $s − t$ *cut* is a cut C such that $s \in C$ and $t \notin C$. The *budgeted minimum $s − t$ cut problem* (BMCP for short) is to find a $s − t$ cut C satisfying the budget constraints $c^h(C) \leqslant b_h$, for $h = 1, \ldots, k − 1$ such that $c^k(C)$ is minimum. This problem has numerous applications in areas such as control of disasters in biological or social networks [10]. While the minimum $s − t$ cut problem can be solved in polynomial time, Papadimitriou and Yannakakis [16] proved that BMCP is strongly NP-complete even for $k = 2$. This implies that for $k > 2$, it is strongly NP-complete even to test if the problem has a feasible solution. Therefore, BMCP is not at all approximable in this case. Chestnut and Zenkluzen [4] give an $O(n)$-approximation algorithm for BMCP with $k = 2$ and show that it cannot be much easier to approximate than Densest k-Subgraph. More precisely, any $n^{o(1)}$-approximation algorithm for BMCP with $k = 2$ gives an $n^{o(1)}$-approximation algorithm for Densest k-Subgraph.

© Springer Nature Switzerland AG 2020
M. Baïou et al. (Eds.): ISCO 2020, LNCS 12176, pp. 81–88, 2020.
https://doi.org/10.1007/978-3-030-53262-8_7

1.1   Related Works

There has been a substantial interest in the study of multicriteria and budgeted versions of several combinatorial optimization problems [1,9,11–13,17]. In general, these versions are shown to be harder than the original ones. One exception is given in [1] where Armon and Zwick showed that the budgeted global minimum cut problem is polynomial-time solvable if graph G is undirected. Ravi and Goemans [17] consider the minimum spanning tree with a single budget constraint and give a polynomial-time approximation scheme. The algorithm exploits the fact that two adjacent spanning trees differ by exactly two edges. Grandoni et al. [9] give polynomial-time approximation scheme algorithms for spanning tree, matroid basis, and bipartite matching problems with a constant number of budget constraints. These algorithms are based on the iterative rounding technique and exploit the structural property of the basic optimal solutions of the linear relaxation of the problems. For instance, the authors show that the support graph of the basic optimal solutions of the linear relaxation of the spanning tree problem with $O(1)$ budget constraints has $O(1)$ edges with fractional values.

Hayrapetyan et al. [10] consider the minimum size bounded cut problem. Given a graph $G = (V, E)$ with edge capacities c_e, source and sink nodes s and t, as well as a bound b, the goal is to find a $s - t$ cut C such that $c(C) \leqslant b$ and minimizing $|C|$. This problem can be reduced to BMCP with $k = 2$ by connecting any node $u \in V \setminus \{s, t\}$ to t. The cost functions are defined as follows: $c_{ut}^2 = 1$ and $c_{ut}^1 = c_{ut}$ and $c_{uv}^2 = 0$ and $c_{uv}^1 = c_{uv}$ for all the remaining edges. Note that only the arcs incident to t have nonzero c^2 cost. The authors give a $(\frac{1}{\varepsilon}, 1)$ or a $(1, \frac{1}{1-\varepsilon})$ pseudo-approximation algorithm for any $0 < \varepsilon < 1$, i.e., the algorithm returns either a super optimal solution violating the budget by a factor $\frac{1}{\varepsilon}$ or a feasible solution with $\frac{1}{1-\varepsilon}$ times more vertices on the s side as the optimal value. However, we don't know a priori which case occurs. This result is based on a combinatorial algorithm for solving the linear relaxation of the problem. The authors reduce the linear program to a special case of the parametric maximum flow problem and use as a black box the algorithm of Gallo et al. [6]. In the parametric maximum flow problem, the arc capacities $c_e(\mu)$ are d-variable affine function of some parameter $\mu \in \mathbb{R}^d$. The goal is to find μ^* which maximizes the maximum flow between s and t. In the problem considered in [10], only the capacities of the arcs incident to t are an affince function of a single parameter. The remaining arcs have static arc capacities (not depending of μ). In this particular case, Gallo et al. show that all the minimum $s - t$ cuts for different μ values can be computed efficiently using a single call to the push relabel algorithm [7]. Hayrapetyan et al. give an optimal solution of the linear relaxation which is a convex combination of two cuts among them. Zhang [18] considers the related problem of the minimum b-size $s - t$ cut. The goal is to find a $s - t$ cut with minimum cost such that the s-side have a size at most b. The author gives a $\frac{b+1}{b+1-b^*}$-approximation algorithm where b^* is the size of the s-side optimal solution. In the worst case, this approximation guarantee could be $O(k)$ if $b^* = b$.

1.2   Our Contributions

The minimum $s - t$ cut problem is one of the few polynomial-time solvable combinatorial problems which become strongly NP-hard by adding a single budget constraint. In order to understand the difficulty of the problem, we investigate the structural properties of the compact linear relaxation of BMCP coming from the dual of the maximum $s - t$ flow problem. We show that the support graph of any fractional basic solution contains few nodes but may have many arcs with fractional values. This contrasts with the simple structure of the basic solution of the minimum spanning tree problem with budget constraints [9,17]. We also give a necessary and sufficient condition for which the linear relaxation of BMCP has an integral optimal basic solution.

2   Structural Properties of the Linear Relaxation

The linear relaxation of BMCP can naturally be formulated by adding the budget constraints to the standard formulation of the minimum $s - t$ cut problem.

$$\min \sum_{(u,v) \in A} c_{uv}^k d_{uv} \tag{1a}$$

$$d_{uv} \geqslant p_u - p_v, \text{for all arcs } (u, v) \in A, \tag{1b}$$

$$p_s = 1, p_t = 0, \tag{1c}$$

$$\sum_{(u,v) \in A} c_{uv}^h d_{uv} \leqslant b_h, \text{for } h = 1, \ldots, k - 1, \tag{1d}$$

$$p_u \geqslant 0 \text{ for all } u \in V \setminus \{s, t\} \tag{1e}$$

$$d_{uv} \geqslant 0 \text{ for all arcs } (u, v) \in A. \tag{1f}$$

In the integer programming problem, of which the above formulation is its linear relaxation, variable p_v (referred to as the potential of v) is set to 1 if $v \in C$ and $p_v = 0$ otherwise, and variable $d_{uv} = 1$ if $u \in C$ and $v \notin C$, and $d_{uv} = 0$ otherwise.

Let $P(G, c, b)$ denote the polyhedron given by inequalities (1b)–(1f), where $c = (c^1, \ldots, c^{k-1})$ denote the cost vectors defining the budget constraints. We consider BMCP under the following assumption.

Assumption: All nodes in $V \setminus \{s, t\}$ are part of an $s - t$ path.

If there exists a node u connected to s but not to t, then there exists an optimal solution of BMCP where u is merged with s. The symmetric case can be handled similarly.

We study in this section structural properties of the linear relaxation of BMCP and give a necessary and sufficient condition for which it has integral optimal basic solution for nonnegative cost vectors $c^1, \ldots, c^k$.

The following example, depicted in Fig. 1, shows that the linear relaxation can have an arbitrary high integrality gap. Assume that $k = 2$ and the cost bound is $b_1 = 2M - 1$ where M is an arbitrary non-negative integer value.

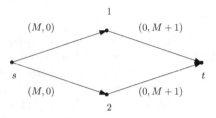

Fig. 1. Instance of BMCP where the linear program has an arbitrary high integrality gap. Each arc (u, v) has two costs (c_{uv}^1, c_{uv}^2).

An optimal integer solution is $\bar{p}_1 = 1, \bar{p}_2 = 0, \bar{d}_{s1} = 0, \bar{d}_{s2} = 1, \bar{d}_{1t} = 1, \bar{d}_{2t} = 0$, with cost values $(M, M+1)$. However, an optimal solution of the linear relaxation is $p_1^* = \frac{1}{M}, p_2^* = 0, d_{s1}^* = 1 - \frac{1}{M}, d_{s2}^* = 1, d_{1t}^* = \frac{1}{M}, d_{2t}^* = 0$, with cost values $(2M - 1, 1 + \frac{1}{M})$. The integrality gap is $M - 1/M$ which may be large.

In order to prove our result, we first give the following definitions and notation.

Definition 1. *A point (d, p) of $P(G, c, b)$ is said to be relevant if the following hold:*

i) $d_{uv} = \max\{p_u - p_v, 0\}$ for all arc $(u, v) \in A$, and
ii) $0 \leqslant p_u \leqslant 1$ for all node $u \in V$.

Definition 2. *Given two points (d^1, p^1) and (d^2, p^2) in $P(G, c, b)$, we say that (d^1, p^1) dominates (d^2, p^2) if $d_{uv}^1 \leqslant d_{uv}^2$ for all arcs $(u, v) \in A$ except for at least one arc $(u_0, v_0) \in A$ where $d_{u_0 v_0}^1 < d_{u_0 v_0}^2$.*

Note that the dominance relation is asymmetric (if (d_1, p_1) dominates (d_2, p_2), then (d_2, p_2) does not dominate (d_1, p_1)) and transitive (if (d_1, p_1) dominates (d_2, p_2) and (d_2, p_2) dominates (d_3, p_3) then (d_1, p_1) dominates (d_3, p_3)). A point (d, p) in $P(G, c, b)$ will be called *minimal* if it is not dominated by any other point in $P(G, c, b)$.

In what follows, we will give a characterization of the linear relaxation of BMCP to be integral by excluding cuts satisfying some specific properties. These cuts are called FRACTIONAL cuts.

Definition 3. *A set $\mathcal{S} = \{C^1, \ldots, C^r\}$ formed by $1 < r \leqslant k$ nested $s - t$ cuts $C^1 \subset C^2 \subset \cdots \subset C^r$ is called FRACTIONAL if the following conditions are met:*

1. *$\mathcal{S}$ contains at least an infeasible cut C^q for some $q \in \{1, \ldots, r\}$, i.e., C^q violates at least one budget constraint,*
2. *there exists a convex combination of these cuts that corresponds to a minimal extreme point (d, p) of $P(G, c, b)$.*

The main result of this paper is the following theorem.

Theorem 1. *Given nonnegative cost vectors $c^1, \ldots, c^k$, the linear relaxation of BMCP has no minimal fractional optimal extreme point if and only if there exists no FRACTIONAL set of nested $s - t$ cuts.*

The proof of Theorem 1 will be given as the consequence of a series of lemmas. A crucial step in the proof is to give a characterization of the minimal points and their relationships with relevant points and Fractional sets of nested $s - t$ cuts.

Lemma 1. *Any minimal point of $P(G, c, b)$ is relevant.*

Proof. Consider a minimal point (d, p) of $P(G, c, b)$ and suppose that it is not relevant. First, assume that Condition i) of Definition 1 does not hold. In this case, there exists an arc $(u_0, v_0) \in A$ such that $d_{u_0 v_0} > \max\{\bar{p}_{u_0} - \bar{p}_{v_0}, 0\}$. Let $(\bar{d}, \bar{p})$ be the solution given by

$$\begin{cases} \bar{p}_u = p_u \text{ for all } u \in V, \\ \bar{d}_{uv} = \max\{\bar{p}_u - \bar{p}_v, 0\} \text{ for all } (u, v) \in A. \end{cases} \tag{2}$$

Since the cost vector c is nonnegative and $\bar{d} \leqslant d$ (with at least a strict inequality), $(\bar{d}, \bar{p})$ satisfies the budget constraints (1d). By construction (2), $(\bar{d}, \bar{p})$ satisfies the constraints (1b). Therefore, $(\bar{d}, \bar{p})$ belongs to $P(G, c, b)$ and $(\bar{d}, \bar{p})$ dominates (d, p). This contradicts that (d, p) is a minimal point.

Now suppose that Condition ii) of Definition 1 does not hold. Therefore, there exists at least one node $\bar{u}_0 \in V$ such that either $p_{\bar{u}_0} > 1$ or $p_{\bar{u}_0} < 0$. Consider solution $(\bar{d}, \bar{p})$ defined as follows:

$$\bar{p}_u = \begin{cases} 1 \text{ if } p_u > 1, \\ 0 \text{ if } p_u < 0, \\ p_u \text{ otherwise,} \end{cases} \tag{3}$$

and

$$\bar{d}_{uv} = \max\{\bar{p}_u - \bar{p}_v, 0\}. \tag{4}$$

By construction (3) and (4) we have $\bar{d} \leqslant d$. Suppose that $p_{\bar{u}_0} > 1$. The case where $p_{\bar{u}_0} < 0$ can be handled similarly. Let V_1 denote the set of nodes $u \in V$ such that $p_u > 1$. By Assumption, any node $u \in V_1$ is part of an $s - t$ path. Since $s, t \notin V_1$, it follows that there exists at least an arc $(u_1, v_1) \in A$ such that $u_1 \in V_1$ and $v_1 \notin V_1$. By (3) and (4), $\bar{d}_{u_1 v_1} < d_{u_1 v_1}$. This shows that $(\bar{d}, \bar{p})$ dominates (d, p) and contradicts again the minimality of (d, p). $\square$

Lemma 2. *For any dominated point (d^0, p^0) of $P(G, c, b)$ there exists a minimal point (d^q, p^q) of $P(G, c, b)$ that dominates it.*

Proof. As (d^0, p^0) is dominated, there exists a point (d^1, p^1) in $P(G, c, b)$ that dominates it. By Definition 2, we have

$$\sum_{(u,v) \in A} d^1_{uv} < \sum_{(u,v) \in A} d^0_{uv}.$$

If (d^1, p^1) is minimal, then we are done. Otherwise, (d^1, p^1) is dominated by a point (d^2, p^2) of $P(G, c, b)$. By transitivity of the dominance relation, (d^2, p^2) dominates (d^0, p^0). Moreover, we have

$$\sum_{(u,v) \in A} d_{uv}^2 < \sum_{(u,v) \in A} d_{uv}^1.$$

Since the optimal value of

$$\min_{(d,p) \in P(G,c,b)} \sum_{(u,v) \in A} d_{uv}$$

is finite, this process must stop at a minimal solution (d^q, p^q) which by transitivity dominates (d^0, p^0).   □

In the following, we show that a minimal optimal basic solution of the linear relaxation of BMCP exists if the cost vectors are nonnegative. Furthermore, we give a structural property satisfied by minimal points.

Lemma 3. *The linear relaxation of BMCP has an optimal extreme point which is minimal for any nonnegative cost vectors $c^1, \ldots, c^k$.*

Lemma 4. *For any minimal point (d, p) of $P(G, c, b)$, there exists an $s - t$ path $P = u_0(= s), u_1, \ldots, u_q(= t)$ such that $p_{u_0} \geqslant p_{u_1} \geqslant \cdots \geqslant p_{u_q}$ and $\sum_{(u_j, u_{j+1}) \in P} d_{u_j u_{j+1}} = 1$.*

We introduce next a reduction operation. Given a vertex (d^*, p^*) of $P(G, c, b)$, consider the following operation:

O: contract an arc $(u, v) \in A$ such that $d_{uv}^* = p_u^* - p_v^*$ and $d_{vu}^* = 0$.

After the operation is over, the potential of the node w (which we call supernode) that results from merging u and v satisfies $p_w^* = p_u^* = p_v^*$. Operation O preserves the d_{uv}^*'s, i.e., for any arc $(i, j) \in A$ such that $i \in \{u, v\}$ and $j \notin \{u, v\}$ we have $d_{wj}^* = d_{ij}^*$. The case where $j \in \{u, v\}$ is handled similarly.

Let $G' = (V', A')$ and $(d^{*'}, p^{*'})$ denote the minor digraph and the solution obtained from (d^*, p^*) by all possible applications of Operation O. By abuse of notation, the supernodes in V' containing s and t are denoted also by s and t, respectively. Let c' denote the restriction of the cost vector c to A'. The following results show that operation O preserves the extremality of the relevant points and yields a reduced graph with at most $k + 1$ nodes.

Lemma 5. *Let (d^*, p^*) be a relevant extreme point of $P(G, c, b)$ and $(d^{*'}, p^{*'})$ denote the point obtained from (d^*, p^*) by all possible application of operation O. Then $(d^{*'}, p^{*'})$ is a relevant extreme point of $P(G', c', b)$.*

Lemma 6. *Let (d^*, p^*) be a relevant extreme point of $P(G, c, b)$. Then the minor G' obtained by all possible applications of Operation O has at most $k + 1$ nodes.*

We investigate in the following the relationship between minimal points and a FRACTIONAL set of nested $s - t$ cuts.

Lemma 7. *For any fractional minimal extreme point (d^*, p^*) of $P(G, c, b)$, there exists a* FRACTIONAL *set $\mathcal{S}$ formed by $r \leqslant k$ nested $s-t$ cuts such that (d^*, p^*) corresponds to a solution satisfying Condition 2 of Definition 3.*

Now we are ready to prove Theorem 1.

Proof of Theorem 1 ($\Leftarrow$) Suppose that there exists no FRACTIONAL set of nested $s-t$ cuts. By Lemma 3, the linear relaxation of BMCP has an optimal minimal extreme point (d^*, p^*). If (d^*, p^*) is fractional, then by Lemma 7 it can be decomposed into a Fractional set of nested $s-t$ cuts. This contradicts that there is no FRACTIONAL set of nested $s-t$ cuts. Therefore, (d^*, p^*) is integral.

($\Rightarrow$) Suppose that there exists a FRACTIONAL set $\mathcal{S} = \{C^1, \ldots, C^r\}$ of nested $s-t$ cuts in G containing an infeasible cut C^q for some $1 \leqslant q \leqslant r$. Let (d^*, p^*) denote the fractional minimal extreme point given in Definition 3 such that $(d^*, p^*) = \sum_{i=1}^{r} \mu_i (d^{C^i}, p^{C^i})$ where $\mu_q > 0$. It is known that any extreme point of a polyhedron is the unique optimal solution of some cost function. In our case, this cost function may not be an arcs cost function (it may associate to some nodes potentials variables p_v nonzero costs). We will construct a cost function $c^k : A \to \{0, 1\}$ such that (d^*, p^*) is an optimal extreme point of BMCP (not necessarily the unique optimal solution).

Let $\mathcal{C}_1(d^*, p^*), \ldots, \mathcal{C}_r(d^*, p^*)$ denote the partition of the nodes into classes according to the decreasing potential values p_i^*. By Lemma 4, there exists an $s-t$ path $P = u_0(=s), u_1, \ldots, u_q = t$ such that $p_{u_0}^* \geqslant \cdots \geqslant p_{u_q}^*$ and $\sum_{(u_j, u_{j+1}) \in P} d_{u_j u_{j+1}}^* = 1$. Define the cost vector c^k such that $c_{uv}^k = 1$ for each arc $(u, v) \in P$ and $c_{uv}^k = 0$ otherwise. Any point (d, p) in $P(G, c, b)$ satisfies

$$\sum_{(u,v) \in A} c_{uv}^k d_{uv} = \sum_{(u_j, u_{j+1}) \in P} d_{u_j u_{j+1}}$$

$$\geqslant \sum_{(u_j, u_{j+1}) \in P} p_{u_j} - p_{u_{j+1}} \text{ (by constraint (1b))}$$

$$= p_{u_0} - p_{u_q} = 1.$$

This shows that (d^*, p^*) is an optimal extreme point of BMCP with cost function c^k. □

References

1. Armon, A., Zwick, U.: Multicriteria global minimum cuts. Algorithmica **46**(1), 15–26 (2006)
2. Carstensen, P.J.: Complexity of some parametric integer and network programming problems. Math. Program. **26**(1), 64–75 (1983)
3. Cohen, E., Megiddo, N.: Maximizing concave function in fixed dimension. In: Pardalos, P.M. (ed.) Complexity in Numerical Optimization, pp. 74–87 (1993)
4. Chestnut, S.R., Zenklusen, R.: Hardness and approximation for network flow interdiction. Networks **69**(4), 378–387 (2017)

5. Ehrgott, M.: Multicriteria Optimization. Springer, Heidelberg (2005). https://doi.org/10.1007/3-540-27659-9
6. Gallo, G., Grigoriadis, M.D., Tarjan, R.E.: A fast parametric maximum flow algorithm and applications. SIAM J. Comput. **18**, 30–55 (1989)
7. Goldberg, A.V., Tarjan, R.E.: A new approach to the maximum-flow problem. J. ACM **35**(4), 921–940 (1988)
8. Garg, N., Vazirani, V.V.: A polyhedron with all s – t cuts as vertices, and adjacency of cuts. Math. Program. **70**, 17–25 (1995)
9. Grandoni, F., Ravi, R., Singh, M., Zenklusen, R.: New approaches to multi-objective optimization. Math. Program. **146**(1–2), 525–554 (2014)
10. Hayrapetyan, A., Kempe, D., Pál, M., Svitkina, Z.: Unbalanced graph cuts. In: Brodal, G.S., Leonardi, S. (eds.) ESA 2005. LNCS, vol. 3669, pp. 191–202. Springer, Heidelberg (2005). https://doi.org/10.1007/11561071_19
11. Hong, S., Chung, S.J., Park, B.H.: A fully polynomial bicriteria approximation scheme for the constrained spanning tree problem. Oper. Res. Lett. **32**(3), 233–239 (2004)
12. Levin, A., Woeginger, G.J.: The constrained minimum weight sum of job completion times. Math. Program. **108**, 115–126 (2006)
13. Martins, E.Q.V.: On a multicriteria shortest path problem. Eur. J. Oper. Res. **16**, 236–245 (1984)
14. Megiddo, N.: Combinatorial optimization with rational objective functions. Math. Oper. Res. **4**(4), 414–424 (1979)
15. Megiddo, N.: Applying parallel computation algorithms in the design of serial algorithms. J. ACM **30**(4), 852–865 (1983)
16. Papadimitriou, C.H., Yannakakis, M.: On the approximability of trade-offs and optimal access of web sources. In: Proceedings of FOCS, pp. 86–92 (2000)
17. Ravi, R., Goemans, M.X.: The constrained minimum spanning tree problem. In: Karlsson, R., Lingas, A. (eds.) SWAT 1996. LNCS, vol. 1097, pp. 66–75. Springer, Heidelberg (1996). https://doi.org/10.1007/3-540-61422-2_121
18. Zhang, P.: A new approximation algorithm for the unbalanced min s–t cut problem. Theor. Comput. Sci. **609**, 658–665 (2016)

An Experimental Study of ILP Formulations for the Longest Induced Path Problem

Fritz Bökler[ID], Markus Chimani[ID], Mirko H. Wagner$^{(\boxtimes)}$[ID], and Tilo Wiedera[ID]

Theoretical Computer Science, Osnabrück University, Osnabrück, Germany
{fboekler,markus.chimani,mirwagner,tilo.wiedera}@uni-osnabrueck.de

Abstract. Given a graph $G = (V, E)$, the LONGESTINDUCEDPATH problem asks for a maximum cardinality node subset $W \subseteq V$ such that the graph induced by W is a path. It is a long established problem with applications, e.g., in network analysis. We propose novel integer linear programming (ILP) formulations for the problem and discuss efficient implementations thereof. Comparing them with known formulations from literature, we prove that they are beneficial in theory, yielding stronger relaxations. Moreover, our experiments show their practical superiority.

1 Introduction

Let $G = (V, E)$ be an undirected graph and $W \subseteq V$. The *W-induced graph* $G[W]$ contains exactly the nodes W and those edges of G whose incident nodes are both in W. If $G[W]$ is a path, it is called an *induced path*. The length of a longest induced path is also referred to as the *induced detour number* which was introduced more than 30 years ago [8]. We denote the problem of finding such a path by LONGESTINDUCEDPATH. It is known to be NP-complete, even on bipartite graphs [18].

The LONGESTINDUCEDPATH problem has applications in molecular physics, analysis of social, telecommunication, and more general transportation networks [3,7,26,33] as well as pure graph and complexity theory. It is closely related to the graph *diameter*—the length of the longest among all shortest paths between any two nodes, which is a commonly analyzed communication property of social networks [30]. A longest induced path witnesses the largest diameter that may occur by the deletion of any node subset in a node failure scenario [30]. The *tree-depth* of a graph is the minimum depth over all of its depth-first-search trees, and constitutes an upper bound on its treewidth [6], which is a well-established measure in parameterized complexity and graph theory. Recently, it was shown that any graph class with bounded degree has bounded induced detour number iff it has bounded tree-depth [32]. Further, the enumeration of induced paths can be used to predict nuclear magnetic resonance [36].

LONGESTINDUCEDPATH is not only NP-complete, but also W[2]-complete [10] and does not allow a polynomial $\mathcal{O}(|V|^{1/2-\epsilon})$-approximation, unless NP =

© Springer Nature Switzerland AG 2020
M. Baïou et al. (Eds.): ISCO 2020, LNCS 12176, pp. 89–101, 2020.
https://doi.org/10.1007/978-3-030-53262-8_8

ZPP [5,25]. On the positive side, it can be solved in polynomial time for several graph classes, e.g., those of bounded mim-width (which includes interval, bi-interval, circular arc, and permutation graphs) [27] as well as k-bounded-hole, interval-filament, and other decomposable graphs [19]. Furthermore, there are NP-complete problems, such as k-COLORING for $k \geq 5$ [23] and INDEPENDENT SET [29], that are polynomial on graphs with bounded induced detour number.

Recently the first non-trivial, general algorithms to solve the LONGESTIN-DUCEDPATH problem exactly were devised by Matsypura et al. [30]. There, three different integer linear programming (ILP) formulations were proposed: the first searches for a subgraph with largest diameter; the second utilizes properties derived from the average distance between two nodes of a subgraph; the third models the path as a walk in which no shortcuts can be taken. Matsypura et al. show that the latter (see below for details) is the most effective in practice.

Contribution. In Sect. 3, we propose novel ILP formulations based on cut and subtour elimination constraints. We obtain strictly stronger relaxations than those proposed in [30] and describe a way to strengthen them even further in Sect. 4. After discussing some algorithmic considerations in Sect. 5, we show in Sect. 6 that our most effective models are also superior in practice.

2 Preliminaries

Notation. For $k \in \mathbb{N}$, let $[k] := \{0, \ldots, k-1\}$. Throughout this paper, we consider a connected, undirected, simple graph $G = (V, E)$ as our input. Edges are cardinality-two subsets of V. If there is no ambiguity, we may write uv for an edge $\{u, v\}$. Given a graph H, we refer to its nodes and edges by $V(H)$ and $E(H)$, respectively. Given a cycle C in G, a *chord* is an edge connecting two nodes of $V(C)$ that are not neighbors along C.

Linear Programming (cf., e.g., [35]). A *linear program* (LP) consists of a cost vector $c \in \mathbb{R}^d$ together with a set of linear inequalities, called *constraints*, that define a polyhedron $\mathcal{P}$ in $\mathbb{R}^d$. We want to find a point $x \in \mathcal{P}$ that maximizes the *objective function* $c^\mathsf{T} x$. This can be done in polynomial time. Unless $\mathsf{P} = \mathsf{NP}$, this is no longer true when restricting x to have integral components; the so-modified problem is an *integer linear program* (ILP). Conversely, the *LP relaxation* of an ILP is obtained by dropping the integrality constraints on the components of x. The optimal value of an LP relaxation is a dual bound on the ILP's objective; e.g., an upper bound for maximization problems. As there are several ways to *model* a given problem as an ILP, one aims for models that yield small dimensions and strong dual bounds, to achieve good practical performance. This is crucial, as ILP solvers are based on a branch-and-bound scheme that relies on iteratively solving LP relaxations to obtain dual bounds on the ILP's objective. When a model contains too many constraints, it is often sufficient to use only a reasonably sized constraint subset to achieve provably optimal solutions. This allows us to add constraints during the solving process, which is called *separation*. We say that model A is *at least as strong* as model B, if for all instances, the LP relaxation's value of model A is no further from the ILP optimum than that of B. If there also exists an instance for which A's LP relaxation yields a tighter bound than that of B, then A is *stronger* than B.

When referring to models, we use the prefix "ILP" with an appropriate subscript. When referring to their respective LP relaxations we write "LP" instead.

Walk-Based Model (State-of-the-Art). Recently, Matsypura et al. [30] proposed an ILP model, ILP_{Walk}, that is the foundation of the fastest known exact algorithm (called A3c therein) for LONGESTINDUCEDPATH. They introduce timesteps, and for every node v and timestep t they introduce a variable that is 1 iff v is visited at time t. Constraints guarantee that nodes at non-consecutive time points cannot be adjacent. We recapitulate details in the arXive version [9]. Unfortunately, ILP_{Walk} yields only weak LP relaxations (cf. [30] and Sect. 4). To achieve a practical algorithm, Matsypura et al. iteratively solve ILP_{Walk} for an increasing number of timesteps until the path found does not use all timesteps, i.e., a non-trivial dual bound is encountered. In contrast to [30], we consider the number of edges in the path (instead of nodes) as the objective value.

3   New Models

We aim for models that exhibit stronger LP relaxations and are practically solvable via single ILP computations. To this end, we consider what we deem a more natural variable space. We start by describing a partial model ILP_{Base}, which by itself is not sufficient but constitutes the core of our new models. To obtain a full model, ILP_{Cut}, we add constraints that prevent subtours.

For notational simplicity, we augment G to $G^* := (V^* := V \cup \{s\}, E^* := E \cup \{sv\}_{v \in V})$ by adding a new node s that is adjacent to all nodes of V. Within G^*, we look for a longest induced cycle through s, where we ignore induced chords incident to s. Searching for a cycle, instead of a path, allows us to homogeneously require that each *selected* edge, i.e., edge in the solution, has exactly two adjacent edges that are also selected. Let $\delta^*(e) \subset E^*$ denote the edges adjacent to edge e in G^*. Each binary x_e-variable is 1 iff edge e is selected. We denote the partial model below by ILP_{Base}:

$$\max \sum_{e \in E} x_e \tag{1a}$$

$$\text{s.t.} \sum_{v \in V} x_{sv} = 2 \tag{1b}$$

$$2x_e \le \sum_{f \in \delta^*(e)} x_f \le 2 \qquad \forall e \in E \tag{1c}$$

$$x_e \in \{0,1\} \qquad \forall e \in E^* \tag{1d}$$

Constraint (1b) requires to select exactly two edges incident with s. To prevent chords, constraints (1c) enforce that any (original) edge $e \in E$, even if not selected itself, is adjacent to at most two selected edges; if e is selected, precisely two of its adjacent edges need to be selected as well.

Establishing Connectivity. The above model is not sufficient: it allows for the solution to consist of multiple disjoint cycles, only one of which contains s. But still, these cycles have no chords in G, and no edge in G connects any two cycles. To obtain a longest single cycle C through s—yielding the longest induced path $G[V(C)\backslash\{s\}]$—we thus have to forbid additional cycles in the solutions that are not containing s. In other words, we want to enforce that the graph induced by the x-variables is connected.

There are several established ways to achieve connectivity: To stay with *compact* (i.e., polynomially sized) models, we could, e.g., augment ILP_{Base} with Miller-Tucker-Zemlin constraints (which are known to be polyhedrally weak [4]) or multi-commodity-flow formulations (ILP_{Flow}; cf. [9]). However, herein we focus on augmenting ILP_{Base} with *cut* or *(generalized) subtour elimination* constraints, resulting in the (non-compact) model we denote by ILP_{Cut}, that is detailed later. Such constraints are a cornerstone of many algorithms for diverse problems where they are typically superior (in particular in practice) than other known approaches [16,17,34]. While ILP_{Cut} and ILP_{Flow} are polyhedrally equally strong (cf. Sect. 4), we know from other problems that the sheer size of the latter typically nullifies the potential benefit of its compactness. Preliminary experiments show that this is indeed the case here as well.

Cut Model (and Generalized Subtour Elimination). Let $\delta^*(W) := \{w\bar{w} \in E^* \mid w \in W, \bar{w} \in V^*\backslash W\}$ be the set of edges in the cut induced by $W \subseteq V^*$. For notational simplicity, we may omit braces when referring to node sets of cardinality one. We obtain ILP_{Cut} by adding *cut constraints* to ILP_{Base}:

$$\sum_{e \in \delta^*(v)} x_e \leq \sum_{e \in \delta^*(W)} x_e \qquad \forall W \subseteq V, v \in W \qquad (2a)$$

These constraints ensure that if a node v is incident to a selected edge (by (1c) there are then two such selected edges), any cut separating v from s contains at least two selected edges, as well. Thus, there are (at least) two edge-disjoint paths between v and s selected. Together with the cycle properties of ILP_{Base}, we can deduce that all selected edges form a common cycle through s.

An alternative view leads to *subtour elimination constraints* $\sum_{e \in E: e \subseteq W} x_e \leq |W| - 1$ for $W \subseteq V$, which prohibit cycles not containing s via counting. It is well known that these constraints can be generalized using binary node variables $y_v := \frac{1}{2}\sum_{e \in \delta^*(v)} x_e$ that indicate whether a node $v \in V$ participates in the solution (in our case: in the induced path) [21]. *Generalized subtour elimination constraints* thus take the form

$$\sum_{e \in E: e \subseteq W} x_e \leq \sum_{w \in W\backslash\{v\}} y_w \qquad \forall W \subseteq V, v \in W. \qquad (2b)$$

One expects ILP_{Cut} and "ILP_{Base} with constraints (2b)" to be equally strong as this is well-known for standard Steiner tree, and other related models [12,13,22]. In fact, there even is a direct one-to-one correspondence between cut constraints (2a) and generalized subtour elimination constraints (2b): By substituting node-variables with their definitions in (2b), we obtain $2\sum_{e \in E: e \subseteq W} x_e \leq$

$\sum_{w\in W\setminus\{v\}}\sum_{e\in\delta^*(v)}x_e$. A simple rearrangement yields the corresponding cut constraint (2a).

Clique Constraints. We further strengthen our models by introducing additional inequalities. Consider any clique (i.e., complete subgraph) in G. The induced path may contain at most one of its edges to avoid induced triangles:

$$\sum_{e\in E:e\subseteq Q}x_e \leq 1 \qquad \forall Q \subseteq V : G[Q] \text{ is a clique} \qquad (3)$$

4 Polyhedral Properties of the LP Relaxations

We compare the above models w.r.t. the strength of their LP relaxations, i.e., the quality of their dual bounds. Achieving strong dual bounds is a highly relevant goal also in practice: one can expect a lower running time for the ILP solvers in case of better dual bounds since fewer nodes of the underlying branch-and-bound tree have to be explored. We defer the proofs of this section to [9].

Since ILP_{Walk} requires *some* upper bound T on the objective value, we can only reasonably compare this model to ours by assuming that we are also given this bound as an explicit constraint. Hence, no dual bound of any of the considered models gives a worse (i.e., larger) bound than T. As it has already been observed in [30], LP_{Walk} in fact *always* yields this worst case bound:

Proposition 1 (Proposition 5 from [30]). *For every instance and every number $T + 1 \leq |V|$ of timesteps LP_{Walk} has objective value T.*

Note that Proposition 1 is independent of the graph. Given that the longest induced path of a complete graph has length 1, we also see that the integrality gap of ILP_{Walk} is unbounded. Furthermore, this shows that ILP_{Base} cannot be weaker than ILP_{Walk}. We show that already the partial model ILP_{Base} is in fact *stronger* than ILP_{Walk}. Let therefore $\theta := T - \text{OPT} \in \mathbb{N}$, where OPT is the instance's (integral) optimum value.

Proposition 2. ILP_{Base} *is stronger than* ILP_{Walk}. *Moreover, for every $\theta \geq 1$ there is an infinite family of instances on which LP_{Base} has objective value at most $\text{OPT} + 1$ and LP_{Walk} has objective value at least $T = \text{OPT} + \theta$.*

Since ILP_{Cut} only has additional constraints compared to ILP_{Base}, this implies that ILP_{Cut} is also stronger than ILP_{Walk}. In fact, since constraints (2a) cut off infeasible integral points contained in ILP_{Base}, LP_{Cut} is clearly even a strict subset of LP_{Base}. As noted before, we can show that using a multi-commodity-flow scheme (cf. [9]) results in LP relaxations equivalent to LP_{Cut}:

Proposition 3. ILP_{Flow} *and* ILP_{Cut} *are equally strong.*

Let $\text{ILP}_{\text{Cut}}^k$ denote ILP_{Cut} with clique constraints added for all cliques on at most k nodes. We show that increasing the clique sizes yields a hierarchy of ever stronger models.

Proposition 4. *For any $k \geq 4$, $\text{ILP}_{\text{Cut}}^k$ is stronger than $\text{ILP}_{\text{Cut}}^{k-1}$.*

5   Algorithmic Considerations

Separation. Since ILP_{Cut} contains an exponential number of cut constraints (2a), it is not practical in its full form. We follow the traditional separation pattern for branch-and-cut-based ILP solvers: We initially omit cut constraints (2a), i.e., we start with model $M := \text{ILP}_{\text{Base}}$. Iteratively, given a feasible solution to the LP relaxation of M, we seek violated cut constraints and add them to M. If no such constraints are found and the solution is integral, we have obtained a solution to ILP_{Cut}. Otherwise, we proceed by branching or—given a sophisticated branch-and-cut framework—by applying more general techniques.

Given an LP solution $\hat{x}$, we call an edge $e \in E$ *active* if $\hat{x}_e > 0$. Similarly, we say that a node is *active*, if it has an active incident edge. These active graph elements yield a subgraph H of G^*. For integral LP solutions, we simply compute the connected components of H and add a cut constraint for each component that does not contain s. We refer to this routine as *integral separation*. For a fractional LP solution, we compute the maximum flow value f_v between s and each active node v in H; the capacity of an edge $e \in E^*$ is equal to $\hat{x}_e$. If $f_v < \sum_{e \in \delta^*(v)} \hat{x}_e$, a cut constraint based on the induced minimum s-v-cut is added. We call this routine *fractional separation*. Both routines manage to find a violated constraint if there is any, i.e., they are *exact* separation routines. In fact, this shows that an optimal solution to LP_{Cut} can be computed in polynomial time [24]. Note that already integral separation suffices to obtain an exact, correct algorithm—we simply may need more branching steps than with fractional separation.

Relaxing Variables. As presented above, our models have $\Theta(|E|)$ binary variables, each of which may be used for branching by the ILP solver. We can reduce this number, by introducing $\Theta(|V|)$ new binary variables y_v, $v \in V$, that allow us to relax the binary x_e-variables, $e \in E$, to continuous ones. The new variables are precisely those discussed w.r.t. generalized subtour elimination, i.e., we require $y_v = \frac{1}{2} \sum_{e \in \delta^*(v)} x_e$. Assuming x_e to be continuous in $[0, 1]$, we have for every edge $e = \{v, w\} \in E$: if $y_v = 0$ or $y_w = 0$ then $x_e = 0$. Conversely, if $y_v = y_w = 1$ then $x_e = 1$ by (1c). Hence, requiring integrality for the y-variables (and, e.g., branching only on them), suffices to ensure integral x values.

Handling Clique Constraints. We use a modified version of the Bron-Kerbosch algorithm [15] to list all maximal cliques. For each such clique we add a constraint during the construction of our model. Recall that there are up to $3^{n/3}$ maximal cliques [31], but preliminary tests show that this effort is negligible compared to solving the ILP. Thus, as our preliminary tests also show, other (heuristic) approaches of adding clique constraints to the initial model are not worthwhile.

6   Computational Experiments

Algorithms. We implement the best state-of-the-art algorithm, i.e., the ILP$_{\text{Walk}}$-based one by Matsypura et al. [30]. We denote this algorithm by "W". For our implementations of ILP$_{\text{Cut}}$, we consider various parameter settings w.r.t. to the algorithmic considerations described in Sect. 5. We denote the arising algorithms by "C" to which we attach sub- and superscripts defining the parameters: the subscript "frac" denotes that we use fractional separation in addition to integral separation. The superscript "n" specifies that we introduce node variables as the sole integer variables. The superscript "c" specifies that we use clique constraints. We consider all eight thereby possible ILP$_{\text{Cut}}$ implementations.

Hard- and Software. Our C++ (GCC 8.3.0) code uses SCIP 6.0.1 [20] as the Branch-and-Cut-Framework with CPLEX 12.9.0 as the LP solver. We use OGDF snapshot-2018-03-28 [11] for the separation of cut constraints. We use igraph 0.7.1 [14] to calculate all maximal cliques. For W, we directly use CPLEX instead of SCIP as the Branch-and-Cut-Framework. This does not give an advantage to our algorithms, since CPLEX is more than twice as fast as SCIP [1] and we confirmed in preliminary tests that CPLEX is faster on ILP$_{\text{Walk}}$. However, we use SCIP for our algorithms, as it allows better parameterizible user-defined separation routines. We run all tests on an Intel Xeon Gold 6134 with 3.2 GHz and 256 GB RAM running Debian 9. We limit each test instance to a single thread with a time limit of 20 min and a memory limit of 8 GB.

Instances. We consider the instances proposed for LONGESTINDUCEDPATH in [30] as well as additional ones. Overall, our test instances are grouped into four sets: RWC, MG, BAS and BAL. The first set, denoted by RWC, is a collection of 22 real-world networks, including communication and social networks of companies and of characters in books, as well as transportation, biological, and technical networks. See [30] for details on the selection. The *Movie Galaxy* (MG) set consists of 773 graphs representing social networks of movie characters [28]. While [30] considered only 17 of them, we use the full set here. The other two sets are based on the Barabási-Albert probabilistic model for scale-free networks [2]. In [30], only the chosen parameter values are reported, not the actual instances. Our set BAS recreates instances with the same values: 30 graphs for each choice $(|V|, d) \in \{(20, 3), (30, 3), (40, 3), (40, 2)\}$, where $|E| = (|V| - d) \cdot d$ describes the density of the graph. As we will see, these small instances are rather easy for our models. We thus also consider a set BAL of graphs on 100 nodes; for each density $d \in \{2, 3, 10, 30, 50\}$ we generate 30 instances. See http://tcs.uos.de/research/lip for all instances, their sources, and detailed experimental results.

Comparison to the State-of-the-Art. We start with the most obvious question: Are the new models practically more effective than the state-of-the-art? See Fig. 1a for BAS and BAL, Fig. 1b for MG, and Table 1 for RWC.

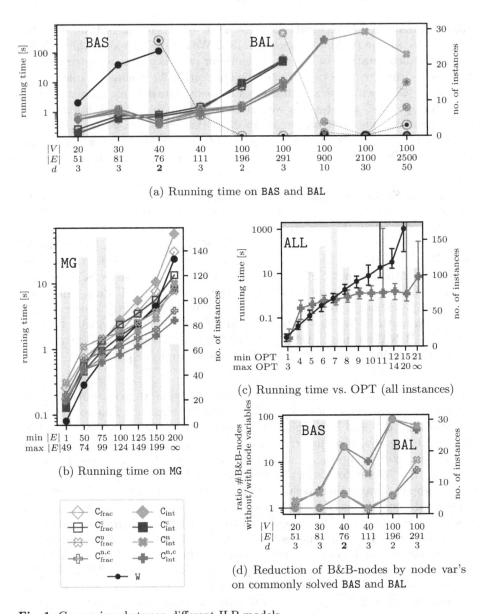

(a) Running time on BAS and BAL

(b) Running time on MG

(c) Running time vs. OPT (all instances)

(d) Reduction of B&B-nodes by node var's on commonly solved BAS and BAL

Fig. 1. Comparison between different ILP models.
(a),(b): Each point is a median, where timeouts are treated as ∞'s. Bars in the background give the number of instances. Gray encircled markers, connected via dotted lines, show the number of solved instances (if not 100%).
(c): Whiskers mark the 20% and 80% percentile. The gray area marks timeouts.

Table 1. Running times [s] on RWC except for **yeast** and **622bus** (solved by none). We denote timeouts by ☉ and mark times within 5% of the minimum in bold.

instance	OPT	$|V|$	$|E|$	W	C_{int}	C_{frac}	C^c_{int}	C^c_{frac}	C^n_{int}	C^n_{frac}	$C^{n,c}_{int}$	$C^{n,c}_{frac}$
high-tech	13	33	91	15.40	0.90	1.11	1.44	3.15	0.51	0.81	**0.41**	2.05
karate	9	34	78	2.98	1.73	1.65	2.12	1.32	1.07	3.71	**0.66**	2.74
mexican	16	35	117	73.30	1.68	2.25	1.12	3.59	1.22	1.34	**0.87**	0.99
sawmill	18	36	62	70.00	0.51	**0.43**	0.50	**0.44**	0.85	3.32	0.82	3.34
tailorS1	13	39	158	83.80	4.78	7.92	4.81	6.45	**1.51**	1.87	3.29	3.55
chesapeake	16	39	170	106.00	**1.84**	13.11	2.11	11.00	2.29	4.88	3.19	4.39
tailorS2	15	39	223	445.00	6.80	21.78	11.92	14.91	3.20	4.31	**2.89**	3.14
attiro	31	59	128	☉	1.76	2.57	2.48	1.75	1.20	1.75	**0.89**	1.19
krebs	17	62	153	522.00	3.86	28.21	18.55	10.03	16.00	11.26	3.90	**2.33**
dolphins	24	62	159	☉	7.95	27.59	22.72	18.33	19.21	**2.99**	3.01	4.70
prison	36	67	142	☉	13.36	5.87	1.09	1.50	3.62	4.05	**1.02**	**1.02**
huck	9	69	297	41.70	☉	144.13	19.46	42.22	114.27	11.63	**5.96**	7.49
sanjuansur	38	75	144	☉	30.67	8.64	24.86	10.33	8.22	**3.65**	3.79	4.71
jean	11	77	254	121.00	464.89	52.89	16.54	9.53	81.03	14.47	**3.88**	5.14
david	19	87	406	☉	666.25	719.46	26.70	45.34	85.88	23.94	**6.93**	10.35
ieeebus	47	118	179	☉	37.10	22.35	39.82	10.60	15.69	**3.13**	22.72	5.61
sfi	13	118	200	44.40	47.41	4.39	4.89	3.77	15.13	2.64	3.31	**2.44**
anna	20	138	493	☉	21.58	296.69	53.21	74.55	439.23	20.27	**7.09**	7.58
usair	46	332	2126	☉	☉	☉	☉	☉	☉	☉	**922.94**	☉
494bus	142	494	586	☉	☉	379.29	☉	379.97	☉	**178.92**	☉	170.74

We observe that on every benchmark set, the various ILP_{Cut} implementations achieve the best running times and success rates. The only exceptions are the instances from MG (cf. Fig. 1b): there, the overhead of the stronger model, requiring an explicit separation routine, does not pay off and W yields comparable performance to the weaker of the cut-based variants. On BAS instances, the cut-based variants dominate (cf. Fig. 1a): while all variants (detailed later) solve all of BAS, W can only solve the instances for $d \in \{20, 30\}$ reliably. On BAL (cf. Fig. 1a) W fails on virtually all instances. The cut-based model, however, allows implementations (detailed later) that solve all of these harder instances. We point out one peculiarity on the BAL instances, visible in Fig. 1a. The instances have 100 nodes but varying density. As the density increases from 2 to 30, the median running times of all algorithmic variants increase and the median success rates decrease. However, from $d = 30$ to $d = 50$ (where only C^n_{int} is successful) the running times drop again and the success rate increases. Interestingly, the number of branch-and-bound (B&B) nodes for $d = 50$ is only roughly 1/7 of those for $d = 30$. This suggests that the denser graphs may allow fewer (near-)optimal solutions and thus more efficient pruning of the search tree.

Comparison of Cut-Based Implementations. Choosing the best among the eight ILP_{Cut} implementations is not as clear as the general choice of ILP_{Cut} over ILP_{Walk}. In Fig. 1a, 1b, and Table 1 we see that, while adding clique constraints

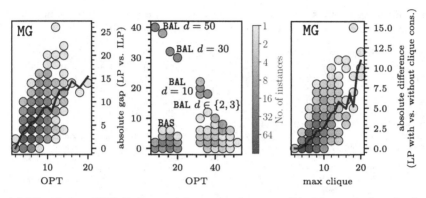

(a) LP value vs. OPT. Left: MG; right: BAS and BAL.   (b) Maximal found clique size vs. LP value on MG.

Fig. 2. Root LP relaxation of cut-based models. The blue line shows the median. (Color figure online)

is clearly beneficial on MG, on BAS and RWC the benefit is less clear. On BAL, we do not see a benefit and for $d \in \{30, 50\}$ we even see a clear benefit of *not* using clique constraints. Each of the graphs from BAL with $d \in \{30, 50\}$ has at least 4541 maximal cliques—and therefore initial clique constraints—, whereas the BAL graphs for $d = 10$ and the RWC graphs yeast and usair have at most 581 maximal cliques and all other graphs have at most 102.

The probably most surprising finding is the choice of the separation routine: while the fractional variant is a quite fast algorithm and yields tighter dual bounds, the simpler integral separation performs better in practice. This is in stark contrast to seemingly similar scenarios like TSP or Steiner problems, where the former is considered by default. In our case, the latter—being very fast and called more rarely—is seemingly strong enough to find effective cutting planes that allow the ILP solver to achieve its computations fastest. This is particularly true when combined with the addition of node variables (detailed later). In fact, C_{int}^n is the only choice that can completely solve all large graphs in BAL.

Adding node variables (and relaxing the integrality on the edge variables) nearly always pays off significantly (cf. Fig. 1a, 1b). Figure 1d shows that the models without node variables require many more B&B-nodes. In fact, looking more deeply into the data, C_{int}^n requires roughly as few B&B-nodes as C_{frac} without requiring the overhead of the more expensive separation routine. Only for BAS with $|V| \in \{20, 30\}$, the configurations without node variables are faster; on these instances, our algorithms only require 2–6.5 B&B-nodes (median).

Dependency of Running Time on the Optimal Value. Since the instances optimal value OPT determines the final size of the ILP_{Walk} instance, it is natural to expect the running time of W to heavily depend on OPT. Figure 1c shows that this is indeed the case. The new models are less dependent on the solution size, as, e.g., witnessed by $C_{int}^{n,c}$ in the same figure.

Practical Strength of the Root Relaxations. For our new models, we may ask how the integer optimal solution value and the value of the LP relaxation (obtained by any cut-based implementation with exact fractional separation) differ, see Fig. 2a. The gap increases for larger values of OPT. Interestingly, we observe that the *density* of the instance seems to play an important role: for BAS and BAL, the plot shows obvious clusters, which—without a single exception—directly correspond to the different parameter settings as labeled. Denser graphs lead to weaker LP bounds in general.

Figure 2b shows the relative improvement to the LP relaxation when adding clique constraints for MG instances. On the other hand for every instance of BAS and BAL the root relaxation did not change by adding clique constraints.

7 Conclusion

We propose new ILP models for LONGESTINDUCEDPATH and prove that they yield stronger relaxations in theory than the previous state-of-the-art. Moreover, we show that they—generally, but also in particular in conjunction with further algorithmic considerations—clearly outperform all known approaches in practice. We also provide strengthening inequalities based on cliques in the graph and prove that they form a hierarchy when increasing the size of the cliques.

It could be worthwhile to separate the proposed clique constraints (at least heuristically) to take advantage of their theoretical properties without overloading the initial model with too many such constraints. As it is unclear how to develop an *efficient* such separation scheme, we leave it as future research.

References

1. Achterberg, T.: SCIP: solving constraint integer programs. Math. Program. Comput. **1**(1), 1–41 (2009)
2. Barabási, A.L., Albert, R.: Emergence of scaling in random networks. Science **286**, 509–512 (1999)
3. Barabási, A.L.: Network Science. Cambridge University Press, Cambridge (2016)
4. Bektaş, T., Gouveia, L.: Requiem for the Miller-Tucker-Zemlin subtour elimination constraints? EJOR **236**(3), 820–832 (2014)
5. Berman, P., Schnitger, G.: On the complexity of approximating the independent set problem. Inf. Comput. **96**(1), 77–94 (1992)
6. Bodlaender, H.L., Gilbert, J.R., Hafsteinsson, H., Kloks, T.: Approximating treewidth, pathwidth, frontsize, and shortest elimination tree. J. Algorithms **18**(2), 238–255 (1995)
7. Borgatti, S.P., Everett, M.G., Johnson, J.C.: Analyzing Social Networks. SAGE Publishing, Thousand Oaks (2013)
8. Buckley, F., Harary, F.: On longest induced paths in graphs. Chin. Quart. J. Math. **3**(3), 61–65 (1988)
9. Bökler, F., Chimani, M., Wagner, M.H., Wiedera, T.: An experimental study of ILP formulations for the longest induced path problem (2020). arXiv:2002.07012 [cs.DS]

10. Chen, Y., Flum, J.: On parameterized path and chordless path problems. In: CCC, pp. 250–263 (2007)
11. Chimani, M., Gutwenger, C., Juenger, M., Klau, G.W., Klein, K., Mutzel, P.: The open graph drawing framework (OGDF). In: Tamassia, R. (ed.) Handbook on Graph Drawing and Visualization, pp. 543–569. Chapman and Hall/CRC (2013). www.ogdf.net
12. Chimani, M., Kandyba, M., Ljubić, I., Mutzel, P.: Obtaining optimal k-cardinality trees fast. J. Exp. Algorithmics **14**, 5:2.5–5:2.23 (2010)
13. Chimani, M., Kandyba, M., Ljubić, I., Mutzel, P.: Strong formulations for 2-node-connected Steiner network problems. In: Yang, B., Du, D.-Z., Wang, C.A. (eds.) COCOA 2008. LNCS, vol. 5165, pp. 190–200. Springer, Heidelberg (2008). https://doi.org/10.1007/978-3-540-85097-7_18
14. Csardi, G., Nepusz, T.: The igraph software package for complex network research. InterJ. Complex Syst. **1695**, 1–9 (2006). http://igraph.sf.net
15. Eppstein, D., Löffler, M., Strash, D.: Listing all maximal cliques in sparse graphs in near-optimal time. In: Cheong, O., Chwa, K.-Y., Park, K. (eds.) ISAAC 2010. LNCS, vol. 6506, pp. 403–414. Springer, Heidelberg (2010). https://doi.org/10.1007/978-3-642-17517-6_36
16. Fischetti, M.: Facets of two Steiner arborescence polyhedra. Math. Program. **51**, 401–419 (1991)
17. Fischetti, M., Salazar-Gonzalez, J.J., Toth, P.: The generalized traveling salesman and orienteering problems. In: Gutin, G., Punnen, A.P. (eds.) The Traveling Salesman Problem and Its Variations. Combinatorial Optimization, vol. 12, pp. 609–662. Springer, Boston (2007). https://doi.org/10.1007/0-306-48213-4_13
18. Garey, M.R., Johnson, D.S.: Computers and Intractability: A Guide to the Theory of NP-Completeness. W. H. Freeman & Co., San Francisco (1979)
19. Gavril, F.: Algorithms for maximum weight induced paths. Inf. Process. Lett. **81**(4), 203–208 (2002)
20. Gleixner, A., et al.: The SCIP optimization suite 6.0. ZIB-Report 18-26, Zuse Institute Berlin (2018). https://scip.zib.de
21. Goemans, M.X.: The steiner tree polytope and related polyhedra. Math. Program. **63**, 157–182 (1994)
22. Goemans, M.X., Myung, Y.S.: A catalog of Steiner tree formulations. Networks **23**, 19–28 (1993)
23. Golovach, P.A., Paulusma, D., Song, J.: Coloring graphs without short cycles and long induced paths. Discrete Appl. Math. **167**, 107–120 (2014)
24. Grötschel, M., Lovász, L., Schrijver, A.: Geometric Algorithms and Combinatorial Optimization. Algorithms and Combinatorics, vol. 2. Springer, Heidelberg (1988)
25. Håstad, J.: Clique is hard to approximate within $n^{1-\epsilon}$. Acta Math. **182**(1), 105–142 (1999)
26. Jackson, M.O.: Social and Economic Networks. Princeton University Press, Princeton (2010)
27. Jaffke, L., Kwon, O., Telle, J.A.: Polynomial-time algorithms for the longest induced path and induced disjoint paths problems on graphs of bounded mim-Width. In: IPEC. LIPIcs, vol. 89, pp. 21:1–13 (2017)
28. Kaminski, J., Schober, M., Albaladejo, R., Zastupailo, O., Hidalgo, C.: Moviegalaxies - Social Networks in Movies. Harvard Dataverse, V3 (2018)
29. Lozin, V., Rautenbach, D.: Some results on graphs without long induced paths. Inf. Process. Lett. **88**(4), 167–171 (2003)
30. Matsypura, D., Veremyev, A., Prokopyev, O.A., Pasiliao, E.L.: On exact solution approaches for the longest induced path problem. EJOR **278**, 546–562 (2019)

31. Moon, J.W., Moser, L.: On cliques in graphs. Israel J. Math. **3**(1), 23–28 (1965)
32. Nesetril, J., de Mendez, P.O.: Sparsity - Graphs, Structures, and Algorithms. Algorithms and Combinatorics, vol. 28. Springer, Heidelberg (2012). https://doi.org/10.1007/978-3-642-27875-4
33. Newman, M.: Networks: An Introduction. Oxford University Press, Oxford (2010)
34. Polzin, T.: Algorithms for the Steiner problem in networks. Ph.D. thesis, Saarland University, Saarbrücken, Germany (2003)
35. Schrijver, A.: Theory of Linear and Integer Programming. Wiley-Interscience Series in Discrete Mathematics and Optimization. Wiley, New York (1999)
36. Uno, T., Satoh, H.: An efficient algorithm for enumerating chordless cycles and chordless paths. In: Džeroski, S., Panov, P., Kocev, D., Todorovski, L. (eds.) DS 2014. LNCS (LNAI), vol. 8777, pp. 313–324. Springer, Cham (2014). https://doi.org/10.1007/978-3-319-11812-3_27

Handling Separable Non-convexities
Using Disjunctive Cuts

Claudia D'Ambrosio[1]([✉]), Jon Lee[2], Daphne Skipper[3],
and Dimitri Thomopulos[4]

[1] LIX CNRS, École Polytechnique, Institut Polytechnique de Paris, Palaiseau, France
dambrosio@lix.polytechnique.fr
[2] Department of IOE, University of Michigan, Ann Arbor, MI, USA
jonxlee@umich.edu
[3] Department of Mathematics, U.S. Naval Academy, Annapolis, MD, USA
skipper@usna.edu
[4] Department of Energy, Process and System Engineering,
Università di Pisa, Pisa, Italy
dimitri.thomopulos@unipi.it

Abstract. D'Ambrosio, Lee, and Wächter (2009, 2012) introduced an algorithmic approach for handling separable non-convexities in the context of global optimization. That algorithmic framework calculates lower bounds (on the optimal min objective value) by solving a sequence of convex MINLPs. We propose a method for addressing the same setting, but employing disjunctive cuts (generated via LP), and solving instead a sequence of convex NLPs. We present computational results which demonstrate the viability of our approach.

Keywords: MINLP · Separable nonconvexity · Disjunctive cuts

Introduction

We consider non-convex global-optimization problems of the form

$$\min \ f_0(\mathbf{x}) + \sum_{j=1}^{n} g_{0,j}(\mathbf{x}_j)$$

subject to:

$$f_i(\mathbf{x}) + \sum_{j=1}^{n} g_{i,j}(\mathbf{x}_j) \leq 0, \text{ for } i = 1, \ldots, m;$$

$$L_j \leq \mathbf{x}_j \leq U_j, \text{ for } j = 1, \ldots, n,$$

C. D'Ambrosio and D. Thomopulos were supported by a public grant as part of the Investissement d'avenir project, reference ANR-11-LABX-0056-LMH, LabEx LMH. This research benefited from the support of the FMJH Program PGMO and from the support of EDF. J. Lee was supported in part by ONR grant N00014-17-1-2296 and LIX, École Polytechnique. D. Skipper was supported in part by ONR grant N00014-18-W-X00709.

© Springer Nature Switzerland AG 2020
M. Baïou et al. (Eds.): ISCO 2020, LNCS 12176, pp. 102–114, 2020.
https://doi.org/10.1007/978-3-030-53262-8_9

where the f_i are convex $(i = 0, \ldots, m)$ and the $g_{i,j}$ are univariate but not necessarily convex $(i = 0, \ldots, m; \ j = 1, \ldots, n)$. So all of the non-convexity is assumed to be separable. We assume that all of the functions are continuous and sufficiently smooth. We may have that some of the variables are restricted to be integers, but this does not directly matter for our approach.

An algorithm for this class of problems was studied in [8,9]. A key aspect of that algorithm is to develop and refine a convex Mixed Integer Nonlinear Program (MINLP) relaxation. The algorithm is based on making a piecewise-convex under-estimator for each $g_{i,j}$, by identifying the concave intervals and using secant under-estimation on them, to get a convex MINLP relaxation — binary variables are used to manage the piecewise functions. Refinement of the piecewise-convex under-estimator is carried out by adding further breakpoints on the concave intervals. After each new breakpoint, a convex MINLP is solved.

Our goal is to use the same starting relaxation, but to relax its integrality, resulting in a continuous, convex Nonlinear Program (NLP). At each iteration, rather than using further breakpoints to tighten the relaxation, we use a Linear Program (LP) to generate a cut, an inequality that is not in the original formulation, but is valid for the formulation and tightens the relaxation. In particular, we iteratively introduce disjunctive cuts as a much more efficient means of improving the NLP relaxation. The efficiency is realized by solving an LP and a convex NLP at each iteration, rather than an expensive convex MINLP.

In Sect. 1, we describe our piecewise-convex under-estimation model. For ease of exposition and economy of notation, we confine our attention to a single univariate function g on domain $[L, U]$. So what we propose applies separately to each of the $g_{i,j}$ defined above. In Sect. 2, we describe a method for tightening our under-estimator using disjunctive cuts, plus some improvements to our basic approach. In Sect. 3, we describe our successful computational results. In Sect. 4, we make some brief conclusions and indicate some plans for future work.

1 A Piecewise-Convex Under-Estimator

Our framework is similar to that of [8,9], with our focus being on optimization models in which all of the non-convexity is separable, with each (univariate) summand being continuous on a finite interval and piecewise sufficiently smooth. In our development, we focus on how to handle each such non-convex summand.

Toward that end, we consider treating a non-convex univariate function $g : [L, U] \to \mathbb{R}$, where $L < U$ are real. We assume that g is continuous on $[L, U]$ and g is piecewise-defined over a finite set of $T \geq 1$ closed subintervals of $[L, U]$ so that g is twice continuously differentiable on each associated open subinterval. To formalize our notation, we assume that g is sub-divided by points p_i,

$$L =: p_0 < p_1 < p_2 < \ldots < p_T := U,$$

so that g is twice continuously differentiable on (p_{i-1}, p_i), for $i = 1, 2, \ldots, T$.

In this section, we develop a convex under-estimator for g (essentially the same as was used in [9]). Toward this end, we assume that g is either convex or concave on each interval $I_i := [p_{i-1}, p_i]$, for $i = 1, 2, \ldots, T$. We note that g could be convex or concave on consecutive intervals. Let $H := \{1, 2, \ldots, T\}$ be the set of indices of these intervals, which we partition into $\check{H} := \{i \in H : g \text{ is convex on } I_i\}$ and $\hat{H} := \{i \in H : g \text{ is not convex on } I_i\}$. Note that g is concave on interval I_i, for $i \in \hat{H}$; and if g is linear on interval I_i, then $i \in \check{H}$.

We employ a set of binary variables z_i, $i = 1, \ldots, T-1$, in order to express g as a separable function of continuous "interval variables" δ_i, for $i = 1, \ldots, T$, using the so-called "delta method" (see [4, pp. 282–283] and [18, Chapter 3], for example). We want to write the variable $x \in [L, U]$ as a function of the interval variables in a certain disciplined manner. Specifically,

$$x = p_0 + \sum_{i=1}^{T} \delta_i, \tag{1}$$

where for each $i = 1, \ldots, T$:

$$0 \le \delta_i \le p_i - p_{i-1}, \text{ and} \tag{2}$$

$$\delta_i > 0 \implies \delta_j = p_j - p_{j-1}, \text{ for } 1 \le j < i. \tag{3}$$

Condition (3) dictates that if δ_i is positive, then all δ_j "to the left" (i.e., $j < i$) should be at their upper bounds (as specified by (2)). In this way, every value of $x \in [L, U]$ is associated with a unique solution of (1).

We accomplish this with the constraints:

$$z_1(p_1 - p_0) \le \delta_1 \le p_1 - p_0; \tag{4}$$

$$z_i(p_i - p_{i-1}) \le \delta_i \le z_{i-1}(p_i - p_{i-1}), \text{ for } i = 2, 3, \ldots, T-1; \tag{5}$$

$$0 \le \delta_T \le z_{T-1}(p_T - p_{T-1}); \tag{6}$$

$$z_i \in \{0, 1\}, \text{ for } i = 1, \ldots, T-1, \tag{7}$$

which drive the correct behavior of the δ_i variables. Note that for a particular $\bar{x} \in [L, U]$, the values of the (ordered) binary z variables form a sequence of 1s followed by a sequence of 0s, where the index $i_{\bar{x}}$ of the first 0 indicates the interval $I_{i_{\bar{x}}}$ that contains $\bar{x}$, so that

$$\delta_i = \begin{cases} p_i - p_{i-1}, & \text{for } i < i_{\bar{x}}; \\ \bar{x} - p_{i-1}, & \text{for } i = i_{\bar{x}}; \\ 0, & \text{for } i > i_{\bar{x}}. \end{cases}$$

Now we can express $g(x)$ as a separable function of the δ-variables,

$$g(x) = \sum_{i=1}^{T} g(p_{i-1} + \delta_i) - \sum_{i=1}^{T-1} g(p_i),$$

giving us access to g on the individual subintervals of $[L, U]$. Our goal now is to give a piecewise-convex under-estimator for g. So we use g as its own convex under-estimator on the subintervals where g is convex, and replace g with a secant under-estimator on the subintervals where g is not convex. Our piecewise-convex under-estimator for g is

$$\underline{g}(x) = \sum_{i=1}^{T} y_i + g(p_0), \tag{8}$$

with

$$y_i \geq \left(\frac{g(p_i) - g(p_{i-1})}{p_i - p_{i-1}} \right) \delta_i, \quad \text{for } i \in \hat{H}, \quad \text{(secant under-estimator)} \quad (9)$$

and

$$y_i \geq g(p_{i-1} + \delta_i) - g(p_{i-1}), \quad \text{for } i \in \check{H}. \quad \text{(under-estimation by } g \text{ itself)} \quad (10)$$

Notice that in the secant under-estimation (9), the right-hand side is linear in δ_i, and is 0 for $\delta_i = 0$ and is $g(p_i) - g(p_{i-1})$ when $\delta_i = p_i - p_{i-1}$.

[10] uses a related approach where concave separable quadratics are extracted from indefinite non-separable quadratics, and then under-estimated by secants.

2 Disjunctive Cuts

Our under-estimator $\underline{g}$ can be quite far from g on the concave subintervals of $[L, U]$. In [9], the algorithmic strategy is to branch at points interior to I_i, for $i \in \hat{H}$, to get tighter lower bounds on the concave segments of g. In doing so, we are faced with solving a convex MINLP at every stage of the overall algorithm.

Our strategy is to use disjunctive cuts to iteratively tighten our formulation, with the goal of reducing the number of branching nodes that would be required for globally optimizing. In contrast to the algorithm of [9], we solve convex NLPs rather than convex MINLPs, thereby decreasing the computational burden as we seek to improve the global lower bound in the overall solution process.

2.1 Disjunctive Cuts in General

We take an aside now to review disjunctive cuts (see [2]). This sub-section is self contained, and the notation is not meant to coincide with its earlier usage. For example, here we have $\mathbf{x} \in \mathbb{R}^n$. Let

$$\mathcal{P} := \{\mathbf{x} \in \mathbb{R}^n : A\mathbf{x} \leq b\},$$
$$\mathcal{D}_1 := \{\mathbf{x} \in \mathbb{R}^n : D^1\mathbf{x} \leq f^1\},$$
$$\mathcal{D}_2 := \{\mathbf{x} \in \mathbb{R}^n : D^2\mathbf{x} \leq f^2\}.$$

In its typical usage, $\mathcal{P}$ would be (a subset of) the inequalities describing a polyhedral relaxation of a non-convex model, and $\mathcal{D}_1 \bigvee \mathcal{D}_2$ would be a disjunction (i.e., $\mathbf{x} \in \mathcal{D}_1$ or $\mathbf{x} \in \mathcal{D}_2$) that is valid for the non-convex model.

It is well known that we can characterize the set of all linear cuts $\alpha^\top \mathbf{x} \leq \beta$ that are valid for $\mathbb{X} := \text{conv}\left((\mathcal{P} \cap \mathcal{D}_1) \cup (\mathcal{P} \cap \mathcal{D}_2)\right)$, via the cone

$$\mathbb{K} := \left\{ \begin{pmatrix} \alpha \\ \beta \end{pmatrix} \in \mathbb{R}^{n+1} : \exists \ \pi_1, \pi_2, \gamma_1, \gamma_2 \geq \mathbf{0}, \text{ with} \right.$$

$$\left. \alpha^\top = \pi_1^\top A + \gamma_1^\top D^1, \ \alpha^\top = \pi_2^\top A + \gamma_2^\top D^2, \ \beta \geq \pi_1^\top b + \gamma_1^\top f^1, \ \beta \geq \pi_2^\top b + \gamma_2^\top f^2 \right\}.$$

Notice how, for $\pi_1, \pi_2, \gamma_1, \gamma_2 \geq \mathbf{0}$, the inequality $\pi_\ell^\top (A\mathbf{x} \leq b) + \gamma_\ell^\top (D^\ell \mathbf{x} \leq f^\ell)$, which we can alternatively write as $(\pi_\ell^\top A + \gamma_\ell^\top D^\ell)\mathbf{x} \leq (\pi_\ell^\top b + \gamma_\ell^\top f^\ell)$, is valid for $\mathcal{P} \cap \mathcal{D}_\ell$, for $\ell = 1, 2$. This implies that if $\alpha^\top = \pi_\ell^\top A + \gamma_1^\top D^\ell$ and $\beta \geq \pi_\ell^\top b + \gamma_\ell^\top f^\ell$, then $\alpha^\top \mathbf{x} \leq \beta$ is valid for $\mathcal{P} \cap \mathcal{D}_\ell$.

The so-called "cut-generating linear program" with respect to an $\bar{\mathbf{x}} \in \mathcal{P}$ is

$$\max \left\{ \alpha^\top \bar{\mathbf{x}} - \beta : \begin{pmatrix} \alpha \\ \beta \end{pmatrix} \in \mathbb{K} \right\}. \tag{CGLP}$$

Because $\mathbb{K}$ is a cone, CGLP either has a maximum value of 0 or is unbounded. If the maximum value is 0, then $\bar{\mathbf{x}} \in X$. Otherwise, the LP is unbounded, and any direction $\begin{pmatrix} \alpha \\ \beta \end{pmatrix}$ with $\alpha^\top \bar{\mathbf{x}} - \beta > 0$ gives a violated valid cut $\alpha^\top \mathbf{x} \leq \beta$.

Typically, we bound $\mathbb{K}$ by using a so-called "normalization constraint", so that CGLP always has a finite optimum. See [11] for an understanding of how to properly solve CGLP using an appropriate normalization. For an example of the use of a CGLP that is very relevant to our setting, see [15–17] for treating indefinite non-separable quadratics, and [3] for an implementation of some of those ideas more broadly.

2.2   Our Disjunction

Now we return to our setting (see Sect. 1). To improve upon the secant under-estimators on the concave portions of g, we generate cuts based on disjunctions of the following form. Choosing $\psi \in (0, p_k - p_{k-1})$ for some $k \in \hat{H}$, we let

$$\mathcal{D}_1^k := \left\{ (y_k, \delta_k, *) : \underbrace{\delta_k \leq \psi}_{\lambda_1}, y_k \geq \underbrace{\left(\frac{g(p_{k-1} + \psi) - g(p_{k-1})}{\psi} \right) \delta_k}_{\omega_{1,k}} \right\} \text{ and}$$

$$\mathcal{D}_2^k := \left\{ (y_k, \delta_k, *) : \underbrace{\delta_k \geq \psi}_{\lambda_2}, \right.$$

$$\left. y_k \geq \underbrace{\left(\frac{g(p_k) - g(p_{k-1} + \psi)}{p_k - (p_{k-1} + \psi)} \right) (\delta_k - \psi) + g(p_{k-1} + \psi) - g(p_{k-1})}_{\omega_{2,k}} \right\},$$

where "$*$" is a place-holder for additional variables in the CGLP (the red annotations define variables that will play the role of the γ-variables in the CGLP)

This disjunction corresponds to making a pair of secant under-estimators, one on $[p_{k-1}, p_{k-1} + \psi]$ and one on $[p_{k-1} + \psi, p_k]$. But note that we are not advocating for refining the set of intervals à la [9]; rather, we want to include some of the lower-bounding power of such a branching refinement (which would come at the substantial cost of solving a further convex MINLP) with a cut (which only leads to a further convex NLP).

An important consideration is which additional variables and constraints to include in our CGLP, playing the role of $\mathcal{P}$ in the previous section. We also need to address the selection of k and ψ.

Suppose that we have a "current solution" to a relaxation, including values

$$\bar{\delta}_1, \ \bar{\delta}_2, \ldots, \bar{\delta}_T; \ \bar{z}_1, \ \bar{z}_2, \ldots, \bar{z}_{T-1}; \ \bar{y}_1, \ \bar{y}_2, \ldots, \bar{y}_T. \tag{11}$$

We choose $k \in \hat{H}$ so that the secant approximation for g on I_k is "bad" at the current solution. This means that $g(p_{k-1} + \bar{\delta}_k) \gg \bar{y}_k$, so we choose $k \in \hat{H}$ so that the difference $g(p_{k-1} + \bar{\delta}_k) - \bar{y}_k$ is maximized. Then we choose $\psi = p_{k-1} + \bar{\delta}_k$, in the context of the disjunction $\mathcal{D}_1^k \vee \mathcal{D}_2^k$.

Now, we turn to the more subtle topic of describing which variable and inequalities comprise $\mathcal{P} = \{\mathbf{x} : A\mathbf{x} \leq b\}$. The variable-space for the CGLP should include all δ-, z-, and y-variables corresponding to g and the following inequalities (the red annotations define variables that will play the role of the π-variables in the CGLP):

$$
\begin{aligned}
z_i(p_i - p_{i-1}) - \delta_i \leq 0, && \text{for } i = 1, 2, \ldots, T-1; && \mu_{\ell,i} \\
-\delta_T \leq 0; && && \mu_{\ell,T} \\
\delta_1 \leq p_1 - p_0; && && \nu_{\ell,1} \\
-z_{i-1}(p_i - p_{i-1}) + \delta_i \leq 0, && \text{for } i = 2, 3, \ldots, T; && \nu_{\ell,i} \\
-z_1 \leq 0; && && \rho_{\ell,0} \\
z_{T-1} \leq 1; && && \rho_{\ell,1} \\
\left(\frac{g(p_i) - g(p_{i-1})}{p_i - p_{i-1}}\right) \delta_i - y_i \leq 0, && \text{for } i \in \hat{H} \setminus \{k\}. && \omega_{\ell,i}
\end{aligned}
$$

Note that we omit the secant inequality for I_k because it is implied by our disjunctive secants.

We also need to include something to represent the convex pieces of g. That is, we would like to use y_i for $i \in \check{H}$ in a constraint similar to (10), but it should be linearized at the point $(\bar{\delta}_i, \bar{y}_i)$ for use in an LP:

$$g(p_{i-1} + \bar{\delta}_i) - g(p_{i-1}) + g'(p_{i-1} + \bar{\delta}_i)(\delta_i - \bar{\delta}_i) - y_i \leq 0, \quad \text{for } i \in \check{H}. \quad \omega_{\ell,i} \tag{12}$$

Note that we use this linearization only in the CGLP. We propose to solve the overall model relaxation as a convex NLP, including (10), directly.

Additional possibilities for inclusion in $\mathcal{P}$ for forming CGLP are:

– linearizations of (10) at other points;
– the variables (δ, z, and y) and constraints corresponding to other univariate functions in the formulation, but maybe only those univariate functions that operate on the same x-variable as g.

The CGLP is designed to seek to separate the current solution (to our relaxation) using a linear inequality. In particular, we have values (11), and so we may seek an inequality of the form

$$\sum_{i=1}^{T} a_i \delta_i + \sum_{i=1}^{T-1} b_i z_i + \sum_{i=1}^{T} c_i y_i \leq \beta \tag{*}$$

that is violated by (11). We do this by solving our version of the CGLP:

$$\max \ \sum_{i=1}^{T} a_i \bar{\delta}_i + \sum_{i=1}^{T-1} b_i \bar{z}_i + \sum_{i=1}^{T} c_i \bar{y}_i - \beta \tag{13}$$

subject to:

$$a_i = -\mu_{\ell,i} + \nu_{\ell,i} + \left(\frac{g(p_i) - g(p_{i-1})}{p_i - p_{i-1}} \right) \omega_{\ell,i}, \quad \text{for } i \in \hat{H} \setminus \{k\}, \ \ell = 1,2; \tag{14}$$

$$a_i = -\mu_{\ell,i} + \nu_{\ell,i} + g'(p_{i-1} + \bar{\delta}_i)\omega_{\ell,i}, \quad \text{for } i \in \check{H}, \ \ell = 1,2; \tag{15}$$

$$a_k = -\mu_{1,k} + \nu_{1,k} + \left(\frac{g(p_{k-1} + \psi) - g(p_{k-1})}{\psi} \right) \omega_{1,k} + \lambda_1; \tag{16}$$

$$a_k = -\mu_{2,k} + \nu_{2,k} + \left(\frac{g(p_k) - g(p_{k-1} + \psi)}{p_k - p_{k-1} - \psi} \right) \omega_{2,k} - \lambda_2; \tag{17}$$

$$b_1 = (p_1 - p_0)\mu_{\ell,1} - (p_2 - p_1)\nu_{\ell,2} - \rho_{\ell,0}, \quad \text{for } \ell = 1,2; \tag{18}$$

$$b_i = (p_i - p_{i-1})\mu_{\ell,i} - (p_{i+1} - p_i)\nu_{\ell,i+1}, \quad \text{for } i = 2,\dots,T-2, \ \ell = 1,2; \tag{19}$$

$$b_{T-1} = (p_{T-1} - p_{T-2})\mu_{\ell,T-1} - (p_T - p_{T-1})\nu_{\ell,T} + \rho_{\ell,1}, \quad \text{for } \ell = 1,2; \tag{20}$$

$$c_i = -\omega_{\ell,i}, \quad \text{for } i = 1,\dots,T, \ \ell = 1,2; \tag{21}$$

$$\beta \geq (p_1 - p_0)\nu_{1,1} + \rho_{1,1} \tag{22}$$
$$- \sum_{i \in \check{H}} \left(g(p_{i-1} + \bar{\delta}_i) - g(p_{i-1}) - g'(p_{i-1} + \bar{\delta}_i)\bar{\delta}_i \right) \omega_{1,i} + \psi\lambda_1;$$

$$\beta \geq (p_1 - p_0)\nu_{2,1} + \rho_{2,1} \tag{23}$$
$$- \sum_{i \in \check{H}} \left(g(p_{i-1} + \bar{\delta}_i) - g(p_{i-1}) - g'(p_{i-1} + \bar{\delta}_i)\bar{\delta}_i \right) \omega_{2,i}$$
$$- \left(-\psi \frac{g(p_k) - g(p_{k-1} + \psi)}{p_k - (p_{k-1} + \psi)} + g(p_{k-1} + \psi) - g(p_{k-1}) \right) \omega_{2,k} - \psi\lambda_2;$$

$$\mu_{\ell,i} \geq 0, \quad \text{for } i = 1,\dots,T, \ \ell = 1,2; \tag{24}$$

$$\nu_{\ell,i} \geq 0, \quad \text{for } i = 1,\dots,T, \ \ell = 1,2; \tag{25}$$

$$\rho_{\ell,j} \geq 0, \quad \text{for } j = 0,1, \ \ell = 1,2; \tag{26}$$

$$\omega_{\ell,i} \geq 0, \quad \text{for } i = 1,\dots,T, \ \ell = 1,2; \tag{27}$$

$$\lambda_\ell \geq 0, \quad \text{for } \ell = 1,2. \tag{28}$$

2.3 Possible Improvements and Another Disjunction: Inviting the z-variables to the Party

In this section, we describe three ideas, the first two of which we have tested, for inviting the z-variables to the disjunctive party.

Improvement to $\mathcal{P}$. Model relaxations similar to ours can be solved much faster when the convex constraints (10) are linearized (and thus weakened, unfortunately) so that we are in the realm of LP rather than NLP. [6,7] suggests that these constraints can be tightened via a "perspective reformulation" (see [12]) and then linearized and thus weakened, but hopefully for a net benefit. In particular, (10) becomes:

$$y_i \geq z_{i-1}\, g(p_{i-1} + \delta_i/z_{i-1}) - z_{i-1}\, g(p_{i-1}), \quad \text{for } i \in \check{H}, \qquad (29)$$

(where $z_0 := 1$ if $1 \in \check{H}$). Inequalities (29) are convex by the perspective-reformulation definition and can be linearized with standard techniques. One could carry this out directly, but we proceed as follows: we linearize (10) as (12) only for the conceptual purpose of using within $\mathcal{P}$ to set up our CGLP; that is, we strengthen (12) via the perspective reformulation, and then linearize only to strengthen our CGLP.

Improvement to $\mathcal{D}_1^k \bigvee \mathcal{D}_2^k$. We can attempt to improve (i.e., tighten) the representation of our disjunction that was based on secant inequalities. We can see that $\delta_k \leq \psi \implies z_k = 0$, and $\delta_k \geq \psi \implies z_{k-1} = 1$. So we could include $z_k = 0$ in $\mathcal{D}_1^k$ and $z_{k-1} = 1$ in $\mathcal{D}_2^k$, which implies a slightly different and potentially stronger CGLP.

Another Disjunction. Related to the last idea, we can make a direct disjunction on a z_k that we could use to generate disjunctive cuts. That is, we could use the disjunction

$$\mathcal{Z}_0^k := \{(z_k, *) : z_k = 0\} \bigvee \mathcal{Z}_1^k := \{(z_k, *) : z_k = 1\},$$

choosing k based on $\bar{z}_k$ being fractional. It is well known that a diversity of cuts can be quite effective, so we propose to use disjunctive cuts based on $\mathcal{Z}_0 \bigvee \mathcal{Z}_1$ as well as ones based on $\mathcal{D}_1 \bigvee \mathcal{D}_2$. Note that in forming the CGLP based on the disjunction $\mathcal{Z}_0 \bigvee \mathcal{Z}_1$, we should include in $\mathcal{P}$ the secant inequality for I_k (i.e., (9) for $i = k$).

3   Computational Experiments

As a proof of concept, we tested our ideas on challenging non-linear continuous knapsack-type problems of the form:

$$\min\ f(x_1, \ldots, x_n) + g(x_0)$$
$$\text{subject to:}$$
$$C_L \leq \sum_{j=0}^{n} r_j x_j \leq C_U;$$
$$0 \leq x_j \leq 1,\ j = 0, 1, \ldots, n,$$

where C_L, C_U, $r_j \in \mathbb{R}^+$ for $j = 0, 1, \ldots, n$, $f : \mathbb{R}^n \mapsto \mathbb{R}$ is convex, and the single univariate g is a highly non-convex function. In particular, we defined $f(x_1, \ldots, x_n)$ as a convex quadratic, namely $\sum_{i=1}^{n} \sum_{j=1}^{n} Q_{ij} x_i x_j$, where $Q := \tilde{Q}^\top \tilde{Q}$ and $\tilde{Q}$ is randomly generated. Note that f is convex by the definition of Q. The convexity of f and the minimization objective drive $(x_1, \ldots, x_n)$ toward $\mathbf{0}$ at optimality. By choosing $0 < C_L < C_U$ appropriately, we can drive the solution into a range where x_0 is in a concave portion of g, thus stressing our relaxation.

We generated ten test instances. For g, we used two types of functions, designed to be difficult (see Fig. 1):

1. $g(x_0) := s \cdot x_0 - \frac{2\cos(h\pi x_0)}{h\pi} - x_0 \sin(h\pi x_0)$, and
2. $g(x_0) := d(\sin((h\pi x_0) + 2e\pi + \sin^{-1}(\frac{m}{d}))) + m((h\pi x_0) + 2e\pi + \sin^{-1}((\frac{m}{d}))^2 + v((h\pi x_0) + 2e\pi + \sin^{-1}(\frac{m}{d})),$

where s is randomly generated (with a uniform distribution) on $[-4, +4]$, h on $[7, 15]$, d on $\{100, 200, 300\}$, e on $\{-3, -2\}$, m on $\{-2, -1\}$, v on $\{10, 15, 20\}$. Entries in the vector r are uniformly randomly generated on $[1, 200]$. C_L and C_U are chosen so that the value of x_0, when $(x_1, \ldots, x_n) = \mathbf{0}$, is in a centrally-located concave interval of g. The first type of example only puts local stress on the relaxation; the second type of example puts global stress on the relaxation, as interpolating across the full domain of x_0 can result in a very poor underestimate of g in the range where x_0 is feasible (when $(x_1, \ldots, x_n) = \mathbf{0}$).

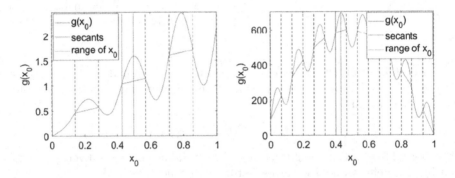

Fig. 1. Example of the function of type 1 (left) and type 2 (right).

We generated 5 instances with $n = 3$ for each of the two types of functions g.

All experiments were performed on a single machine equipped with an Intel Xeon E5649 processor clocked at 2.53 GHz and 50 GB RAM. We used open-source solvers like `IPOPT 3.12.8`, `Bonmin 1.8.6`, and `Couenne 0.5` to solve convex NLPs, convex MINLPs, and non-convex (MI)NLPs, respectively. As for the MILPs (solved within `Bonmin` and `Couenne`), we used `IBM CPLEX 12.6`.

We tested and compared three strategies: the basic approach we have outlined (called "Alg" in the tables), the improvement of the relaxation with the Perspective Reformulation (29) ("Alg+PC"), and the improvement using $\mathcal{D}_1^k \vee \mathcal{D}_2^k$ ("Alg+IDC"). We have not yet tested the modifications in tandem.

Table 1 summarizes the performance of the algorithm when employing each of the three cut-generating strategies, and applied to each of our ten test instances. In particular, the algorithm iteratively adds violated cuts to the convex NLP relaxation, then solves the strengthened convex MINLP. For each instance and strategy, we present the objective-function value of the convex MINLP, along with the number of iterations of cut generation (an iteration limit of 300 was set). The strategy Alg+PC most frequently requires the fewest iterations among the three strategies, making it the fastest. Among the five instances for which the Alg+PC strategy reaches the iteration limit, the final solution obtained with this strategy is better (i.e., greater lower bound) in three cases, and equivalent in one of the two remaining instances. Strategy Alg+IDC sometimes finds the best solution, but often requires the most iterations.

Table 1. Results on the solution of the convex MINLP obtained after strengthening the convex NLP relaxation by adding iteratively the cuts based on the different strategies

Inst.	Strategy	CMINLP	#iter	Inst.	Strategy	CMINLP	#iter
1	Alg	1.07	18	6	Alg	502.44	300
1	Alg+PC	1.07	12	6	Alg+PC	506.85	300
1	Alg+IDC	1.11	69	6	Alg+IDC	502.44	300
2	Alg	2.09	36	7	Alg	502.48	300
2	Alg+PC	2.15	22	7	Alg+PC	505.81	300
2	Alg+IDC	2.09	38	7	Alg+IDC	502.92	300
3	Alg	2.56	221	8	Alg	246.46	300
3	Alg+PC	2.57	60	8	Alg+PC	252.14	300
3	Alg+IDC	2.66	300	8	Alg+IDC	246.46	300
4	Alg	−2.10	300	9	Alg	504.40	300
4	Alg+PC	−2.10	29	9	Alg+PC	504.40	300
4	Alg+IDC	−2.13	26	9	Alg+IDC	504.40	300
5	Alg	2.70	73	10	Alg	587.70	300
5	Alg+PC	2.72	63	10	Alg+PC	587.70	300
5	Alg+IDC	2.70	300	10	Alg+IDC	589.60	300

In Table 2 we compare the strategies by defining a measure of the impact of disjunctive cuts. In particular, for each instance (one per row) and each strategy (one per block of 3 columns), we display the values of GAP1 := $100 \cdot \frac{GO-CMINLP}{GO-NLP}$, GAP2 := $100 \cdot \frac{GO-MINLP}{GO-NLP}$, and GAP3 := $100 \cdot \frac{GO-CNLP}{GO-NLP}$, where GO is the global optimum value of the non-convex MINLP problem, NLP is the optimal value of the convex NLP relaxation, MINLP is the optimal value of the convex MINLP relaxation, CNLP is the optimal value of the convex NLP relaxation after applying the disjunctive cuts, and CMINLP is the optimal value of the convex MINLP relaxation after applying disjunctive cuts.

Most importantly, GAP3 measures the effectiveness of disjunctive cuts on closing the gap in the convex NLP relaxation (initial gap is 100). We can see that with all three strategies substantial gap is often closed using the disjunctive cuts. Comparing across the approaches, Alg+IDC usually gives the greatest effect from the disjunctive cuts.

By imposing integrality and solving a single convex MINLP, we can compare gaps obtained via convex MINLP versus convex NLP: GAP2 vs 100 (no disjunctive cuts) and GAP1 vs GAP3 (with disjunctive cuts). Generally, we obtain much smaller gaps in the first of each pair, but at the cost of solving a convex MINLP.

Table 2. Results per instance using the basic algorithm (Alg), the algorithm with perspective cuts (Alg+PC), and the algorithm with improved disjunctive cuts (Alg+IDC).

Inst.	Alg			Alg+PC			Alg+IDC		
	GAP1	GAP2	GAP3	GAP1	GAP2	GAP3	GAP1	GAP2	GAP3
1	4.25	4.25	81.43	8.38	8.38	79.01	2.42	4.25	29.67
2	15.62	15.62	80.78	22.70	27.06	79.72	15.62	15.62	79.54
3	4.57	4.57	98.28	8.27	8.88	79.39	0.83	4.57	18.82
4	0.34	1.96	80.37	0.69	3.96	74.37	1.66	1.96	76.91
5	88.34	88.34	99.38	94.36	94.47	99.21	88.34	88.34	93.10
6	7.34	7.34	93.26	12.33	13.63	92.24	7.34	7.34	41.12
7	8.20	8.20	89.62	14.31	14.88	92.74	8.16	8.20	42.62
8	9.77	9.77	93.16	17.02	17.61	90.73	9.77	9.77	50.65
9	7.32	7.32	92.70	15.19	15.19	86.48	7.32	7.32	74.71
10	4.08	4.08	93.85	8.31	8.31	88.87	3.93	4.08	47.93

Note that GAP2 is the same for Alg and Alg+IDC, because they use the same NLP relaxation; while Alg+PC uses a different relaxation, so GAP2 is different. Comparing GAP1 vs GAP2, we see several instances where the disjunctive cuts close much of the remaining gap when included in the single convex MINLP; e.g., instance 4 using Alg or Alg+PC, and instances 1 and 3, using Alg+IDC. Again, comparing across the approaches, Alg+IDC is usually the most effective strategy of the three.

To conclude, strategy Alg+PC seems to be the most promising both on the speed of convergence and on the strengthening of the convex MINLP problem, while strategy Alg+IDC gives the strongest disjunctive cuts.

As a future direction, we plan to explore a hybrid Alg+PC+IDC strategy, and to experiment with including a diversity of disjunctive cuts (e.g., those produced from the disjunction described at the end of Sect. 2). Finally, we have concentrated on lower bounds, but we plan to experiment with incorporating branching to get a complete algorithm for reaching global optimality.

4   Conclusions and Outlook

We have introduced a technique for using disjunctive cuts so as to improve a framework for handling mathematical-optimization models where the non-convexities are separable. The efficiency is realized by solving an LP and a convex NLP at each iteration, rather than an expensive convex MINLP. We have presented preliminary computational work to demonstrate the promise of our methodology. Further work will center on enhancing and tuning the method to efficiently handle models coming from real-world applications as the minimization of the ripple effect (see, e.g., [13]) or the optimization of the design of bandpass and low-pass filters (see, e.g., [14]). Optimization models with non-convexities are indeed typical of electromagnetic problems where the outputs are often characterized by a strongly sinusoidal or generally periodic trend (see [1,5]).

References

1. Alexander, C.K., Sadiku, M.N.: Fundamentals of Electric Circuits. McGraw-Hill Education, Boston (2000)
2. Balas, E.: Disjunctive Programming. Springer, Cham (2018). https://doi.org/10.1007/978-3-030-00148-3
3. Belotti, P.: Disjunctive cuts for nonconvex MINLP. In: Lee, J., Leyffer, S. (eds.) Mixed Integer Nonlinear Programming. IMA, vol. 154, pp. 117–144. Springer, New York (2012). https://doi.org/10.1007/978-1-4614-1927-3_5
4. Bradley, S.P., Hax, A.C., Magnanti, T.L.: Applied Mathematical Programming. Addison-Wesley, Reading (1977)
5. Ceraolo, M., Poli, D.: Fundamentals of Electric Power Engineering: From Electromagnetics to Power Systems. Wiley, New York (2014)
6. D'Ambrosio, C., Frangioni, A., Gentile, C.: Strengthening convex relaxations of mixed integer non linear programming problems with separable non convexities. In: Rocha, A., Costa, M., Fernandes, E. (eds.) Proceedings of the XIII Global Optimization Workshop (GOW 2016), pp. 49–52 (2016)
7. D'Ambrosio, C., Frangioni, A., Gentile, C.: Strengthening the sequential convex MINLP technique by perspective reformulations. Optim. Lett. 13(4), 673–684 (2018). https://doi.org/10.1007/s11590-018-1360-9
8. D'Ambrosio, C., Lee, J., Wächter, A.: A global-optimization algorithm for mixed-integer nonlinear programs having separable non-convexity. In: Fiat, A., Sanders, P. (eds.) ESA 2009. LNCS, vol. 5757, pp. 107–118. Springer, Heidelberg (2009). https://doi.org/10.1007/978-3-642-04128-0_10
9. D'Ambrosio, C., Lee, J., Wächter, A.: An algorithmic framework for MINLP with separable non-convexity. In: Lee, J., Leyffer, S. (eds.) Mixed Integer Nonlinear Programming. IMA, vol. 154, pp. 315–347. Springer, New York (2012). https://doi.org/10.1007/978-1-4614-1927-3_11
10. Fampa, M., Lee, J., Melo, W.: On global optimization with indefinite quadratics. EURO J. Comput. Optim. 5(3), 309–337 (2016). https://doi.org/10.1007/s13675-016-0079-6
11. Fischetti, M., Lodi, A., Tramontani, A.: On the separation of disjunctive cuts. Math. Program. 128(1), 205–230 (2011)
12. Frangioni, A., Gentile, C.: Perspective cuts for a class of convex 0–1 mixed integer programs. Math. Program. 106(2), 225–236 (2006)
13. Mahmoudi, H., Aleenejad, M., Ahmadi, R.: Torque ripple minimization for a permanent magnet synchronous motor using a modified quasi-Z-source inverter. IEEE Trans. Power Electron. 34(4), 3819–3830 (2019)
14. Quendo, C., Rius, E., Person, C., Ney, M.: Integration of optimized low-pass filters in a bandpass filter for out-of-band improvement. IEEE Trans. Microw. Theory Tech. 49(12), 2376–2383 (2001)
15. Saxena, A., Bonami, P., Lee, J.: Disjunctive cuts for non-convex mixed integer quadratically constrained programs. In: Lodi, A., Panconesi, A., Rinaldi, G. (eds.) IPCO 2008. LNCS, vol. 5035, pp. 17–33. Springer, Heidelberg (2008). https://doi.org/10.1007/978-3-540-68891-4_2

16. Saxena, A., Bonami, P., Lee, J.: Convex relaxations of non-convex mixed integer quadratically constrained programs: extended formulations. Math. Program. **124**(1–2), 383–411 (2010)
17. Saxena, A., Bonami, P., Lee, J.: Convex relaxations of non-convex mixed integer quadratically constrained programs: projected formulations. Math. Program. **130**(2), 359–413 (2011)
18. Wilson, D.: Polyhedral methods for piecewise-linear functions. Ph.D. thesis, University of Kentucky (1998)

Improving Proximity Bounds Using Sparsity

Jon Lee[1], Joseph Paat[2(✉)], Ingo Stallknecht[2], and Luze Xu[1]

[1] Department of Industrial and Operations Engineering, University of Michigan, Ann Arbor, USA
[2] Department of Mathematics, ETH Zürich, Zürich, Switzerland
joseph.paat@ifor.math.ethz.ch

Abstract. We refer to the distance between optimal solutions of integer programs and their linear relaxations as *proximity*. In 2018 Eisenbrand and Weismantel proved that proximity is independent of the dimension for programs in standard form. We improve their bounds using results on the *sparsity* of integer solutions. We first bound proximity in terms of the largest absolute value of any full-dimensional minor in the constraint matrix, and this bound is tight up to a polynomial factor in the number of constraints. We also give an improved bound in terms of the largest absolute entry in the constraint matrix, after efficiently transforming the program into an equivalent one. Our results are stated in terms of general sparsity bounds, so any new sparsity results immediately improve our work. Generalizations to mixed integer programs are also discussed.

Keywords: Proximity · Sparsity · Mixed integer programming

1 Introduction

Let $A \in \mathbb{Z}^{m \times n}$ with $\text{rank}(A) = m$, $c \in \mathbb{Z}^n$, and $b \in \mathbb{Z}^m$. Denote the largest absolute value of a minor of A of order $k \in \{1, \ldots, m\}$ by

$$\Delta_k := \Delta_k(A) := \max\{|\det(B)| : B \text{ is a } k \times k \text{ submatrix of } A\}.$$

Note that $\Delta_1 = \|A\|_\infty$ is the largest absolute entry of A. For simplicity, we set $\Delta := \Delta_m$. We consider the standard form integer program

$$\max\{c^\mathsf{T} z : Az = b, \ z \in \mathbb{Z}^n_{\geq 0}\} \tag{IP}$$

and its linear relaxation

$$\max\{c^\mathsf{T} x : Ax = b, \ x \in \mathbb{R}^n_{\geq 0}\}. \tag{LP}$$

We assume that (IP) is both feasible and bounded.

Given an optimal vertex solution x^* to (LP), we investigate the question of (LP) to (IP) proximity: can we bound the distance from x^* to some optimal solution z^* of (IP)? We refer to any bound τ on $\|x^* - z^*\|_1$ as a *proximity*

© Springer Nature Switzerland AG 2020
M. Baïou et al. (Eds.): ISCO 2020, LNCS 12176, pp. 115–127, 2020.
https://doi.org/10.1007/978-3-030-53262-8_10

bound. Proximity bounds have a variety of implications in the theory of integer programming. For example, a proximity bound of τ translates into a bound of $\tau \cdot \|c\|_\infty$ on the so-called integrality gap [7,12,13]. Furthermore, strong proximity bounds reduce the time needed for a local search algorithm to find an optimal (IP) solution starting from an optimal (LP) solution, see, e.g. [10].

One of the first seminal results on proximity is by Cook et al. [7], who established that there exists an optimal solution z^* to (IP) satisfying

$$\|z^* - x^*\|_\infty \leq n \cdot \max\{\Delta_k : k \in \{1, \ldots, m\}\}. \tag{1}$$

Cook et al. actually consider problems in inequality form, i.e., with constraints $Ax \leq b$ rather than $Ax = b$ and $x \geq 0$, but their results easily translate to the standard form setting. A closer analysis reveals that Δ suffices for the standard form problem rather than $\max\{\Delta_k : k \in \{1, \ldots, m\}\}$ stated in (1). Furthermore, if we naively extend (1) to a bound on proximity in terms of the ℓ_1-norm, then we obtain $\|z^* - x^*\|_1 \leq n^2\Delta$. Another closer analysis gives us the bound

$$\|z^* - x^*\|_1 \leq (m+1)n\Delta. \tag{2}$$

See the proof of Lemma 3 for the two 'closer analyses' referred to above. Cook et al.'s bound has been generalized to various problems including those with separable convex objective functions [12,13,23] or with mixed integer constraints [19], and extended to alternative data parameters such as k-regularity [16,24] and the magnitude of Graver basis elements [11].

The proximity bound in (2) depends on the dimension n. In 2018 Eisenbrand and Weismantel [10] proved that proximity is independent of the dimension by establishing the bound

$$\|z^* - x^*\|_1 \leq m(2m \cdot \|A\|_\infty + 1)^m. \tag{3}$$

Eisenbrand and Weismantel use the so-called Steinitz Lemma with the ℓ_∞-norm [22] in their proof of (3). Their proof can be modified using the norm $\|x\|_* = \|B^{-1}x\|_\infty$, where B is an $m \times m$ submatrix of A with $|\det(B)| = \delta$, to obtain the bound

$$\|z^* - x^*\|_1 \leq m(2m + 1)^m \cdot \Delta. \tag{4}$$

The proximity bounds (3) and (4) also hold for standard form problems with additional upper bound constraints on the variables. Oertel et al. established that the upper bound $\|z^* - x^*\|_1 \leq (m+1) \cdot (\Delta - 1)$ holds for most problems, where 'most' is defined parametrically with b treated as input [17].

As for lower bounds on proximity, it is not difficult to come up with examples demonstrating $\|z^* - x^*\|_1 \geq m \cdot (\Delta - 1)$ and $\|z^* - x^*\|_1 \geq \|A\|_\infty^m$. Aliev et al. [3] give a tight lower bound $\Delta - 1$ on proximity in terms of the ℓ_∞-norm when $m = 1$. However, it remains an open question if (2), (3), or (4) is tight in general.

1.1   Statement of Results and Overview of Proof Techniques

The focus of this paper is to create stronger proximity bounds. Recall that $\Delta := \Delta_m$. Our first main result is an improvement over (4) for fixed Δ. We always consider the logarithm $\log(\cdot)$ to have base two.

Theorem 1. *For every optimal* (LP) *vertex solution* x^*, *there exists an optimal* (IP) *solution* z^* *such that*

$$\|z^* - x^*\|_1 < 3m^2 \log(\sqrt{2m} \cdot \Delta^{1/m}) \cdot \Delta.$$

Theorem 1 demonstrates that proximity in the ℓ_1-norm between (IP) solutions and (LP) *vertex* solutions is bounded by a polynomial in m and Δ. We focus on vertex solutions because proximity may depend on n for general non-vertex (LP) solutions. For example, suppose that $c = 0$ and take any feasible solution to (LP), which is optimal in this case, such that each of the n components is in $1/2 + \mathbb{Z}$.

Our second main proximity result is in terms of $\|A\|_\infty$ *after A is transformed by a suitable unimodular matrix*. Recall that a unimodular matrix $U \in \mathbb{Z}^{m \times m}$ satisfies $|\det(U)| = 1$, so the $m \times m$ minors of UA have the same magnitudes as those of A. Moreover, the optimal solutions of (IP) are the same as the optimal solutions to

$$\max\{c^\mathsf{T} z : UAz = Ub, \ z \in \mathbb{Z}_{\geq 0}^n\}. \qquad (U \text{ - IP})$$

Given an $m \times m$ submatrix B of A, we can find a unimodular matrix U in polynomial time such that UB is upper triangular. The Hermite normal form provides one method for computing such U, see [20]. If B satisfies $|\det(B)| = \Delta$, then $\Delta \leq \|UB\|_\infty^m$, and we can apply Theorem 1 to obtain the bound

$$\|z^* - x^*\|_1 < 3m^2 \log(\sqrt{2m} \cdot \|UB\|_\infty) \cdot \|UB\|_\infty^m. \qquad (5)$$

The previous bound is predicated on the knowledge of an $m \times m$ submatrix of maximum absolute determinant, which is NP-hard to find [15]. However, Di Summa et al. [8] established that this submatrix can be approximated in polynomial time. In particular, they demonstrated that there exists an $m \times m$ submatrix B of A satisfying

$$\Delta \leq |\det(B)| \cdot (2 \log(m + 1))^{m/2} \qquad (6)$$

that can be found in time polynomial in m and n.[1] We can use this approximate largest absolute determinant to derive our second main result. We denote the linear relaxation of (U - IP) by (U - LP).

Theorem 2. *Let B be an $m \times m$ submatrix B of A satisfying* (6) *and* $U \in \mathbb{Z}^{m \times m}$ *a unimodular matrix such that UB is upper triangular. Then for every optimal* (U - LP) *vertex solution* x^*, *there exists an optimal* (U - IP) *solution* z^* *satisfying*

$$\|z^* - x^*\|_1 < 3m^2 \log(2\sqrt{m \log(m + 1)} \cdot \|UB\|_\infty) \cdot (2 \log(m + 1))^{m/2} \cdot \|UB\|_\infty^m.$$

---

[1] The approximation result of Di Summa et al. involves an ϵ factor of precision and the running time is polynomial in $m, n, 1/\epsilon$. For the sake of presentation, we have fixed this ϵ to $1/m$ and obtain a polynomial time algorithm in m, n.

It is worth reemphasizing that the proximity bound in Theorem 2 can be determined in polynomial time, which is in contrast to the bound in (5), and the dependence on m is significantly less than the bound in (3).

The proofs of Theorems 1 and 2 are based on combining proof techniques of Cook et al. [7] with results on the sparsity of optimal solutions to (IP). The *support* of a vector $x \in \mathbb{R}^n$ is defined as

$$\text{supp}(x) := \{i \in \{1, \ldots, n\} : x_i \neq 0\}.$$

A classic theorem of Carathéodory states that $|\text{supp}(x^*)| \leq m$ for every vertex solution of (LP). It turns out that the minimum support of an optimal solution to (IP) is not much larger; denote this value by

$$S := \min \{|\text{supp}(z^*)| : z^* \text{ is an optimal solution for (IP)}\}.$$

Aliev et al. [2,3] established that

$$S \leq m + \log\left(\sqrt{\det(AA^{\mathsf{T}})}\right) \leq 2m \log(2\sqrt{m} \cdot \|A\|_\infty). \tag{7}$$

For other results regarding sparsity, see [9,18] for general A and [4–6,21] for matrices that form a Hilbert basis. See also the manuscript of Aliev et al. [1], who give improved sparsity bounds for *feasible solutions* to special classes of integer programs and provide efficient algorithms for finding such solutions. Using sparsity we derive the following proximity bound, which forms the basis for Theorems 1 and 2.

Lemma 3. *For every optimal* (LP) *vertex solution x^*, there exists an optimal* (IP) *solution z^* such that*

$$\|z^* - x^*\|_1 < (m + 1) \cdot S \cdot \Delta.$$

Lemma 3 improves (2) by replacing the dependence on n to S. Lemma 3 is stated for a generic sparsity bound, so one could use it together with (7) to achieve a proximity bound in terms of Δ and $\|A\|_\infty$. In order to provide a bound for proximity that is uniform in the data parameter, we prove a new sparsity result in terms of Δ. A bound in terms of Δ is also of interest because it is invariant under unimodular transformations of A.

Theorem 4. *There exists an optimal* (IP) *solution z^* such that*

$$|\text{supp}(z^*)| < 2m \log(\sqrt{2m} \cdot \Delta^{1/m}).$$

Our proximity bounds can be generalized to mixed integer programs. Given an index set $\mathcal{I} \subseteq \{1, \ldots, n\}$, the mixed integer program with integrality constraints indexed by $\mathcal{I}$ is

$$\max\{c^\mathsf{T} z : Az = b, \ z \geq 0, \ z_i \in \mathbb{Z} \ \forall \ i \in \mathcal{I}\}. \tag{MIP}$$

Similarly to [2, Corollary 4], we establish the extension of Theorem 4 to (MIP).

Corollary 5. *There exists an optimal* (MIP) *solution* z^* *satisfying*

$$|\operatorname{supp}(z^*)| < m + 2m \log\left(\sqrt{2m} \cdot \Delta^{1/m}\right) = 2m \log\left(2\sqrt{m} \cdot \Delta^{1/m}\right).$$

We obtain the following proximity result by applying Corollary 5.

Corollary 6. *For every optimal* (LP) *vertex solution* x^*, *there exists an optimal* (MIP) *solution* z^* *such that*

$$\|z^* - x^*\|_1 < 3m^2 \log(2\sqrt{m} \cdot \Delta^{1/m}) \cdot \Delta.$$

Our results also extend to integer programs in general form. Let $A \in \mathbb{Z}^{m \times n}$ and $B \in \mathbb{Z}^{m \times d}$ be matrices satisfying $\operatorname{rank}([A, B]) = m$. Note that it is not necessary to assume that $\operatorname{rank}(A) = m$ in our general form results. Let $C \in \mathbb{Z}^{t \times d}$, $c \in \mathbb{Z}^{n+d}$, $b_1 \in \mathbb{Z}^m$, and $b_2 \in \mathbb{Z}^t$. The *general form integer program* is

$$\max\left\{c^\mathsf{T} z : \begin{matrix}[A,\ B]\ z = b_1 \\ [0,\ C\ ]\ z \le b_2\end{matrix},\ z \in \mathbb{Z}^{n+d},\ z_i \ge 0\ \forall\ i \in \{1, \ldots, n\}\right\}. \qquad \text{(GIP)}$$

We define the general form linear program (GLP) similarly. Previously cited bounds on proximity hold for (GIP). However, our analysis reveals that proximity for (GIP) depends on the potentially smaller data parameter

$$\delta := \max\left\{|\det(E)| : \begin{matrix}E \text{ is any submatrix of } \begin{pmatrix}A\ B \\ 0\ C\end{pmatrix} \\ \text{defined using the first } m \text{ rows}\end{matrix}\right\}.$$

If $t = 0$ and $d = 0$, then (GIP) is a standard form problem and $\delta = \Delta_m(A)$. If $m = 0$ and $n = 0$, then (GIP) is an inequality form problem and $\delta = \max\{\Delta_k(C) : k \in \{1, \ldots, d\}\}$.

Corollary 7. *For every optimal* (GLP) *vertex solution* x^*, *there exists an optimal* (GIP) *solution* z^* *such that*

$$\|z^* - x^*\|_1 < \min\{m+t+1, n+d\} \cdot \left(\min\{n-m, 2m \cdot \log(\sqrt{2m} \cdot \delta^{1/m})\} + d\right) \cdot \delta.$$

The proximity bound in Corollary 7 matches the best known bounds in both the standard form setting and the inequality form setting.

Going beyond integer linear optimization problems, it would be ideal for proximity bounds in terms of sparsity to extend to integer programs with separable convex objective functions (see [12,13] for similarities between the linear and separable convex setting). However, for separable convex *maximization problems*, strong proximity bounds do not exist for exact solutions, in general, even though sparsity results apply. In contrast, for separable convex *minimization problems*, strong proximity bounds do not exist for exact solutions, in general, even though the classic proximity techniques apply.

The paper is structured as follows. In Sect. 2, we present our proofs of the proximity bound derived from a generic sparsity bound (Lemma 3) and the sparsity bounds (Theorem 4 and Corollary 5). Then in Sect. 3, we provide the proofs of the proximity results for the standard form integer programs (Theorems 1 and 2). The proof of the mixed integer case (Corollary 6) is omitted because it is the same as the proof of the pure integer case except that a different sparsity bound is applied. Additionally, we provide a proof of the proximity result in the general form setting (Corollary 7).

2   Proofs Regarding Sparsity

Given $A \in \mathbb{R}^{m \times n}$ and $I \subseteq \{1, \ldots, n\}$, we let $A_I \in \mathbb{R}^{m \times |I|}$ denote the columns of A indexed by I. If $I = \{i\}$ for some $i \in \{1, \ldots, n\}$, then $A_i := A_I$. Similarly, given $u \in \mathbb{R}^n$, we let $u_I \in \mathbb{R}^{|I|}$ denote the components of u indexed by I.

Proof (of Lemma 3). We prove the result by projecting the optimization problems onto the union of the supports of x^* and an optimal (IP) solution with minimal support. Let $\bar{z} \in \mathbb{Z}_{\geq 0}^n$ be an optimal (IP) solution with minimum support. By the definition of S we have $|\operatorname{supp}(\bar{z})| = S$. As x^* is an optimal vertex solution of (LP) we also have $|\operatorname{supp}(x^*)| \leq m$. Define

$$H := \operatorname{supp}(x^*) \cup \operatorname{supp}(\bar{z}),$$

and note that

$$|H| = |\operatorname{supp}(x^*) \cup \operatorname{supp}(\bar{z})| \leq |\operatorname{supp}(x^*)| + |\operatorname{supp}(\bar{z})| \leq |\operatorname{supp}(x^*)| + S. \quad (8)$$

If $n = m$, then A is invertible and there exists a unique solution $A^{-1}b$ to the system $Ax = b$. In this case $x^* = \bar{z} = A^{-1}b$. Therefore, $\|x^* - \bar{z}\|_1 = 0$. For the rest of the proof, we assume that $n > m$ and $H = \{1, \ldots, |H|\}$.

Consider the optimization problems

$$\max \left\{ c_H^\mathsf{T} z : A_H z = b, \ z \in \mathbb{Z}_{\geq 0}^{|H|} \right\} \quad \text{(IP2)}$$

and

$$\max \left\{ c_H^\mathsf{T} x : A_H x = b, \ x \in \mathbb{R}_{\geq 0}^{|H|} \right\}. \quad \text{(LP2)}$$

Observe that x_H^* is an optimal vertex solution for (LP2), and $\bar{z}_H$ is an optimal solution for (IP2).

Rewrite (LP2) in inequality form:

$$\max \left\{ c_H^\mathsf{T} x : A_H x = b, -I_H x \leq 0, \ x \in \mathbb{R}^{|H|} \right\},$$

where I_H is the $|H| \times |H|$ identity matrix. Partition the rows of $-I_H$ into D_1 and D_2 such that $D_1 \bar{z}_H < D_1 x_H^*$ and $D_2 \bar{z}_H \geq D_2 x_H^*$. Define the pointed polyhedral cone

$$K := \left\{ u \in \mathbb{R}^{|H|} : A_H u = 0, \ D_1 u \leq 0, \ D_2 u \geq 0 \right\}. \quad (9)$$

Observe that $\bar{z}_H - x_H^* \in K$. By (8) we see that the rank of A_H is at least $|\operatorname{supp}(x^*)|$ because the columns of A corresponding to the support x^* are linearly independent. Thus, the dimension of K, which we denote by $\dim(K)$, is at most $|\operatorname{supp}(x^*)| + S - \operatorname{rank}(A_H) \leq S$.

Let $U := \{u^1, \ldots, u^t\} \subseteq \mathbb{R}^{|H|} \setminus \{0\}$ be a set of vectors that generate the extreme rays of K, i.e.,

$$K = \left\{ \sum_{i=1}^{t} \lambda_i u^i : \lambda_i \geq 0 \; \forall \; i \in \{1, \ldots, t\} \right\}$$

and u^i satisfies $|H| - 1$ linearly independent constraints in (9) at equality for each $i \in \{1, \ldots, t\}$.

Claim 8. *For each $\tilde{u} \in U$ we have*

$$|\operatorname{supp}(\tilde{u})| \leq m + 1. \tag{10}$$

Also, each $\tilde{u} \in U$ can be scaled to have integer components and satisfy $\|\tilde{u}\|_\infty \leq \Delta$.

Proof. Set $T := \operatorname{supp}(\tilde{u})$ and without loss of generality assume $T = \{1, \ldots, |T|\}$. Recall that $\tilde{u}$ satisfies a set of $|H| - 1$ linearly independent constraints in (9) at equality. One such set is composed of $|H| - |T|$ constraints from the system $D_1 \tilde{u} \leq 0$, $D_2 \tilde{u} \geq 0$ and $|T| - 1$ constraints from $0 = A_H \tilde{u} = A_T \tilde{u}_T$. By this choice of constraints it follows that

$$|\operatorname{supp}(\tilde{u})| = |T| \leq m + 1$$

and $|T| - 1 \leq \operatorname{rank}(A_T) \leq \min\{|T|, m\}$. Recalling $n > m$ and $\operatorname{rank}(A) = m$, the latter inequalities imply that there exists an index set $\overline{T}$ satisfying $T \subseteq \overline{T} \subseteq \{1, \ldots, n\}$, $A_{\overline{T}} \in \mathbb{Z}^{m \times (m+1)}$ and $\operatorname{rank}(A_{\overline{T}}) = m$.

Let $\bar{u} \in \mathbb{R}^{m+1}$ denote the vector obtained by appending $m + 1 - |T|$ zeros to $\tilde{u}_T$. There exists an index set $I \subseteq \overline{T}$ with $|I| = m$ and A_I invertible. Let i denote the singleton in $\overline{T} \setminus I$. Because $A_{\overline{T}} \bar{u} = 0$, we have

$$A_I \bar{u}_I = -A_i \bar{u}_i.$$

If $\bar{u}_i = 0$, then $\bar{u} = 0$ and so $\tilde{u} = 0$. However, this contradicts that $\tilde{u} \in U$. Hence, $\bar{u}_i \neq 0$. Scale $\bar{u}$ such that $|\bar{u}_i| = |\det(A_I)|$. Applying Cramer's rule demonstrates that

$$|\bar{u}_j| = |\det(A_{I \cup \{i\} \setminus \{j\}})| \qquad \forall \; j \in I.$$

Hence, $\bar{u}$, and consequently $\tilde{u}$, can be scaled to have integer components with $\|\tilde{u}\|_\infty \leq \Delta$. $\qquad \square$

For the rest of the proof we assume that each $\tilde{u} \in U$ is scaled such that the conclusions of Claim 8 hold.

Recall $\bar{z}_H - x_H^* \in K$. By Carathéodory's theorem, there exists an index set $I \subseteq \{1, \ldots, t\}$ with $|I| \leq \dim(K) \leq S$ and $\lambda_i \in \mathbb{R}_{\geq 0}$ for each $i \in I$ such that

$$\bar{z}_H - x_H^* = \sum_{i \in I} \lambda_i u^i.$$

Set $w := \sum_{i \in I} \lfloor \lambda_i \rfloor u^i$. Using standard techniques in proximity proofs (see, e.g., [7, Theorem 1]), it can be verified that $\tilde{z} := \bar{z}_H - w$ is a feasible solution to (IP2), $\tilde{x} := x_H^* + w$ is a feasible solution to (LP2), and

$$c_H^\mathsf{T} \bar{z}_H + c_H^\mathsf{T} x_H^* = c_H^\mathsf{T} \tilde{z} + c_H^\mathsf{T} \tilde{x}. \tag{11}$$

Because x_H^* is optimal for (LP2), we have $c_H^\mathsf{T} \tilde{x} \le c_H^\mathsf{T} x_H^*$. Combining this with (11) proves that $c_H^\mathsf{T} \tilde{z} \ge c_H^\mathsf{T} \bar{z}_H$. Because $\bar{z}_H$ is optimal for (IP2), we have $c_H^\mathsf{T} \tilde{z} = c_H^\mathsf{T} \bar{z}_H$ and $\tilde{z}$ is also an optimal solution to (IP2).

Define $z^* \in \mathbb{Z}_{\ge 0}^n$ component-wise to be

$$z_i^* := \begin{cases} \tilde{z}_i & \text{if } i \in \{1, \dots, |H|\} \\ 0 & \text{otherwise.} \end{cases}$$

By construction z^* is an optimal solution to (IP) because $Az^* = A_H \tilde{z} = A_H \bar{z}_H = b$ and $c^\mathsf{T} z^* = c_H^\mathsf{T} \tilde{z} = c_H^\mathsf{T} \bar{z}_H = c^\mathsf{T} \bar{z}$. By (8) and (10), we arrive at the final result:

$$\|z^* - x^*\|_1 = \|\tilde{z} - x_H^*\|_1 \le \sum_{i \in I} (\lambda_i - \lfloor \lambda_i \rfloor) \|u^i\|_1$$
$$< (m+1) \cdot S \cdot \|u^i\|_\infty \le (m+1) \cdot S \cdot \Delta.$$

$\square$

Proof (of Theorem 4). Let z^* be an optimal solution of (IP) with minimum support. The definition of S states that $S := |\operatorname{supp}(z^*)|$. Define $\tilde{A} \in \mathbb{Z}^{m \times S}$ as the submatrix of A corresponding to the support of z^*. If $S \le 2m$, then the result holds. Thus, assume that $S > 2m$, which implies $\log(S/m) < (S/m) - 1$. Theorem 1 in Aliev et al. [2] states that

$$S < m + \log\left(\sqrt{\det(\tilde{A}\tilde{A}^\mathsf{T})}\right).$$

The Cauchy-Binet formula for $\det(\tilde{A}\tilde{A}^\mathsf{T})$ states that

$$\det(\tilde{A}\tilde{A}^\mathsf{T}) = \sum_{\substack{B \text{ is an } m \times m \\ \text{submatrix of } \tilde{A}}} \det(B)^2.$$

See, e.g., [14]. Combining the previous inequalities yields

$$S < m + \log\left(\sqrt{\det(\tilde{A}\tilde{A}^\mathsf{T})}\right) \le m + \log\left(\sqrt{\binom{S}{m} \Delta^2}\right)$$
$$\le m + \log\left(S^{m/2} \Delta\right) = m + \frac{m}{2} \log\left(\frac{S}{m}\right) + \frac{m}{2} \log(m) + \log(\Delta)$$
$$< m + \frac{m}{2}\left(\frac{S}{m} - 1\right) + \frac{m}{2} \log(m) + \log(\Delta) = \frac{S}{2} + \frac{m}{2} + \frac{m}{2} \log(m) + \log(\Delta).$$

Therefore,

$$| \operatorname{supp}(z^*)| = S < m + m \log(m) + 2 \log(\Delta) \leq 2m \log\left(\sqrt{2m} \cdot \Delta^{1/m}\right).$$

$\square$

Proof (of Corollary 5). Let z^* be an optimal (MIP) solution with minimal support. By applying Theorem 4 to the standard form integer program with constraint matrix $A_{\mathcal{I}}$ and right hand side $b - A_{\mathcal{J}} z_{\mathcal{J}}^* \in \mathbb{Z}^m$, where $\mathcal{J} := \{1, \ldots, n\} \setminus \mathcal{I}$, we see that

$$| \operatorname{supp}(z_{\mathcal{I}}^*)| < 2m \log\left(\sqrt{2m} \cdot \Delta^{1/m}\right).$$

Similarly, by considering the standard form linear program with constraint matrix $A_{\mathcal{J}}$ and right hand side $b - A_{\mathcal{I}} z_{\mathcal{I}}^*$, we see that $| \operatorname{supp}(z_{\mathcal{J}}^*)| \leq m$. Hence,

$$| \operatorname{supp}(z^*)| \leq | \operatorname{supp}(z_{\mathcal{J}}^*)| + | \operatorname{supp}(z_{\mathcal{I}}^*)| \leq m + 2m \log\left(\sqrt{2m} \cdot \Delta^{1/m}\right).$$

$\square$

3   Results on Proximity

Proof (of Theorem 1). For now assume that $m \geq 2$. Combining Lemma 3 with Theorem 4 demonstrates that there exists an optimal (IP) solution z^* satisfying

$$\begin{aligned}
\|z^* - x^*\|_1 &< (m+1) \cdot S \cdot \Delta \\
&\leq (m+1) \cdot 2m \log(\sqrt{2m} \cdot \Delta^{1/m}) \cdot \Delta \\
&\leq 3m^2 \log(\sqrt{2m} \cdot \Delta^{1/m}) \cdot \Delta.
\end{aligned}$$

This completes the proof when $m \geq 2$.

It is left to consider the case $m = 1$. Here we have $\Delta = \|A\|_\infty$. If $\Delta = 1$, then x^* is integral and there is nothing to prove. Thus, assume $\Delta \geq 2$. The proximity bound (3) states that there exists an optimal solution z^* to (IP) satisfying

$$\|z^* - x^*\|_1 \leq 2 \cdot \|A\|_\infty + 1 = 2\Delta + 1 < 3\Delta \leq 3m^2 \log(\sqrt{2m} \cdot \Delta^{1/m}) \cdot \Delta.$$

This completes the proof. $\square$

We use the following result to prove Theorem 2.

Lemma 9 (Theorem 1 in [8]). *For every $\epsilon > 0$, there exists an $m \times m$ submatrix B of A satisfying*

$$\Delta \leq | \det(B)| \cdot (e \cdot \ln ((1 + \epsilon) \cdot m))^{m/2} .$$

The matrix B can be found in time that is polynomial in m, n, and $1/\epsilon$.

Setting $\epsilon = 1/m$ in Lemma 9 yields the approximation factor

$$\Delta \leq |\det(B)| \cdot (e \cdot \ln(m+1))^{m/2}$$

$$= |\det(B)| \cdot \left(\frac{e}{\log(e)} \cdot \log(m+1) \right)^{m/2}$$

$$\leq |\det(B)| \cdot (2 \cdot \log(m+1))^{m/2},$$

which is precisely (6).

Proof (of Theorem 2). Let B be an $m \times m$ submatrix of A satisfying (6). There exists a unimodular matrix $U \in \mathbb{Z}^{m \times m}$ such that UB is an upper triangular matrix with non-negative diagonal entries d_i.

Unimodular matrices preserve the absolute value of $m \times m$ determinants, so $|\det(B)| = |\det(UB)|$. By (6) we see that

$$\Delta \leq |\det(B)| \cdot (2 \cdot \log(m+1))^{m/2} = |\det(UB)| \cdot (2 \cdot \log(m+1))^{m/2}$$

$$= \left(\prod_{i=1}^{m} d_i \right) \cdot (2 \cdot \log(m+1))^{m/2} \leq \|UB\|_\infty^m \cdot (2 \cdot \log(m+1))^{m/2}.$$

By applying Theorem 1 we obtain the bound

$$\|z^* - x^*\|_1 < 3m^2 \log(\sqrt{2m} \cdot \Delta^{1/m}) \cdot \Delta$$

$$< 3m^2 \log(2\sqrt{m \log(m+1)} \|UB\|_\infty) \cdot (2\log(m+1))^{m/2} \cdot \|UB\|_\infty^m.$$

This completes the proof. □

Next, we present a proof of Corollary 7. We advise the reader that the proof is similar to the proof of Lemma 3.

Proof (of Corollary 7). Let z^* be an optimal solution to (GIP) with minimal support on the first n components. Consider the n-dimensional integer program obtained by fixing the last d variables of (GIP) to the last d components of z^*. Similarly, consider the n-dimensional linear program obtained by fixing the last d variables of (GLP) to the last d components of x^*. These lower dimensional problems are in standard form. Therefore, by applying Theorem 4 to these lower dimensional problems and recalling $\Delta_m(A) \leq \delta$ we see that

$$\bar{S} := |\operatorname{supp}(z^*) \cap \{1, \ldots, n\}| \leq 2m \cdot \log(\sqrt{2m} \cdot \delta^{1/m})$$

and

$$|\operatorname{supp}(x^*) \cup \operatorname{supp}(z^*)| \leq \min\{n, \bar{S} + m\} + d.$$

We project the original optimization problems onto the variables corresponding to $|\operatorname{supp}(x^*) \cup \operatorname{supp}(z^*)|$ to bound proximity. We complete the proof of the corollary by showing

$$\|z^* - x^*\|_1 < \min\{m+t+1, n+d\} \cdot \min\{n+d-m, \bar{S}+d\} \cdot \delta.$$

As in the proof of Lemma 3, we create a pointed cone from the constraints defining (GIP). In order to assure a pointed cone, we introduce redundant constraints. Let $b_3 \in \mathbb{Z}^d$ be the vector where every component is $\lceil \|x^*\|_\infty \rceil + \|z^*\|_\infty$, and define

$$D := \begin{pmatrix} 0 & C \\ -I_n & 0 \\ 0 & I_d \end{pmatrix} \in \mathbb{Z}^{(t+n+d) \times (n+d)} \quad \text{and} \quad f = \begin{pmatrix} b_2 \\ 0 \\ b_3 \end{pmatrix} \in \mathbb{Z}^{t+n+d}.$$

By construction, z^* is an optimal solution to the integer program

$$\max \left\{ c^\mathsf{T} z : \begin{array}{l} [A,B]z = b_1, \ Dz \leq f, \ z \in \mathbb{Z}^{n+d}, \\ z_i = 0 \ \forall \ i \in \{1,\ldots,d+n\} \setminus (\operatorname{supp}(x^*) \cup \operatorname{supp}(z^*)) \end{array} \right\},$$

and x^* is an optimal vertex solution to the corresponding linear relaxation.

Subdivide the rows of D such that $D_1 z^* < D_1 x^*$ and $D_2 z^* \geq D_2 x^*$. Define the polyhedral cone

$$K := \left\{ u \in \mathbb{R}^{n+d} : \begin{array}{l} [A,B]u = 0, \ \begin{pmatrix} D_1 \\ -D_2 \end{pmatrix} u \leq 0, \\ u_i = 0 \ \forall \ i \in \{1,\ldots,d+n\} \setminus (\operatorname{supp}(x^*) \cup \operatorname{supp}(z^*)) \end{array} \right\}.$$

Observe that $z^* - x^* \in K$. Moreover, the introduction of b_3 and I_d ensures that $\operatorname{rank}(D) = n+d$ and that K is pointed. We bound $\dim(K)$ in two ways by counting the number of linearly independent equations in the definition of K. First, there are m linearly independent constraints of the form $[A,B]u = 0$ because $\operatorname{rank}([A,\ B]) = m$. Second, there are $n+d-|\operatorname{supp}(x^*) \cup \operatorname{supp}(z^*)|$ many linearly independent constraints of the form $u_i = 0$. Additional linearly independent equations that are independent from this second set are $|\operatorname{supp}(x^*) \cap \{1,\ldots,n\}|$ many rows from $[A,B]u = 0$ corresponding to the independent columns of $\operatorname{supp}(x^*) \cap \{1,\ldots,n\}$. Hence, $\dim(K)$ can be upper bounded as follows:

$$\dim(K) \leq \min\{n+d-m, |[\operatorname{supp}(x^*) \cup \operatorname{supp}(z^*)] \cap \{n+1,\ldots,n+d\}| + \bar{S}\}$$
$$\leq \min\{n+d-m, d+\bar{S}\}.$$

Let U be a finite set generating the extreme rays of K. The proof of Claim 8 can be used to demonstrate that for each $\tilde{u} \in U$ we have

$$|\operatorname{supp}(\tilde{u})| \leq \min\{m+t+1, n+d\},$$

and each $\tilde{u} \in U$ can be scaled such that $\tilde{u} \in \mathbb{Z}^{n+d}$ and $\|\tilde{u}\|_\infty \leq \delta$. The main difference that arises when repeating this proof is that the matrices D_1 and D_2 contain rows of $[0,C]$ rather than simply rows from the identity matrix as was the case in the proof of Lemma 3. It is this difference that necessitates the choice of the data parameter δ and dictates its definition.

By Carathéodory's theorem there exist $k \leq \dim(K)$ vectors $u^1,\ldots,u^k \in U$ and coefficients $\lambda_1,\ldots,\lambda_k \in \mathbb{R}_{\geq 0}$ such that $z^* - x^* = \sum_{i=1}^k \lambda_i u^i$. Because

$z^* - \sum_{i=1}^{k} \lfloor \lambda_i \rfloor u^i$ is also an optimal solution to (GIP), we can assume without loss of generality that $\lambda_1, \ldots, \lambda_k < 1$ (the reasoning is similar to that in the proof of Lemma 3). This implies that

$$\|z^* - x^*\|_1 \leq \sum_{i=1}^{k} \lambda_i \|u^i\|_1 < \min\{m+t+1, n+d\} \cdot \dim(K) \cdot \delta$$

$$\leq \min\{m+t+1, n+d\} \cdot \min\{n+d-m, \bar{S}+d\} \cdot \delta.$$

This completes the proof. □

Acknowledgements. J. Lee was supported in part by ONR grant N00014-17-1-2296.

References

1. Aliev, I., Averkov, G., De Loera, J.A., Oertel, T.: Optimizing sparsity over lattices and semigroups. In: Bienstock, D., Zambelli, G. (eds.) IPCO 2020. LNCS, vol. 12125, pp. 40–51. Springer, Cham (2020). https://doi.org/10.1007/978-3-030-45771-6_4
2. Aliev, I., De Loera, J., Oertel, T., O'Neil, C.: Sparse solutions of linear diophantine equations. SIAM J. Appl. Algebra Geom. **1**, 239–253 (2017)
3. Aliev, I., Henk, M., Oertel, T.: Distances to lattice points in knapsack polyhedra. To appaear in Mathematical Programming (2019)
4. Bruns, W., Gubeladze, J.: Normality and covering properties of affine semigroups. J. Reine Angew. Math. **510**, 151–178 (2004)
5. Bruns, W., Gubeladze, J., Henk, M., Martin, A., Weismantel, R.: A counterexample to an integer analogue of Carathéodory's theorem. J. Reine Angew. Math. **510**, 179–185 (1999)
6. Cook, W., Fonlupt, J., Schrijver, A.: An integer analogue of Carathéodory's theorem. J. Combin. Theory Ser. B **40**(1), 63–70 (1986)
7. Cook, W., Gerards, A.M.H., Schrijver, A., Tardos, É.: Sensitivity theorems in integer linear programming. Math. Program. **34**, 251–264 (1986)
8. Di Summa, M., Eisenbrand, F., Faenza, Y., Moldenhauer, C.: On largest volume simplices and sub-determinants. In: Proceedings of SODA 2015, pp. 315–323 (2015)
9. Eisenbrand, F., Shmonin, G.: Carathéodory bounds for integer cones. Oper. Res. Lett. **34**, 564–568 (2006)
10. Eisenbrand, F., Weismantel, R.: Proximity results and faster algorithms for integer programming using the Steinitz lemma. In: Proceedings of SODA 2018, pp. 808–816 (2018)
11. Eisenbrand, F., Hunkenschröder, C., Klein, K.M., Koutecký, M., Levin, A., Onn, S.: An algorithmic theory of integer programming. preprint arXiv:1904.01361 (2019)
12. Granot, F., Skorin-Kapov, J.: Some proximity and sensitivity results in quadratic integer programming. Math. Program. **47**, 259–268 (1990)
13. Hochbaum, D.S., Shanthikumar, J.G.: Convex separable optimization is not much harder than linear optimization. J. ACM **37**, 843–862 (1990)
14. Horn, R.A., Johnson, C.R.: Matrix Analysis, 2nd edn. Cambridge University Press, New York (2012)
15. Khachiyan, L.: On the complexity of approximating extremal determinants in matrices. J. Complex. **11**, 138–153 (1995)

16. Lee, J.: Subspaces with well-scaled frames. Linear Algebra Appl. **114**(115), 21–56 (1989)
17. Oertel, T., Paat, J., Weismantel, R.: The distributions of functions related to parametric integer optimization. preprint arXiv:1907.07960 (2019)
18. Oertel, T., Paat, J., Weismantel, R.: Sparsity of integer solutions in the average case. In: Lodi, A., Nagarajan, V. (eds.) IPCO 2019. LNCS, vol. 11480, pp. 341–353. Springer, Cham (2019). https://doi.org/10.1007/978-3-030-17953-3_26
19. Paat, J., Weismantel, R., Weltge, S.: Distances between optimal solutions of mixed-integer programs. Math. Program. **179**, 1–14 (2018)
20. Schrijver, A.: Theory of Linear and Integer Programming. Wiley, New York (1986)
21. Sebő, A.: Hilbert bases, Carathéodory's theorem and combinatorial optimization. In: Proceedings of IPCO 1990, pp. 431–455 (1990)
22. Steinitz, E.: Bedingt konvergente reihen und konvexe systeme. J. Angew. Math. **143**, 128–176 (1913)
23. Werman, M., Magagnosc, D.: The relationship between integer and real solutions of constrained convex programming. Math. Program. **51**, 133–135 (1991)
24. Xu, L., Lee, J.: On proximity for k-regular mixed-integer linear optimization. In: Le Thi, H.A., Le, H.M., Pham Dinh, T. (eds.) WCGO 2019. AISC, vol. 991, pp. 438–447. Springer, Cham (2020). https://doi.org/10.1007/978-3-030-21803-4_44

Cut and Flow Formulations for the Balanced Connected k-Partition Problem

Flávio K. Miyazawa[1], Phablo F. S. Moura[2(✉)], Matheus J. Ota[1], and Yoshiko Wakabayashi[3]

[1] Institute of Computing, University of Campinas, Campinas, Brazil
fkm@ic.unicamp.br, matheus.ota@students.ic.unicamp.br
[2] Department of Computer Science, Federal University of Minas Gerais, Belo Horizonte, Brazil
phablo@dcc.ufmg.br
[3] Institute of Mathematics and Statistics, University of São Paulo, São Paulo, Brazil
yw@ime.usp.br

Abstract. For a fixed integer $k \geq 2$, the *balanced connected k-partition problem* (BCP_k) consists in partitioning a graph into k mutually vertex-disjoint connected subgraphs of similar weight. More formally, given a connected graph G with nonnegative weights on the vertices, find a partition $\{V_i\}_{i=1}^k$ of $V(G)$ such that each class V_i induces a connected subgraph of G, and the weight of a class with the minimum weight is as large as possible. This problem, known to be $\mathcal{NP}$-hard, is used to model many applications arising in image processing, cluster analysis, operating systems and robotics. We propose an ILP and a MILP formulation for BCP_k. The first one contains only binary variables and a potentially large number of constraints that can be separated in polynomial time. We also present polyhedral results on the polytope associated with this formulation, introduce new valid inequalities and design separation algorithms. The other formulation is based on flows and has a polynomial number of constraints and variables. Computational experiments show that our formulations achieve better results than the other formulations presented in the literature.

Keywords: Connected partition · Integer linear programming · Facet · Polyhedra · Branch-and-cut

1 Introduction

Let $G = (V, E)$ be a connected graph, $n := |V|$ and $m := |E|$. For an integer $k \geq 1$, as usual, the symbol $[k]$ denotes the set $\{1, 2, \ldots, k\}$. A *k-partition* of G

Research partially supported by grant #2015/11937-9, São Paulo Research Foundation (FAPESP). Miyazawa is supported by CNPq (Proc. 314366/2018-0 and 425340/2016-3) and FAPESP (Proc. 2016/01860-1). Moura is supported by FAPESP grants #2016/21250-3 and #2017/22611-2, CAPES, and Pró-Reitoria de Pesquisa da Universidade Federal de Minas Gerais. Ota is supported by CNPq. Wakabayashi is supported by CNPq (Proc. 306464/2016-0 and 423833/2018-9).

© Springer Nature Switzerland AG 2020
M. Baïou et al. (Eds.): ISCO 2020, LNCS 12176, pp. 128–139, 2020.
https://doi.org/10.1007/978-3-030-53262-8_11

is a collection $\{V_i\}_{i \in [k]}$ of nonempty subsets of V such that $\bigcup_{i=1}^{k} V_i = V$, and $V_i \cap V_j = \emptyset$ for all $i, j \in [k]$, $i \neq j$. We refer to each set V_i as a *class* of the partition. We say that a k-partition $\{V_i\}_{i \in [k]}$ of G is *connected* if $G[V_i]$, the subgraph of G induced by V_i, is connected for each $i \in [k]$.

Let $w \colon V \to \mathbb{Q}_>$ be a function that assigns weights to the vertices of G. For every subset $V' \subseteq V$, we define $w(V') = \sum_{v \in V'} w(v)$. In the *balanced connected k-partition problem* (BCP_k), we are given a vertex-weighted connected graph, and we seek a connected k-partition such that the weight of a lightest class of this partition is maximized. In other words, for each fixed positive integer k, an input of BCP_k is given by a pair (G, w), with $n \geq k$, and the objective is to find a connected k-partition $\{V_i\}_{i \in [k]}$ of V such that $\min_{i \in [k]}\{w(V_i)\}$ is maximized.

There are several problems in police patrolling, image processing, data base, operating systems, cluster analysis, and robotics that can be modeled as a balanced connected partition problem [2,17,19,23]. These different real-world applications indicate the importance of designing algorithms for BCP_k, and reporting on the computational experiments with their implementations. Not less important are the theoretical studies of the rich and diverse mathematical formulations and the polyhedral investigations BCP_k leads to.

2 Some Known Results and Our Contributions

Problems on partitioning a vertex-weighted graph into a fixed number of connected subgraphs with similar weights have been largely investigated in the literature since early eighties. Such partitions are generally called balanced, and a number of different functions have been considered to measure this feature.

The *balanced connected k-partition problem* (BCP_k), to be focused here, is one of these problems. It is closely related to another problem, referred to as MIN-MAX BCP_k, whose objective function is to minimize the weight of the heaviest class. When $k = 2$, for any instance, an optimal 2-partition for BCP_k is also an optimal solution for MIN-MAX BCP_k; but it is easy to see that when $k > 2$ the corresponding optimal k-partitions may differ.

The *unweighted* BCP_k, restricted version in which all vertices have unit weight, has also attracted much attention. Dyer and Frieze [10] showed that it is $\mathscr{NP}$-hard on bipartite graphs. In special, when the input graph is k-connected, polynomial-time algorithms and other interesting structural results have been obtained [14,16,18] for BCP_k. For the weighted case, Becker et al. [4] proved that BCP_k is already $\mathscr{NP}$-hard on (nontrivial) grid graphs. Chlebíková [8] designed a $4/3$-approximation algorithm for BCP_2. For BCP_3 (resp. BCP_4) on 3-connected (resp. 4-connected) graphs approximation algorithms were proposed by Chataigner et al. [7]. Wu [22] proved that BCP_k is $\mathscr{NP}$-hard on interval graphs; and designed an FPTAS for BCP_2 on such graphs. More recently, Borndörfer et al. [6] designed approximation algorithms for BCP_k and MIN-MAX BCP_k. Both BCP_k and MIN-MAX BCP_k can be solved in polynomial-time on trees [5,21]. Many other results on both problems have appeared in the literature.

Mixed integer linear programming formulations for BCP_2 were proposed by Matić [20] and for MIN-MAX BCP_k by Zhou et al. [23]. Matić also presented a VNS-based heuristic for BCP_2, and Zhou et al. devised a genetic algorithm for MIN-MAX BCP_k. Both authors reported on the computational results obtained with the proposed formulations and heuristics, but presented no further polyhedral study of their formulations.

In this work we advance the state of the art for exact algorithms for BCP_k (and also for MIN-MAX BCP_k). In Sect. 3, we introduce a cut-based ILP formulation for BCP_k, and show two stronger valid inequalities for this formulation. A further polyhedral study is presented in Sect. 4. In Sect. 5, we propose a compact flow-based MILP formulation for BCP_k. In Sect. 6 we discuss polynomial-time separation routines for the inequalities in the cut formulation. Both of our formulations for BCP_k can be used to solve MIN-MAX BCP_k (and some other variants) just by changing the objective function. Lastly, the computational experiments in Sect. 7 show that our formulations outperform the previous formulations in the literature.

3   Cut Formulation

In this section, we consider that (G, w), where $G = (V, E)$, is the input for BCP_k. The ILP formulation we propose for BCP_k, called $C_k(G, w)$, or simply C, is based on the following central concept. Let u and v be two non-adjacent vertices in a graph $G = (V, E)$. We say that a set $S \subseteq V \setminus \{u, v\}$ is a (u, v)-separator if u and v belong to different components of $G - S$. We denote by $\Gamma(u, v)$ the collection of all minimal (u, v)-separators in G. In the formulation, we use a binary variable $x_{v,i}$, for every $v \in V$ and $i \in [k]$, that is set to one if and only if v belongs to the ith class.

$$\max \sum_{v \in V} w(v)\, x_{v,1}$$

$$\text{s.t.} \sum_{v \in V} w(v)\, x_{v,i} \le \sum_{v \in V} w(v)\, x_{v,i+1} \qquad \forall i \in [k-1], \qquad (1)$$

$$\sum_{i \in [k]} x_{v,i} \le 1 \qquad \forall v \in V, \qquad (2)$$

$$x_{u,i} + x_{v,i} - \sum_{z \in S} x_{z,i} \le 1 \qquad \forall uv \notin E, S \in \Gamma(u,v), i \in [k], \qquad (3)$$

$$x_{v,i} \in \{0, 1\} \qquad \forall v \in V \text{ and } i \in [k]. \qquad (4)$$

Inequalities (1) impose a non-decreasing weight ordering of the classes. Inequalities (2) require that every vertex is assigned to at most one class. Inequalities (3) guarantee that every class induces a connected subgraph. The objective function maximizes the weight of the first class. Thus, in an optimal solution no class will be empty, and therefore it will always correspond to a connected k-partition of G.

In Sect. 6 we show that the separation problem for inequalities (3) can be solved in polynomial time. Thus, the linear relaxation of $\mathcal{C}$ can be solved in polynomial time (see [13]).

Since feasible solutions of formulation $\mathcal{C}_k(G, w)$ may have empty classes and nodes not assigned to any particular class, to refer to these solutions we introduce the following concept. We say that $\mathcal{V} = \{V_i\}_{i=1}^k$ is a *connected k-subpartition* of G, if it is a connected k-partition of a subgraph (not necessarily proper) of G, and additionally, $w(V_i) \leq w(V_{i+1})$ for all $i \in [k-1]$. For such a k-subpartition $\mathcal{V}$, we denote by $\xi(\mathcal{V}) \in \mathbb{B}^{nk}$ the binary vector such that its non-null entries are precisely $\xi(\mathcal{V})_{v,i} = 1$ for all $i \in [k]$ and $v \in V_i$ (that is, $\xi(\mathcal{V})$ denotes the incidence vector of $\mathcal{V}$). To show our next results, let us define the polytope

$$\mathcal{P}_k(G, w) = \text{convex-hull}\,\{x \in \mathbb{B}^{nk} : x \text{ satisfies inequalities } (1)-(3) \text{ of } \mathcal{C}_k(G, w)\}.$$

We prove that formulation $\mathcal{C}_k(G, w)$ correctly models BCP_k. Then, we present two classes of valid inequalities that strengthen formulation $\mathcal{C}_k(G, w)$.

Proposition 1

$$\mathcal{P}_k(G, w) = \text{convex-hull}\,\{\xi(\mathcal{V}) \in \mathbb{B}^{nk} : \mathcal{V} \text{ is a connected k-subpartition of } G\}.$$

Proof. Consider first an extreme point $x \in \mathcal{P}_k(G, w)$. For each $i \in [k]$, we define the set of vertices $U_i = \{v \in V : x_{v,i} = 1\}$. It follows from inequalities (1) and (2) that $\mathcal{U} := \{U_i\}_{i=1}^k$ is a k-subpartition of G such that $w(U_i) \leq w(U_{i+1})$ for all $i \in [k-1]$.

To prove that $\mathcal{U}$ is a connected k-subpartition, we suppose to the contrary that there exists $i \in [k]$ such that $G[U_i]$ is not connected. Hence, there exist vertices u and v belonging to two different components of $G[U_i]$. Moreover, there is a minimal set of vertices S that separates u and v and such that $S \cap U_i = \emptyset$. Thus, vector x violates inequalities (3), a contradiction. To show the converse, consider now a connected k-subpartition $\mathcal{V} = \{V_i\}_{i=1}^k$ of G. Clearly $\xi(\mathcal{V})$, satisfies inequalities (1) and (2). Take a fixed $i \in [k]$. For every pair u,v of non-adjacent vertices in V_i, and every (u, v)-separator S in G, it holds that $S \cap V_i \neq \emptyset$, because $G[V_i]$ is connected. Therefore, $\xi(\mathcal{V})$ satisfies inequalities (3). $\qquad\square$

Proposition 2. *Let u and v be two non-adjacent vertices of G, and let S be a minimal (u, v)-separator. Let $i \in [k]$, and let $L = \{z \in S : w(P_z) > w(V)/(k - i+1)\}$, where P_z is a minimum-weight path between u and v in G that contains z. The following inequality is valid for $\mathcal{P}_k(G, w)$:*

$$x_{u,i} + x_{v,i} - \sum_{z \in S \setminus L} x_{z,i} \leq 1. \tag{5}$$

Proof. Consider an extreme point x of $\mathcal{P}_k(G, w)$, and define $V_i = \{v \in V : x_{v,i} = 1\}$ for each $i \in [k]$. Suppose to the contrary that there is $j \in [k]$ such that $w(V_j) > \frac{w(V)}{k-j+1}$. Since x satisfies inequalities (1), it holds that $\sum_{i \in [k] \setminus [j-1]} w(V_i) > w(V)$, a contradiction. Thus, if u and v belong to V_i, then there exists a vertex $z \in S \setminus L_i$ such that z also belongs to V_i. Therefore, x satisfies inequality (5). $\qquad\square$

The next class of inequalities was inspired by a result proposed by de Aragão and Uchoa [1] for a connected assignment problem.

Proposition 3. *Let q be a fixed integer such that $q \geq 2$, and let S be a subset of vertices of G containing q distinct pairs of vertices (s_i, t_i), $i \in [q]$, all mutually disjoint. Let $N(S)$ be the set of neighbors of S in $V \setminus S$. Moreover, let $\sigma \colon [q] \to [k]$ be an injective function, and let $I = \{\sigma(i) \in [k] \colon i \in [q]\}$. If there is no collection of q vertex-disjoint (s_i, t_i)-paths in $G[S]$, then the following inequality is valid for $\mathcal{P}_k(G, w)$:*

$$\sum_{i \in [q]} \left(x_{s_i, \sigma(i)} + x_{t_i, \sigma(i)}\right) + \sum_{v \in N(S)} \sum_{i \in [k] \setminus I} x_{v,i} \leq 2q + |N(S)| - 1. \tag{6}$$

Proof. Suppose to the contrary that there exists an extreme point x of $\mathcal{P}_k(G, w)$ that violates inequality (6). Let us define $A = \sum_{i \in [q]} \left(x_{s_i, \sigma(i)} + x_{t_i, \sigma(i)}\right)$ and $B = \sum_{v \in N(S)} \sum_{i \in [k] \setminus I} x_{v,i}$. From inequalities (2), we have that $A \leq 2q$. Since x violates (6), it follows that $B > |N(S)| - 1$. Thus $B = |N(S)|$ (because x satisfies inequalities (2)). Hence, every vertex in $N(S)$ belongs to a class that is different from those indexed by I. This implies that every class indexed by I contains precisely one of the q distinct pairs $\{s_i, t_i\}$. Therefore, there exists a collection of q vertex-disjoint (s_i, t_i)-paths in $G[S]$, a contradiction. ☐

Kawarabayashi et al. [15] proved that, given an n-vertex graph G and a set of q pairs of terminals in G, the problem of deciding whether G contains q vertex-disjoint paths linking the given pairs of terminals can be solved in time $\mathcal{O}(n^2)$. Hence, inequalities (6) can be separated in polynomial time when $S = V$.

4   Polyhedral Results for Unweighted BCP$_k$

This section is devoted to 1-BCP$_k$, the special case of BCP$_k$ in which all vertices have unit weight. In this case, the polytope $\mathcal{P}_k(G, w)$ is simply denoted as $\mathcal{P}_k(G)$.

We assume that G has a matching of size k, otherwise it is easy to find an optimal solution. Thus, from now on we assume that $n \geq 2k$.

Due to space constraints, we omit the proof of the results stated in this section.

Theorem 1. *Let G be an input for 1-BCP$_k$. Then, the following hold.*

(a) The polytope $\mathcal{P}_k(G)$ is full-dimensional, that is, $\dim(\mathcal{P}_k(G)) = kn$;
(b) For every $v \in V$ and $i \in [k]$, the inequality $x_{v,i} \geq 0$ defines a facet of $\mathcal{P}_k(G)$;
(c) For every $v \in V$, the inequality $\sum_{i \in [k]} x_{v,i} \leq 1$ defines a facet of $\mathcal{P}_k(G)$.

The next result characterizes when the inequalities (3) define facets of $\mathcal{P}_k(G)$. To present it, we introduce some concepts and notation. Let u and v be non-adjacent vertices in G, and let S be a minimal (u, v)-separator. Denote by H_u and H_v the components of $G - S$ that contain u and v, respectively. For any vertex z in G, denote by G_z (when S is clear from the context) a minimum size connected subgraph of G containing z, with the following property: if $z \in V(H_v)$, then G_z contains v (and is contained in H_v); if $z \in V(H_u)$, then G_z contains u (and is contained in H_u); otherwise, G_z contains u and v. Note that, in the latter case, G_z contains exactly one vertex of S. Clearly, G_z exists (and may not be unique).

Let $i \in [k]$ and $z \in V(G)$. We say that $\mathcal{V}_z$ is a *robust connected* $(k-i)$-*partition* of $G - G_z$ if $\mathcal{V}_z = \{V_j\}_{j \in [k] \setminus [i]}$, and $|V(G_z)| \le |V_j|$ for all $j \in [k] \setminus [i]$. We say that G is (u,v,S,i)-*robust* if for every $z \in V(G)$, there is a graph G_z such that $G - G_z$ has a robust connected $(k-i)$-partition.

We are now ready to present the mentioned characterization.

Theorem 2. *Let u and v be non-adjacent vertices in G, let S be a minimal (u,v)-separator, and let $i \in [k]$. Then, the inequality $x_{u,i} + x_{v,i} - \sum_{s \in S} x_{s,i} \le 1$ defines a facet of $\mathcal{P}_k(G)$ if and only if G is (u,v,S,i)-robust.*

5   Flow Formulation

We present in this section a mixed integer linear programming formulation for BCP_k, called $\mathcal{F}(G,w)$, or simply $\mathcal{F}$, that is based on flows in a digraph.

Given a graph G, with $G = (V, E)$, we construct a digraph $D = (V \cup S, A)$, where S is a set $\{s_1, \ldots, s_k\}$ of k new vertices. The arc set A is created by replacing every edge of G with two arcs with the same endpoints and opposite directions. Finally, we add an arc from each vertex in S to each vertex of V.

For each arc $a \in A$, we associate a real variable f_a that represents the amount of flow passing through arc a. Then, to relate the flow on the arcs with the weights on the vertices, for each vertex $v \in V$, we impose that the amount of flow entering v minus the amount of flow leaving v be precisely $w(v)$ (that is, we force vertex v to consume $w(v)$ of the flow that enters it). In this way, we can control that the total amount of flow that each source s_i sends and spreads to other vertices corresponds precisely to the total weight of the vertices in the ith class of the desired partition. We also use binary variables y_a (such that $y_a = 1$ if $f_a > 0$) that allow us to impose that flows from different sources do not mix. In the MILP formulation given below, the abbreviated form $f(T)$ and $y(T)$ stands for $\sum_{a \in T} f(a)$ and $\sum_{a \in T} y(a)$, respectively.

$$\max f(\delta^+(s_1))$$

$$\text{s.t. } f(\delta^+(s_i)) \le f(\delta^+(s_{i+1})) \qquad \forall i \in [k-1], \qquad (7)$$

$$f(\delta^-(v)) - f(\delta^+(v)) = w(v) \qquad \forall v \in V, \qquad (8)$$
$$f_a \le w(V)y_a \qquad \forall a \in A, \qquad (9)$$
$$y(\delta^+(s_i)) \le 1 \qquad \forall i \in [k], \qquad (10)$$
$$y(\delta^-(v)) \le 1 \qquad \forall v \in V, \qquad (11)$$
$$y_a \in \{0,1\} \qquad \forall a \in A \qquad (12)$$
$$f_a \in \mathbb{R}_{\ge} \qquad \forall a \in A \qquad (13)$$

Inequalities (7) impose that the flows sent by sources $s_1, s_2, \ldots, s_k$ are in a non-decreasing order. This explains the objective function. Inequalities (8) guarantee that each vertex $v \in V$ consumes $w(v)$ of the flow that it receives. By (9), a positive flow can only pass through arcs that are chosen (arcs a for which $y(a) = 1$). Inequalities (10) impose that from every source s_i at most one arc leaving it transports a positive flow to a single vertex in V.

Inequalities (11) require that every non-source vertex receives a positive flow from at most one vertex of D. This guarantees that the flows sent by any two distinct sources do not pass through a same vertex. That is, if a source s_i sends an amount of flow, say F_i, this amount F_i is distributed and consumed by a subset of vertices, say V_i (whose total weight is precisely F_i). Moreover, this "distribution" guarantees that $G[V_i]$ is connected, and all subsets V_i are mutually disjoint.

It follows from these remarks that formulation $\mathcal{F}_k(G, w)$ correctly models BCP_k. Moreover, it has a polynomial number of variables and constraints.

6   Separation Algorithms

We implemented a Branch-and-Cut approach based on the cut formulation that is introduced in Sect. 3. In what follows, we describe the separation routines for inequalities (3) and (6) that we designed and implemented.

6.1   Connectivity Inequalities

We focus first on the class of inequalities (3) of Sect. 3, henceforth called *connectivity inequalities*. We address here its corresponding separation problem: given a vector $x' \in \mathbb{R}^{nk}$, find connectivity inequalities that are violated by x' or prove that this vector satisfies all such inequalities.

To tackle this problem, given the input graph $G = (V, E)$, for each $i \in [k]$, we define a digraph D_i with capacities $c_i : A(D_i) \to \mathbb{Q}_{\ge} \cup \{\infty\}$ assigned to its arcs, in the following manner. We set $V(D_i) = \{v_1, v_2 : v \in V\}$ and $A(D_i) = A_1 \cup A_2$, where $A_1 = \{(u_2, v_1), (v_2, u_1) : \{u, v\} \in E\}$ and $A_2 = \{(v_1, v_2) : v \in V\}$. We define $c_i(a) = x'_{v,i}$ if $a \in A_2$; and $c_i(a) = \infty$, otherwise. Note that each arc in D_i with a finite capacity (i.e. each arc in A_2) is associated with a vertex of G. Now, for every pair of non-adjacent vertices $u, v \in V$ such that $x'_{u,i} + x'_{v,i} > 1$, we find in D_i a minimum (u_1, v_2)-separating cut. If the weight of such a cut is smaller than $x'_{u,i} + x'_{v,i} - 1$, then it is finite and the vertices of G associated with the arcs in this cut give a (u, v)-separator S in G that violates the connectivity inequality $x'_{u,i} + x'_{v,i} - \sum_{z \in S} x'_{z,i} \leq 1$.

The time complexity to separate the connectivity inequalities depends on the algorithm used to find a minimum cut. We use Goldberg's preflow algorithm [12] for maximum flow, whose time complexity is $\mathcal{O}(\tilde{n}^2 \sqrt{\tilde{m}})$, for a digraph with $\tilde{n}$ vertices and $\tilde{m}$ arcs. Thus, in the worst-case, checking for every $i \in [k]$, and candidate pairs u, v in D_i, the time complexity of this separation algorithm is $\mathcal{O}(kn^4 \sqrt{n + m})$.

Given a (u, v)-separator S, let H_u (resp. H_v) be the connected component of $G - S$ containing u (resp. v). We now describe a procedure for performing the lifting of the connectivity inequalities by removing iteratively unnecessary vertices from S. First we remove every vertex z from S such that the neighborhood of z does not intersect with H_u and H_v. Since removing a vertex from S changes the components of $G - S$, we use a Union-Find data structure to update the components. Next, we use Dijkstra's algorithm to remove from S the set L, as described in Proposition 2.

6.2   Cross Inequalities

Now we turn to the separation of inequalities (6) on planar graphs $G = (V, E)$, restricted to the case $S = V$. Consider a plane embedding of G, and let F be the boundary of a face with at least 4 vertices. Take four different vertices, say s_1, s_2, t_1, t_2, that appear in this order in F. Since G is planar, it does not contain vertex-disjoint paths P_1 and P_2, with endpoints s_1, t_1 and s_2, t_2, respectively. For $S = V$, inequalities (6) simplifies to $x_{s_1,\sigma(1)} + x_{s_2,\sigma(2)} + x_{t_1,\sigma(1)} + x_{t_2,\sigma(2)} \leq 3$. We refer to these inequalities as *cross inequalities*.

For the separation problem of the cross inequalities induced by the vertices in F, where $|F| = f$, we implemented an $\mathcal{O}(fk^2)$ time complexity algorithm (the same complexity mentioned by Barboza [3], without much detail; the algorithms may possibly be different). Next, we give more details on this separation algorithm.

Let $x' \in \mathbb{R}^{nk}$ be a fractional solution of formulation $\mathcal{C}$ and consider a cyclic ordering of the vertices in F. For every $j \in [f]$, let $F[j]$ be the j-th vertex in this ordering. Furthermore, we define matrices L_1 and R_1 such that, for each $j \in [f]$ and each $i \in [k]$, $L_1[j][i] = \max_{j' \in [j]}\{x'_{F[j'],i}\}$ and $R_1[j][i] = \max_{j' \in [f] \setminus [j-1]}\{x'_{F[j'],i}\}$. In other words, $L_1[j][i]$ (resp. $R_1[j][i]$) corresponds to the maximum value in an entry of x' indexed by i and by a vertex that appears before (resp. after) $F[j]$ in the ordering. Clearly, L_1 and R_1 can be created in $\mathcal{O}(fk)$.

For every $j \in [f] \setminus \{1\}$, and every $i_1, i_2 \in [k]$ with $i_1 \neq i_2$, we define:

$$L_2[j][i_1][i_2] = \begin{cases} x'_{F[1],i_1} + x'_{F[2],i_2}, & \text{if } j = 2, \\ \max\left\{L_2[j-1][i_1][i_2];\ L_1[j-1][i_1] + x'_{F[j],i_2}\right\}, & \text{otherwise.} \end{cases}$$

Note that, given $j \geq 2$ and $i_1, i_2 \in [k]$, $L_2[j][i_1][i_2]$ is the maximum value of $x'_{F[j'],i_1} + x'_{F[j''],i_2}$ for all $j', j'' \in [j]$ with $j' < j''$. Our algorithm works as follows: for every $j \in \{3, \ldots, f-1\}$ and every $i_1, i_2 \in [k]$ with $i_1 \neq i_2$, it checks whether $L_2[j-1][i_1][i_2] + x'_{F[j],i_1} + R_1[j+1][i_2] > 3$, that is, whether there is a violated cross inequality (w.r.t. F) such that $\sigma(1) = i_1$, $\sigma(2) = i_2$ and $t_1 = F[j]$. Clearly, one may also keep track of the violated inequalities (if any).

7   Computational Results

To compare the performance of our algorithms with the exact algorithms that have been proposed before [20,23], we ran our experiments on the same classes of graphs that were considered in the literature, namely grid graphs and random connected graphs. Our algorithms are based on the two formulations (and further results) that we described in the previous sections.

To make clear the size of the instances, we use the format *gg_height_width_[a|b]* for the grid graphs, and the format *rnd_n_m_[a|b]* for the random connected graphs with n vertices and m edges. The character a (resp. b) indicates that the weights are integers uniformly distributed in the interval $[1, 100]$ (resp. $[1, 500]$).

The computational experiments were carried out on a PC with Intel(R) Xeon(R) CPU E5-2630 v4 @ 2.20 GHz, 40 cores, 64 GB RAM and Ubuntu 18.04.2 LTS. The code was written in C++ using the graph library Lemon [9]. We implemented a Branch-and-Cut algorithm based on the cut formulation $\mathcal{C}$ using SCIP [11] and Gurobi 9.0 as the LP solver. We also implemented Branch-and-Bound algorithms (using only Gurobi 9.0) based on the flow formulation $\mathcal{F}$, and on the models previously proposed by Matić [20] and Zhou et al. [23]. We used SCIP in our branch-and-cut implementation because, unlike Gurobi, it allows multiple rounds of cut generation when obtaining fractional solutions in non-root nodes of the branch-and-bound tree. In order to evaluate strictly the performance of the described formulations, we deactivated all standard cuts used by SCIP and Gurobi.

Henceforth, CUT refers to the Branch-and-Cut algorithm based on formulation $\mathcal{C}$, while FLOW refer to the Branch-and-Bound algorithm based on $\mathcal{F}$.

The execution time limit for each instance was set to 1800 s. For each graph type, we generated 10 instances. In the tables, the column *Sol* indicates the number of instances solved, and the column *Nodes* shows the average number of nodes explored in the branch-and-bound tree. The column *Time* either indicates the average time (in seconds) spent to solve some instances (ignoring the unsolved instances), or indicates, with a dash (-) that none of the 10 instances could be solved. The minimum time in a row is indicated in boldface, and the minimum number of nodes explored is underlined.

First we tested the efficiency of the separation routine for the cross inequalities. Using such separation, the algorithm solved all 240 grid instances, but without separation, only 183 of these instances could be solved within the time limit. For all these 183 instances, the separation produced a reduction of the average execution time by a factor greater than 10. From now on, we assume that CUT on grid graphs always separates the cross inequalities.

Table 1. Computational results for BCP_2.

Instance	CUT			FLOW			Matić			Zhou et al.		
	Sol	Nodes	Time	Sol	Nodes	Time	Sol	Nodes	Time	Sol	Nodes	Time
gg_05_05_a	10	91	0.20	10	919	**0.09**	10	4,659	0.52	10	2,564	0.45
gg_05_05_b	10	1,136	1.30	10	320,162	19.17	10	540,447	73.77	10	7,341	**1.22**
gg_05_06_a	10	245	0.49	10	460	**0.07**	10	75,140	9.38	10	1,843	0.43
gg_05_06_b	10	692	1.12	10	1,592	**0.13**	10	32,185	5.83	10	4,454	0.91
gg_05_10_a	10	164	0.72	10	500	**0.15**	10	368,184	57.52	10	12,542	3.27
gg_05_10_b	10	254	0.87	10	806	**0.17**	10	144,469	27.60	10	18,258	5.13
gg_05_20_a	10	240	2.71	10	454	**0.25**	1	844,935	228.56	2	116,374	42.32
gg_05_20_b	10	678	6.45	10	1,146	**0.34**	0	–	–	4	2,221,308	522.95
gg_07_07_a	10	200	0.77	10	645	**0.14**	10	211,299	29.04	10	10,346	2.66
gg_07_07_b	10	740	1.99	10	784	**0.18**	9	1,360,869	177.56	10	15,657	3.60
gg_07_10_a	10	136	0.85	10	366	**0.16**	8	871,395	164.53	10	529,336	116.87
gg_07_10_b	10	773	3.22	10	1,304	**0.29**	5	2,977,315	520.85	9	400,859	66.10
gg_10_10_a	10	179	1.77	10	186	**0.20**	2	336,860	74.81	6	1,472,514	396.10
gg_10_10_b	10	537	3.24	10	905	**0.36**	1	404,182	97.20	5	544,515	141.69
gg_15_15_a	10	116	7.05	10	184	**0.40**	0	–	–	0	–	–
gg_15_15_b	10	564	17.45	10	696	**0.59**	0	–	–	0	–	–
rnd_100_150_a	10	204	1.81	10	206	**0.20**	10	558,004	129.67	10	1,918	2.06
rnd_100_150_b	10	1,238	9.08	10	531	**0.25**	10	230,281	50.88	10	1,711	1.77
rnd_100_300_a	10	18	**0.16**	10	69	0.21	8	370,546	47.67	10	247	1.44
rnd_100_300_b	10	70	0.32	10	57	**0.23**	8	250,040	43.69	10	565	2.32
rnd_100_800_a	10	3	**0.04**	10	1	0.30	10	104,655	62.19	10	29	3.29
rnd_100_800_b	10	41	**0.17**	10	83	0.35	10	228,209	221.98	10	71	3.01
rnd_200_300_a	10	61	2.85	10	519	**0.41**	7	61,372	56.53	10	1,951	7.04
rnd_200_300_b	10	1,374	28.23	10	849	**0.49**	7	9,413	11.20	10	2,857	10.71
rnd_200_600_a	10	16	**0.40**	10	356	0.56	6	87,323	34.40	10	728	9.65
rnd_200_600_b	10	123	1.21	10	515	**0.62**	6	64,853	70.62	10	995	9.81
rnd_200_1500_a	10	1	**0.05**	10	1	0.64	6	42,437	91.31	10	3	7.34
rnd_200_1500_b	10	8	**0.11**	10	303	1.05	5	347,862	380.35	10	241	14.85
rnd_300_500_a	10	40	4.71	10	732	**0.72**	6	48,026	40.73	10	2,811	28.61
rnd_300_500_b	10	1,079	25.86	10	1,029	**0.84**	5	8,562	17.01	10	3,629	21.84
rnd_300_1000_a	10	8	**0.51**	10	614	1.07	6	53,284	58.70	10	982	20.34
rnd_300_1000_b	10	54	**1.06**	10	1,113	1.39	5	23,831	36.13	10	1,030	19.75
rnd_300_2000_a	10	1	**0.09**	10	1	1.05	4	234,170	371.89	10	35	25.69
rnd_300_2000_b	10	28	**0.48**	10	754	1.90	5	96,513	195.54	10	54	80.63

Table 1 shows that for BCP_2 the formulations we presented in this work substantially outperform the previous formulations in the literature. On most of the instances, FLOW had the best average execution time. On the other hand, CUT explored a smaller number of nodes in the branch-and-bound tree.

The experiments that we carried out for $k > 2$ are shown on Table 2. To compare with the formulation proposed by Zhou et al. [23], which is for MIN-MAX BCP_k, we considered the cut and flow formulations with min-max objective (i.e. minimizing the weight of the kth class), denoted by CUT (MIN-MAX) and FLOW (MIN-MAX). We remark that the specific instance that Zhou et al.'s branch-and-bound solved in 508.93 s, FLOW solved in 388.67 s.

Table 2. Computational results for MIN-MAX BCP_k when $k \in \{2, 3, 4, 5, 6\}$.

Instance	k	CUT (MIN-MAX)			FLOW (MIN-MAX)			Zhou et al.		
		Sol	Nodes	Time	Sol	Nodes	Time	Sol	Nodes	Time
gg_07_10_a	3	0	–	–	10	4,177	**0.93**	2	1,072,107	603.97
gg_07_10_a	4	0	–	–	10	25,242	**3.86**	1	99,880	68.95
gg_07_10_a	5	0	–	–	10	1,578,669	**247.61**	0	–	–
gg_07_10_a	6	0	–	–	3	3,077,826	**482.45**	0	–	–
rnd_100_150_a	3	10	9,922	223.21	10	1,933	**0.74**	9	109,799	70.04
rnd_100_150_a	4	0	–	–	10	13,404	**3.14**	4	1,019,886	641.36
rnd_100_150_a	5	0	–	–	10	627,636	**149.27**	0	–	–
rnd_100_150_a	6	0	–	–	6	2,268,233	682.23	1	205,247	508.93

8 Conclusion and Further Research

We proposed two formulations for Balanced Connected k-Partition Problem: an ILP and a MILP. While the first one has possibly an exponential amount of constraints, the second one is a compact formulation based on flows in a digraph. We reported on the computational results obtained with the implementation of a Branch-and-Cut algorithm (named CUT) for the first formulation and a Branch-and-Bound algorithm (FLOW) for the second formulation.

We introduced a new class of valid inequalities for the polytope associated with the ILP formulation, and implemented a polynomial-time separation routine for a special subclass on planar graphs that improved greatly the performance of the algorithm. For the cardinality version (unit weight), we presented some further polyhedral results. To the best of our knowledge, a polyhedral approach to BCP_k has not been used before.

Both CUT and FLOW outperform the previous MILP models for BCP_2 in the literature. For BCP_k, $k > 2$, FLOW has shown to be superior to CUT. Moreover, preliminary computational results for MIN-MAX BCP_k with $k > 2$ suggest that FLOW with min-max objective function is also superior to other formulations in the literature. Further experiments are necessary to confirm the efficiency of the proposed algorithm when $k > 2$, but the compact flow formulation seems to be very effective. Additionally, we plan to carry out experiments on real-world instances. On the theoretical side, it would be interesting to find new strong valid inequalities, and design efficient separation algorithms for them.

References

1. de Aragão, M.P., Uchoa, E.: The γ-connected assignment problem. Eur. J. Oper. Res. **118**(1), 127–138 (1999)
2. Assunção, T., Furtado, V.: A heuristic method for balanced graph partitioning: an application for the demarcation of preventive police patrol areas. In: Geffner, H., Prada, R., Machado Alexandre, I., David, N. (eds.) IBERAMIA 2008. LNCS (LNAI), vol. 5290, pp. 62–72. Springer, Heidelberg (2008). https://doi.org/10.1007/978-3-540-88309-8_7

3. Barboza, E.U.: Problemas de classificação com restrições de conexidade flexibilizadas. Master's thesis, Universidade Estadual de Campinas (1997)
4. Becker, R.I., Lari, I., Lucertini, M., Simeone, B.: Max-min partitioning of grid graphs into connected components. Networks **32**(2), 115–125 (1998)
5. Becker, R.I., Schach, S.R., Perl, Y.: A shifting algorithm for min-max tree partitioning. J. ACM **29**(1), 58–67 (1982)
6. Borndörfer, R., Elijazyfer, Z., Schwartz, S.: Approximating balanced graph partitions. Technical report 19–25, ZIB, Takustr. 7, 14195 Berlin (2019)
7. Chataigner, F., Salgado, L.R.B., Wakabayashi, Y.: Approximation and inapproximability results on balanced connected partitions of graphs. Discrete Math. Theor. Comput. Sci. **9**(1) (2007)
8. Chlebíková, J.: Approximating the maximally balanced connected partition problem in graphs. Inf. Process. Lett. **60**(5), 225–230 (1996)
9. Dezső, B., Jüttner, A., Kovács, P.: Lemon-an open source C++ graph template library. Electron. Notes Theor. Comput. Sci. **264**(5), 23–45 (2011)
10. Dyer, M., Frieze, A.: On the complexity of partitioning graphs into connected subgraphs. Discrete Appl. Math. **10**(2), 139–153 (1985)
11. Gleixner, A., Bastubbe, M., Eifler, L., et al.: The SCIP optimization suite 6.0. T. Report, optimization online, July 2018. http://www.optimization-online.org/DB_HTML/2018/07/6692.html
12. Goldberg, A.V., Tarjan, R.E.: A new approach to the maximum-flow problem. J. ACM **35**(4), 921–940 (1988)
13. Grötschel, M., Lovász, L., Schrijver, A.: Geometric Algorithms and Combinatorial Optimization, vol. 2. Springer, Heidelberg (2012)
14. Györi, E.: On division of graph to connected subgraphs. In: Combinatoris (Proceedings of Fifth Hungarian Colloquium, Koszthely, 1976), vol. I, Colloq. Math. Soc. János Bolyai, vol. 18, North-Holland, Amsterdam, New York, pp. 485–494 (1978)
15. Kawarabayashi, K., Kobayashi, Y., Reed, B.: The disjoint paths problem in quadratic time. J. Combin. Theory Ser. B **102**(2), 424–435 (2012)
16. Lovász, L.: A homology theory for spanning tress of a graph. Acta Math. Acad. Sci. Hungarica **30**, 241–251 (1977)
17. Lucertini, M., Perl, Y., Simeone, B.: Most uniform path partitioning and its use in image processing. Discrete Appl. Math. **42**(2), 227–256 (1993)
18. Ma, J., Ma, S.: An $O(k^2n^2)$ algorithm to find a k-partition in a k-connected graph. J. Comput. Sci. Technol. **9**(1), 86–91 (1994)
19. Maravalle, M., Simeone, B., Naldini, R.: Clustering on trees. Comput. Stat. Data Anal. **24**(2), 217–234 (1997)
20. Matić, D.: A mixed integer linear programming model and variable neighborhood search for maximally balanced connected partition problem. Appl. Math. Comput. **237**, 85–97 (2014)
21. Perl, Y., Schach, S.R.: Max-min tree partitioning. J. ACM **28**(1), 5–15 (1981)
22. Wu, B.Y.: Fully polynomial-time approximation schemes for the max-min connected partition problem on interval graphs. Discrete Math. Algorithms Appl. **04**(01), 1250005 (2012)
23. Zhou, X., Wang, H., Ding, B., Hu, T., Shang, S.: Balanced connected task allocations for multi-robot systems: an exact flow-based integer program and an approximate tree-based genetic algorithm. Expert Syst. Appl. **116**, 10–20 (2019)

Scheduling

Polynomial Scheduling Algorithm for Parallel Applications on Hybrid Platforms

Massinissa Ait Aba[1,2]([⊠]), Lilia Zaourar[1], and Alix Munier[2]

[1] CEA, LIST, Computing and Design Environment Laboratory,
91191 Gif Sur Yvette Cedex, France
massinissa.aitaba@gmail.com
[2] LIP6-UPMC, 4 place Jussieu, 75005 Paris, France

Abstract. This work addresses the problem of scheduling parallel applications into hybrid platforms composed of two different types of resources. We focus on finding a generic approach to schedule applications represented by directed acyclic graphs that minimises makespan with performance guarantee. A three-phase algorithm is proposed; the first two phases consist in solving linear formulations to find the type of processor assigned to execute each task. In the third phase, we compute the start execution time of each task to generate a feasible schedule. Finally, we test our algorithm on a large number of instances. These tests demonstrate that the proposed algorithm achieves a close-to-optimal performance.

Keywords: Scheduling · DAG applications · Makespan · Hybrid platform · CPU · GPU · Approximation algorithm

1 Introduction

Nowadays, High Performance Computers (HPC) are popular and powerful commercial platform due to the increasing demand for developing efficient computing resources to execute large parallel applications. In order to increase the computing power of these platforms while keeping a reasonable level of energy consumption, the heterogeneous platforms have appeared. It is possible to integrate several types of material resources such that each one is specialised for certain types of calculations. Thus we have to take into account that the execution time for any task of the application depends on the type of resource used to execute it. However, using these platforms efficiently became very challenging. Consequently, more and more attention has been focused on scheduling techniques for solving the problem of optimizing the execution of parallel applications on heterogeneous computing systems [1,2].

This work addresses the problem of scheduling parallel applications onto a particular case of HPC composed of two different types of resources: CPU (Central Processing Unit) and GPU (Graphics Processing Unit). These platforms are

© Springer Nature Switzerland AG 2020
M. Baïou et al. (Eds.): ISCO 2020, LNCS 12176, pp. 143–155, 2020.
https://doi.org/10.1007/978-3-030-53262-8_12

often called hybrid platform. The number of platforms of the $TOP500$[1] equipped with accelerators has significantly increased during the last years. TGCC Curie supercomputer[2] is an example of these platforms.

We focus here in finding a generic approach to schedule applications presented by DAG (Directed Acyclic Graph) into a hybrid platform that minimises the completion time of the application by considering communication delays. An algorithm with three phases has been proposed; the first phase consists in solving a mathematical formulation (P') and then define a new formulation using the solution obtained. The second phase solves an assignment problem to find the type of processor affected to execute the tasks (processing element type 1 or 2) using a linear formulation. In the last phase, we compute the starting execution time of each task to generate a feasible schedule. Our algorithm has been experimented on a large number of instances and evaluated compared to the exact solution.

The rest of the paper is organised as follows: Sect. 2 gives a quick overview of previous research in scheduling strategies on hybrid platforms. Section 3 presents the detailed problem with mathematical formulation. In Sect. 4, we describe the proposed algorithm for our problem and the approximation ratio we obtain for the scheduling problem. Section 5 shows some preliminary numerical results. Finally, we conclude and provide insights for future work in Sect. 6.

2   Related Work

The problem of scheduling tasks on hybrid parallel platforms has attracted a lot of attention. In the case where all processors have the same processing power and there is a cost for any communication $(P|prec, com|C_{max})$, the problem has been shown to be NP-hard [3].

Several works have studied the problem of scheduling independent tasks on ℓ (resp. k) processors of type $\mathcal{A}$ (resp. $\mathcal{B}$) which is represented by $(P\ell, Pk)||C_{max}$. Imreh [4] proves that the greedy algorithm provides a solution with a performance guarantee of $(2 + \frac{\ell-1}{k})$, where $k \leqslant \ell$. Recently, a 2-approximation algorithm has been proposed in [5]. For the same problem, Kedad-Sidhoum et al. [6] proposed two families of approximation algorithms that can achieve an approximation ratio smaller than $(\frac{3}{2}+\epsilon)$. By considering precedence constraints without communication delays $(P\ell, Pk)|prec|C_{max}$, Kedad-Sidhoum et al. [7] developed a tight 6-approximation algorithm for general structure graphs on hybrid parallel multi-core machines. This work was later revisited in [8] who showed that by separating the allocation phase and the scheduling phase, they could obtain algorithms with a similar approximation ratio but that performs significantly better in practice.

In term of heuristic strategies, the most famous one is Heterogeneous Earliest Finish Time algorithm (HEFT) [9], which is developed for the problem of DAG scheduling on heterogeneous platforms considering communication delays

---

[1] Top500.org ranking. URL https://www.top500.org/lists/2017/11/.

[2] Tgcc curie supercomputer, http://www-hpc.cea.fr/en/complexe/tgcc-curie.htm.

($Rm|prec, com|C_{max}$). It could also be applied for hybrid platforms. It has no performance guarantee, but performs particularly well. Other heuristics for this problem can be roughly partitioned into two classes: clustering and list scheduling algorithms.

Clustering algorithms [10,11] usually provide good solutions for communication-intensive graphs by scheduling heavily communicating tasks onto the same processor. After grouping tasks into a set of clusters using different clustering policies. Clusters are mapped onto processors using communication sensitive or insensitive heuristics.

List scheduling algorithms [12] are often used to handle a limited number of processors. Most of them [13,14] can be decomposed in two main phases. The first one assigns priorities based on certain task properties, typically run time and/or communication delays. The second phase assigns tasks to processors following a priority list. Experimentally, a comparison of different list scheduling algorithms can be found in the work of Kushwaha and Kumar [14].

Our problem was first treated in [15], a non polynomial-time two-phase approach was proposed with a performance guarantee of 6. Numerical evaluations demonstrate that the proposed algorithm achieves a close-to-optimal performance. However, the running time of this method can be important for large instances. We focus here in finding a polynomial-time approach which is able to maintain an interesting performance with reasonable complexity.

3 Problem Definition

We consider in this work a hybrid platform composed of 2 unrelated Processing elements Pe_1 and Pe_2 (1 CPU and 1 GPU, or 2 different GPUs or CPUs, ...).

An application A of n tasks is represented by a Directed Acyclic Graph (DAG) oriented $G(V, E)$, each vertex represents a task t_i. Each arc $e = (t_i, t_j)$ represents a precedence constraint between two tasks t_i and t_j. We associate it with the value $ct_{i,j}$ which represents the communication delay between t_i and t_j if they are executed on two different resource types. The exact formula to evaluate $ct_{i,j}$ which takes into consideration latencies and available bandwidth between processors is provided in [16]. We denote by $\Gamma^-(i)$ (resp. $\Gamma^+(i)$) the sets of the predecessors (resp. successors) of task t_i. Any task t_i can be executed by both processing elements. Executing the task t_i on Pe_1 (resp. Pe_2) generates an execution time equal to $w_{i,0}$ (resp. $w_{i,1}$). A task t_i can be executed only after the complete execution of its predecessors $\Gamma^-(i)$. We do not allow duplication of tasks and preemption. We denote by C_{max} the completion time of the application A (makespan). The aim is to minimise C_{max}.

Our problem can be modelled by a Mixed Integer formulation (Opt). Let x_i be the decision variable which is equal to 1 if the task t_i is assigned to a Pe_1 and 0 otherwise. Let C_i be the finish time of the task t_i. To manage overlapping tasks on the same processing element, we add an intermediary variable $o_{i,j}$ for each two different tasks t_i and t_j. If t_i and t_j are executed in the same processing element and t_j is executed after the finish execution time of t_i, then $o_{i,j} = 1$, otherwise $o_{i,j} = 0$. Finally, for each two successive tasks $(t_i, t_k) \in E$, we add an intermediary variable $\zeta_{i,k}$ to manage communication delays.

$$
(Opt)
\begin{cases}
C_i + x_j w_{j,0} + (1 - x_j) w_{j,1} + \zeta_{i,j} ct_{i,j} \leq C_j \ \forall (t_i, t_j) \in E & (1) \\
x_i - x_j \leq \zeta_{i,j}, \ \forall (t_i, t_j) \in E & (2) \\
x_j - x_i \leq \zeta_{i,j}, \ \forall (t_i, t_j) \in E & (3) \\
x_i w_{i,0} + (1 - x_i) w_{i,1} \leq C_i, \forall i \in \{1, \dots, n\}, \Gamma^-(i) = \emptyset & (4) \\
0 \leq C_i \leq C_{max}, \forall i \in \{1, \dots, n\}, \Gamma^+(i) = \emptyset & (5) \\
C_i + x_j w_{j,0} \leq C_j + B \times (3 - x_i - x_j - o_{i,j}) \ \forall t_i \neq t_j & (6) \\
C_j + x_i w_{i,0} \leq C_i + B \times (2 - x_i - x_j + o_{i,j}) \ \forall t_i \neq t_j & (7) \\
C_i + (1 - x_j) w_{j,1} \leq C_j + B \times (1 + x_i + x_j - o_{i,j}) \ \forall t_i \neq t_j & (8) \\
C_j + (1 - x_i) w_{i,1} \leq C_i + B \times (x_i + x_j + o_{i,j}) \ \forall t_i \neq t_j & (9) \\
x_i, \zeta_{i,j}, o_{i,j} \in \{0, 1\}, \ \forall i \in \{1, \dots, n\}, \ B = Cte \\
Z(min) = C_{max}
\end{cases}
$$

Constraints (1 to 3) describe the critical path, such as if task t_i precedes t_j, and these two tasks are assigned to two different processors, we obtain two cases: either $x_i = 1$ and $x_j = 0$ or $x_i = 0$ and $x_j = 1$. In the two cases, we obtain $\zeta_{i,j} \geq 1$. If tasks t_i and t_j are assigned to the same processor, $x_i = 0$ and $x_j = 0$ or $x_i = 1$ and $x_j = 1$. In the two cases, $\zeta_{i,j} \geq 0$. Since it is a minimisation problem and without loss of generality, $\zeta_{i,j}$ should take the smallest possible value. Tasks without predecessors (respectively successors) are considered in the constraint (4) (resp. (5)). Overlapping tasks on Pe_1 (resp. Pe_2) is avoided by constraints (6) and (7) (resp. (8) and (9)) by using a large constant B (upper bound for example), such that if two tasks t_i and t_j are executed on the same processor, then either t_i starts after the completion time of the task t_j or t_j starts after the completion time of the task t_i. We have two cases:

1. t_i and t_j are executed on Pe_1, then $x_i = 1$ and $x_j = 1$:

$$
\begin{cases}
C_i + x_j w_{j,0} \leq C_j + B(1 - o_{i,j}) & (6) \\
C_j + x_i w_{i,0} \leq C_i + B(o_{i,j}) & (7)
\end{cases}
\quad
\begin{cases}
C_i + x_j w_{j,0} \leq C_j + B(3 - o_{i,j}) & (8) \\
C_j + x_i w_{i,0} \leq C_i + B(2 + o_{i,j}) & (9)
\end{cases}
$$

If $o_{i,j} = 1$ (resp. $o_{i,j} = 0$), only constraint (6) (resp. (7)) becomes relevant, with $C_i + x_j w_{j,0} \leq C_j$ (resp. $C_j + x_i w_{i,0} \leq C_i$), then t_j (resp. t_i) starts after the finish execution time of task t_i (resp. t_j). Other constraints will remain valid no matter the execution order of t_i and t_j.

2. t_i and t_j are executed on Pe_2, then $x_i = 0$ and $x_j = 0$:

$$
\begin{cases}
C_i + x_j w_{j,0} \leq C_j + B(3 - o_{i,j}) & (6) \\
C_j + x_i w_{i,0} \leq C_i + B(2 - o_{i,j}) & (7)
\end{cases}
\quad
\begin{cases}
C_i + x_j w_{j,0} \leq C_j + B(1 - o_{i,j}) & (8) \\
C_j + x_i w_{i,0} \leq C_i + B(o_{i,j}) & (9)
\end{cases}
$$

If $o_{i,j} = 1$ (resp. $o_{i,j} = 0$), only constraint (8) (resp. (9)) becomes relevant, with $C_i + (1 - x_j) w_{j,1} \leq C_j$ (resp. $C_j + (1 - x_i) w_{i,1} \leq C_i$), then t_j (resp. t_i) starts after the finish execution time of task t_i (resp. t_j). Other constraints will remain valid no matter the execution order of t_i and t_j.

The formulation (Opt) can be used to obtain an optimal solution for only small instances with limited number of tasks using solvers like $CPLEX$ [17]. To solve larger instances, a polynomial method is proposed in the following.

4   Solution Method

In this section, we develop a three-phase algorithm. In Phase 1, we start by proposing a new formulation (P) then we solve its relaxation (P'). After that, we use in Phase 2 the solution obtained by this formulation to define another formulation $(P1)$. Finally, after rounding the fractional solution of the formulation $(P1)$ to obtain a feasible assignment of the tasks, in Phase 3 we use a list scheduling algorithm to find a feasible schedule. Details of each phase are described in the following.

4.1   Phase 1: Formulation (P) and Its Relaxation (P')

We solve here a linear formulation with continuous variables. From the formulation (Opt), we define a more simplified formulation (P) which is more useful for the next phase. The first 5 constraints of (Opt) are thus taken up again, but the non-overlapping constraints (6) and (7) are replaced by two workload constraints. Thus, (P) is defined as follow, where Constraint (6) (resp. (7)) simply expresses that the makespan should be be larger than the average Pe_1 (resp. Pe_2) workload. The aim is to minimise C_{maxp}.

$$(P) \begin{cases} C_i + x_j w_{j,0} + (1 - x_j) w_{j,1} + \zeta_{i,j} ct_{i,j} \leq C_j, \ \forall (t_i, t_j) \in E & (1) \\ x_i - x_j \leq \zeta_{i,j}, \ \forall (t_i, t_j) \in E & (2) \\ x_j - x_i \leq \zeta_{i,j}, \ \forall (t_i, t_j) \in E & (3) \\ x_i w_{i,0} + (1 - x_i) w_{i,1} \leq C_i, \forall i \in \{1, \ldots, n\}, \Gamma^-(i) = \emptyset & (4) \\ 0 \leq C_i \leq C_{maxp}, \forall i \in \{1, \ldots, n\}, \Gamma^+(i) = \emptyset & (5) \\ \sum_{i=1}^n x_i w_{i,0} \leq C_{maxp} & (6) \\ \sum_{i=1}^n (1 - x_i) w_{i,1} \leq C_{maxp} & (7) \\ x_i, \zeta_{i,j} \in \{0, 1\}, \ \forall i \in \{1, \ldots, n\} \\ Z(min) = C_{maxp} \end{cases}$$

Remark 1. The optimal solution $C_{maxp}^\star$ of this formulation does not take into account non-overlapping constraints, so it represents a lower bound for our problem, $C_{maxp}^\star \leq C_{max}^\star$.

Lemma 1. *For each two successive tasks* $(t_i, t_j) \in E$, *constraints (2) and (3) can be written as* $max(x_i, x_j) - min(x_i, x_j) \leq \zeta_{i,j}$. *Furthermore,* $max(x_i, x_j) - min(x_i, x_j) = (1 - min(x_i, x_j)) + (max(x_i, x_j) - 1) = max(1 - x_i, 1 - x_j) - min(1 - x_i, 1 - x_j)$. *Thus, constraints (2) and (3) can also be written as* $max(1 - x_i, 1 - x_j) - min(1 - x_i, 1 - x_j) \leq \zeta_{i,j}$.

Remark 2. In each feasible solution of (P), for each couple of tasks $(t_i, t_j) \in E$, we have always $max(x_i, x_j) = 1$ or $max(1 - x_i, 1 - x_j) = 1$ (or both), $\forall (x_i, x_j) \in \{0, 1\} \times \{0, 1\}$.

Lemma 2. *In the optimal solution of (P), for each couple of tasks $(t_i, t_j) \in E$, from Lemma 1 we have at least $max(x_i, x_j) = 1$ or $max(1 - x_i, 1 - x_j) = 1$, such that:*

- *If $max(x_i, x_j) = 1$, then constraints (2) and (3) can be represented by $\widetilde{Con}_{i,j}^{1}$: $1 - min(x_i, x_j) \leqslant \zeta_{i,j}$.*
- *If $max(1 - x_i, 1 - x_j) = 1$, then constraints (2) and (3) can be represented by $\widetilde{Con}_{i,j}^{2}$: $1 - min(1 - x_i, 1 - x_j) \leqslant \zeta_{i,j}$.*

The optimal solution obtained by the formulation (P) without constraints (6) and (7) represents the optimal solution of the scheduling problem on platforms with unlimited resources. This problem has been proven to be NP-hard [18]. Thus, the problem of finding the optimal mapping using (P) is also NP-complete. In order to simplify the problem, we relax the integer variables x_i for $i \in \{1, \ldots, n\}$ and we obtain the relaxed formulation (P'). We denote by $\widetilde{x}_i' \in [0, 1]$, the value of x_i in the optimal solution of the formulation (P').

4.2 Phase 2: Formulation $(P1)$

Based on Lemma 1 and using the solution of (P'), we define another formulation $(P1)$. The decision variables are x_i', and an intermediary variable $y_{i,j}' \in [0, 1]$, with $i \in \{1, \ldots, n\}$ and $j \in \{1, \ldots, n\}$. For all $(t_i, t_j) \in E$, we define the constraint $Con_{i,j}$ as follows:

- If $min\{\widetilde{x}_i', \widetilde{x}_j'\} > min\{1 - \widetilde{x}_i', 1 - \widetilde{x}_j'\}$, then $Con_{i,j} = \begin{cases} y_{i,j}' \leqslant x_i' & (1) \\ y_{i,j}' \leqslant x_j' & (2) \\ \zeta_{i,j}' = (1 - y_{i,j}') & (3) \end{cases}$

From $Con_{i,j}$, we have $y_{i,j}' \leqslant min\{x_i', x_j'\}$. Then, $\zeta_{i,j}' = 1 - y_{i,j}' \geqslant (1 - min\{x_i', x_j'\})$, which is equivalent to constraint $\widetilde{Con}_{i,j}^{1}$. Since it is a minimisation problem, we can set $\zeta_{i,j}' = (1 - min\{x_i', x_j'\})$.

- If $min\{\widetilde{x}_i', \widetilde{x}_j'\} \leqslant min\{1 - \widetilde{x}_i', 1 - \widetilde{x}_j'\}$ then $Con_{i,j} = \begin{cases} y_{i,j}' \leqslant 1 - x_i' & (1) \\ y_{i,j}' \leqslant 1 - x_j' & (2) \\ \zeta_{i,j}' = (1 - y_{i,j}') & (3) \end{cases}$

From $Con_{i,j}$, we have $y_{i,j}' \leqslant min\{1 - x_i', 1 - x_j'\}$. Then, $\zeta_{i,j}' = 1 - y_{i,j}' \geqslant (1 - min\{1 - x_i', 1 - x_j'\})$, which is equivalent to constraint $\widetilde{Con}_{i,j}^{2}$. Since it is a minimisation problem, we can set $\zeta_{i,j}' = (1 - min\{1 - x_i', 1 - x_j'\})$.

The formulation $(P1)$ is then given by:

$$(P1) \begin{cases} C'_i + x'_j w_{j,0} + (1 - x'_j) w_{j,1} + \zeta'_{i,j} ct_{i,j} \leqslant C'_j, \forall (t_i, t_j) \in E & (1) \\ Con_{i,j}, \forall (t_i, t_j) \in E & (2) \\ x'_i w_{i,0} + (1 - x'_i) w_{i,0} \leqslant C'_i, \forall i \in \{1, \ldots, n\}, \Gamma^-(i) = \emptyset & (3) \\ 0 \leqslant C'_i \leqslant C_{max1'}, \forall i \in \{1, \ldots, n\}, \Gamma^+(i) = \emptyset & (4) \\ \sum_{i=1}^{n} x'_i w_{i,0} \leqslant C_{max1'} & (5) \\ \sum_{i=1}^{n} (1 - x'_i) w_{i,0} \leqslant C_{max1'} & (6) \\ x'_i, y_{i,j}, \zeta'_{i,j} \in [0,1], \forall i \in \{1, \ldots, n\}, j \in \{1, \ldots, n\} \\ Z(min) = C_{max1'} \end{cases}$$

We can notice that constraints (1, 3, 4, 5, 6) of the formulation $(P1)$ are equivalent to constraints (1, 4, 5, 6, 7) of the formulation (P'). We denote by $C^{\star}_{max1'}$ the optimal solution of the formulation $(P1)$ and $C^{\star}_{maxp'}$ the optimal solution of (P').

Theorem 1. *If the optimal solution $\tilde{x}^{\star}_i$ obtained by (P') is an integer for all $i \in \{1, \ldots, n\}$, then $C^{\star}_{max1'} = C^{\star}_{maxp'}$.*

Proof. By setting the value of $x'_i = \tilde{x}^{\star}_i$, for all $i \in \{1, \ldots, n\}$, then for each two successive tasks $(t_i, t_j) \in E$, we have two cases:

1. $\min\{\tilde{x}^{\star}_i, \tilde{x}^{\star}_j\} > \min\{1 - \tilde{x}^{\star}_i, 1 - \tilde{x}^{\star}_j\}$, then $Con_{i,j} = \begin{cases} y'_{i,j} \leqslant x'_i & (1) \\ y'_{i,j} \leqslant x'_j & (2) \\ \zeta'_{i,j} = (1 - y'_{i,j}) & (3) \end{cases}$

   Furthermore, $\min\{\tilde{x}^{\star}_i, \tilde{x}^{\star}_j\} = 1$, then $\tilde{x}^{\star}_i = 1$ and $\tilde{x}^{\star}_j = 1$, follows $\tilde{x}^{\star}_i - \tilde{x}^{\star}_j = 0 \leqslant \tilde{\zeta}^{\star}_{i,j}$ ($\tilde{\zeta}^{\star}_{i,j} = 0$ since it is a minimisation problem). Furthermore, $\zeta'_{i,j} = (1 - \min\{x'_i, x'_j\}) = 1 - 1 = 0$, then $\zeta'_{i,j} = \tilde{\zeta}^{\star}_{i,j}$.

2. $\min\{\tilde{x}^{\star}_i, \tilde{x}^{\star}_j\} \leqslant \min\{1 - \tilde{x}^{\star}_i, 1 - \tilde{x}^{\star}_j\}$, then $Con_{i,j} = \begin{cases} y'_{i,j} \leqslant 1 - x'_i & (1) \\ y'_{i,j} \leqslant 1 - x'_j & (2) \\ \zeta'_{i,j} = (1 - y'_{i,j}) & (3) \end{cases}$

   - if $\min\{1 - \tilde{x}^{\star}_i, 1 - \tilde{x}^{\star}_j\} = 1$, then $1 - \tilde{x}^{\star}_i = 1$ and $1 - \tilde{x}^{\star}_j = 1$, follows $\tilde{x}^{\star}_i - \tilde{x}^{\star}_j = 0 \leqslant \tilde{\zeta}^{\star}_{i,j}$ with $\tilde{x}^{\star}_i = 0$ and $\tilde{x}^{\star}_j = 0$. Furthermore, $\zeta'_{i,j} = (1 - \min\{1 - x'_i, 1 - x'_j\}) = 1 - 1 = 0$.
   - if $\min\{1 - \tilde{x}^{\star}_i, 1 - \tilde{x}^{\star}_j\} = 0$, then we suppose that $\tilde{x}^{\star}_i = 1$ and $\tilde{x}^{\star}_j = 0$, follows $\tilde{x}^{\star}_i - \tilde{x}^{\star}_j = 1 \leqslant \tilde{\zeta}^{\star}_{i,j}$. Furthermore, $\zeta'_{i,j} = (1 - \min\{1 - x'_i, 1 - x'_j\}) = 1 - 0 = 1$.
   In both cases, we have $\zeta'_{i,j} = \tilde{\zeta}^{\star}_{i,j}$.

Thus, the finish execution time of each task in the formulation (P') is the same in $(P1)$. Furthermore, since $\sum_{i=1}^{n} x'_i w_{i,0} = \sum_{i=1}^{n} x^{\star}_i w_{i,0}$ and $\sum_{i=1}^{n} (1 - x'_i) w_{i,1} = \sum_{i=1}^{n} (1 - x^{\star}_i) w_{i,1}$, then constraints (5) and (6) of the formulation (P') are the same in $(P1)$. Finally, $C^{\star}_{max1'} = C^{\star}_{maxp'}$.

However, finding the ratio between $C^{\star}_{max1'}$ and $C^{\star}_{maxp'}$ for the general case is difficult. In the following, we suppose that $C^{\star}_{max1'} \leqslant \alpha C^{\star}_{maxp'}$, with $\alpha \in \mathbb{R}^+$.

Table 1 shows the standard deviation between $C^\star_{max1'}$ and $C^\star_{maxp'}$ for 20 randomly generated instances of different sizes (DAG graphs). For each instance I_i, we compute $\alpha_i = \frac{C^\star_{max1'}(I_i)}{C^\star_{maxp'}(I_i)}$. Then, we calculate **Average GAP**$= \frac{\sum_{i=1}^{20} \alpha_i}{20}$ and **Standard deviation**$= \sqrt{\frac{\sum_{i=1}^{20} \alpha_i^2}{20}}$.

From table 1, we can notice that the value of α tends towards 1 when we increase the size of the instances. What can be said, is that the solution of $C^\star_{max1'}$ is very close to the solution of $C^\star_{maxp'}$ in the general case.

Table 1. GAP and standard deviation

Instances	Number of tasks	Average GAP	Standard deviation
test_1	10	1.12457	1.12871
test_2	30	1.11632	1.12065
test_3	60	1.00046	1.00046
test_4	100	1.00007	1.00007
test_5	200	1.00124	1.00126
test_6	400	1	1
test_7	500	1	1
test_8	600	1	1
test_9	800	1	1
test_10	1000	1	1
Average	/	1,024266	1,025115

Lemma 3. *The ratio between the optimal solution $C^\star_{max1'}$ of the formulation* (P1) *and the optimal scheduling solution $C^\star_{max}$ of our main problem is given by* $C^\star_{max1'} \leqslant \alpha C^\star_{max}$.

Proof. From Remark 1, we have $C^\star_{maxp} \leqslant C^\star_{max}$. Then, $C^\star_{max1'} \leqslant \alpha C^\star_{maxp'} \leqslant \alpha C^\star_{maxp} \leqslant \alpha C^\star_{max}$.

Rounding strategy: If x_i' is integer for $i \in \{1, \ldots, n\}$, the solution obtained is feasible and optimal for (P1), otherwise the fractional values are rounded. We denote by x_i^r the rounded value of the fractional value of the assignment variable of task t_i in the optimal solution of (P1). We set $x_i^r = 0$ if $x_i' < \frac{1}{2}$, $x_i^r = 1$ otherwise.

Let θ_1 be the mapping obtained by this rounding. Each task t_i is mapped in either Pe_1 or Pe_2. Thus, $\theta_1(t_i) \longrightarrow \{Pe_1, Pe_2\}$.

4.3   Phase 3: Scheduling Algorithm

Using the mapping θ_1, the following algorithm determines for a task order given by a priority list L, the corresponding scheduling by executing the first task ready of the list as long as there are free processing elements.

The priority list L can be defined in different ways. To achieve good scheduling, the most important and influential tasks must be executed first. For this purpose, the following list is particularly interesting for this problem because it takes into account the critical path of the graph. First, we start by defining graph $G'(V, E)$, with $V = \{t_1, t_2, ..., t_n\}$ and E represents the set of graph edges. The vertices are labelled by the execution time of each task according to their assignments. The edges are labelled by the communication costs if t_i precedes t_j and $x_i^r \neq x_j^r$, 0 otherwise. Then, we can calculate the longest path $LP(t_i)$ from each task t_i to its last successor. The list LLP is given by $LLP = \{t_1, t_2, ..., t_n\}$, such that $LP(t_1) \geqslant LP(t_2) \geqslant \ldots \geqslant LP(t_n)$. The following algorithm executes task by task, executing first the task t_i with the highest $LP(t_i)$ from ready tasks. It uses an insertion policy that tries to insert a task at the earliest idle time between two already scheduled tasks on the processing element, if the slot is large enough to execute the task.

Algorithm 1: PLS (Polynomial List Scheduling) algorithm

Data: mapping θ_1, list LPP.
Result: Feasible scheduling.
begin

    Create an empty list ready-list;
    ready-list= $\{t_j,\ \Gamma^-(j) = \emptyset,\ j \in \{1, \ldots, n\}\}$;
    while *ready-list* $\neq \emptyset$ **do**

        $t_i \longleftarrow$ task with highest $LP(t_i)$ from ready-list;
        Execute t_i on $\theta_1(t_i)$ using insertion-based scheduling policy;
        Update ready-list;

The three steps of PLS (Polynomial List Scheduling) algorithm can be summarized as follows. Solve the relaxed formulation (P') then us its solution to define another formulation $(P1)$, then solve $(P1)$. Finally, After rounding the solutions obtained by $(P1)$, use Algorithm 1 with the obtained mapping θ_1 and the priority list LLP.

Complexity: Mapping θ_1 is based on solving two linear formulations $((P')$ and $(P1))$ with continuous variables, which are two polynomial problems. This gives polynomial-time solving methods for the first two phases of PLS algorithm. In the last phase, ready-list is calculated with $\mathcal{O}(n^2)$ time complexity. The insertion policy is verified on a processing element by checking the non-overlapping with at most $(n-1)$ tasks. This makes a complexity of $\mathcal{O}(n^2)$ for the last phase. Thus, the complexity time of PLS algorithm is polynomial.

Algorithm Analysis: In the following, we study the performance of PLS algorithm in the worst case compared to the optimal solution. We look for the ratio between the solution $\widehat{C}_{max}$ obtained by PLS algorithm and the optimal scheduling solution $C^\star_{max}$ of our main problem.

Lemma 4. *The rounding θ_1 previously defined satisfies the following inequalities:* $x_i^r \leqslant 2x_i'$ *and* $(1 - x_i^r) \leqslant 2(1 - x_i')$.

Proof. If $0 \leqslant x_i' < \frac{1}{2}$, then $x_i^r = 0 \leqslant 2x_i'$. Furthermore, $2x_i' \leqslant 1$, then $0 \leqslant 1 - 2x_i'$, follows $-x_i^r = 0 \leqslant 1 - 2x_i'$, then $1 - x_i^r \leqslant 2(1 - x_i')$. If $\frac{1}{2} \leqslant x_i'$ then $1 \leqslant 2x_i'$, follows $x_i^r = 1 \leqslant 2x_i'$. Furthermore, $x_i' \leqslant 1$ then $-2x_i' \geqslant -2$, follows $1 - 2x_i' \geqslant -1$, then $-x_i^r = -1 \leqslant 1 - 2x_i'$, then $1 - x_i^r \leqslant 2(1 - x_i')$.

Let w_i^r be the execution time of the task t_i by considering the rounding θ_1, where $w_i^r = w_{i,0}$ if $x_i^r = 1$, $w_i^r = w_{i,1}$ otherwise. Let $w_i^{'}$ be the execution time of the task t_i by considering the solution of $(P1)$, where $w_i^{'} = x_i^{'} w_{i,0} + (1 - x_i^{'}) w_{i,1}$.

Proposition 1. *The relation between w_i^r and $w_i^{'}$ of each task t_i is given by $w_i^r \leqslant 2w_i^{'}$, $i \in \{1, \ldots, n\}$.*

Proof. From Lemma 4, $2w_i^{'} = 2x_i^{'} w_{i,0} + 2(1 - x_i^{'}) w_{i,1} \geqslant x_i^r w_{i,0} + (1 - x_i^r) w_{i,1} = w_i^r$.

Lemma 5. *For two successive tasks $(t_i, t_j) \in E$, if t_i and t_j are executed by two different processing elements, then $\zeta_{i,j}^{'} > \frac{1}{2}$.*

Proof. We have two cases:

1. If $\min\{\tilde{x}_i, \tilde{x}_j\} > \min\{1 - \tilde{x}_i, 1 - \tilde{x}_j\}$, then from $Con_{i,j}$ constraint, we have $\zeta_{i,j}^{'} = (1 - \min\{x_i^{'}, x_j^{'}\})$:
   a. If $x_i^{'} < \frac{1}{2}$ and $x_j^{'} \geqslant \frac{1}{2}$, then $\zeta_{i,j}^{'} = (1 - x_i^{'}) > \frac{1}{2}$.
   b. If $x_i^{'} \geqslant \frac{1}{2}$ and $x_j^{'} < \frac{1}{2}$, then $\zeta_{i,j}^{'} = (1 - x_j^{'}) > \frac{1}{2}$.
2. If $\min\{\tilde{x}_i, \tilde{x}_j\} \leqslant \min\{1 - \tilde{x}_i, 1 - \tilde{x}_j\}$, then from $Con_{i,j}$ constraint, we have $\zeta_{i,j}^{'} = (1 - \min\{1 - x_i^{'}, 1 - x_j^{'}\})$:
   a. If $1 - x_i^{'} < \frac{1}{2}$ and $1 - x_j^{'} \geqslant \frac{1}{2}$, then $\zeta_{i,j}^{'} = x_i^{'} > \frac{1}{2}$.
   b. If $1 - x_i^{'} \geqslant \frac{1}{2}$ and $1 - x_j^{'} < \frac{1}{2}$, then $\zeta_{i,j}^{'} = x_j^{'} > \frac{1}{2}$.

For each couple of tasks $(t_i, t_j) \in E$, we denote by $Cost_{i,j}^r$ the value given by $Cost_{i,j}^r = 0$ if $x_i^r = x_j^r$, $Cost_{i,j}^r = ct_{i,j}$ otherwise. Let $Cost_{i,j}^{'}$ be the value given by $Cost_{i,j}^{'} = \zeta_{i,j}^{'} ct_{i,j}$.

Proposition 2. *For each couple of tasks $(t_i, t_j) \in E$, the relation between $Cost_{i,j}^r$ and $Cost_{i,j}^{'}$ is given by $Cost_{i,j}^r < 2Cost_{i,j}^{'}$.*

Proof. If t_j and t_j are executed by the same processing element, $Cost_{i,j}^r = 0 \leqslant 2\zeta_{i,j}^{'} ct_{i,j}$, because $\zeta_{i,j}^{'} \geqslant 0$. If t_j and t_j are executed by two different processing elements, then $Cost_{i,j}^r = ct_{i,j}$. Then, from Lemma 5, $\zeta_{i,j}^{'} > \frac{1}{2}$, then $2\zeta_{i,j}^{'} > 1$, follows $Cost_{i,j}^r = ct_{i,j} \leqslant 2\zeta_{i,j}^{'} ct_{i,j} = 2Cost_{i,j}^{'}$.

Proposition 3. *For each two successive tasks $(t_i, t_j) \in E$, let be $l_{i,j}^r = w_i^r + Cost_{i,j}^r + w_j^r$ (resp. $l_{i,j}^{'} = w_i^{'} + Cost_{i,j}^{'} + w_j^{'}$) the length of (t_i, t_j) in PLS solution (resp. $(P1)$ solution). Then, we have $l_{i,j}^r < 2l_{i,j}^{'}$.*

Proof. From Proposition 1 and Proposition 2, $l_{i,j}^r = w_i^r + Cost_{i,j}^r + w_j^r < 2w_i^{'} + 2Cost_{i,j}^{'} + 2w_j^{'} = 2l_{i,j}^{'}$. Thus, $l_{i,j}^r < 2l_{i,j}^{'}$.

Theorem 2. *Let $\widehat{C}_{max}$ be the solution obtained by using PLS algorithm, then $\widehat{C}_{max} < 6C_{max1'}^{\star}$.*

Proof. From Proposition 3, the length of each path L from $G(V, E)$ is given by $length(L)^r = \sum_{(t_i, t_{i+1}) \in L} l^r_{i,i+1} \leqslant 2 \sum_{(t_i, t_{i+1}) \in L} l'_{i,i+1} = 2length(L)'$, where $length(L)^r$ (resp. $length(L)'$) is the length of L in PLS solution (resp. ($P1$) solution). Furthermore, the workload of the tasks assigned to Pe_1 (resp. Pe_2) is given by $\sum_{i=1}^{n} x_i^r w_{i,0} = 2 \sum_{i=1}^{n} x_i' w_{i,0}$ (resp. $\sum_{i=1}^{n} (1 - x_i^r) w_{i,0} = 2 \sum_{i=1}^{n} (1 - x_i') w_{i,0}$). Thus, the value of the critical path and the workloads on Pe_1 and Pe_2 will be at most doubled compared to the lower bounds. Finally, the interaction between the longest paths and the workload on each processing element has been studied in [15], such that if we have these properties, then $\widehat{C}_{max} < 6C^\star_{max1'}$.

Theorem 3. *The ratio between the solution $\widehat{C}_{max}$ obtained by PLS algorithm and the optimal scheduling solution $C^\star_{max}$ is given by $\dfrac{\widehat{C}_{max}}{C^\star_{max}} < 6\alpha$.*

Proof. From Lemma 3, we have $C^\star_{max1'} \leqslant \alpha C^\star_{max}$. Then, $\dfrac{\widehat{C}_{max}}{C^\star_{max}} < \dfrac{6C^\star_{max1'}}{C^\star_{max}} \leqslant \dfrac{6\alpha C^\star_{max}}{C^\star_{max}} \leqslant 6\alpha$.

5   Numerical Results

We compare here the performance of PLS (Polynomial List Scheduling) algorithm to HEFT (Heterogeneous Earliest Finish Time) and LS (List Scheduling) algorithm using benchmarks generated by Turbine [19].

The benchmark is composed of ten parallel DAG applications. We denote by $test\_i$ instance number i. We generate 10 different applications for each $test\_i$ with $i \in \{1, \ldots, 10\}$. The execution times of the tasks are generated randomly over an interval $[w_{min}, w_{max}]$, w_{min} has been fixed at 5 and w_{max} at 70. The number of successors of each task is generated randomly over an interval $[d_{min}, d_{max}]$, d_{min} has been fixed at 1 and d_{max} at 10. The communication rate for each arc was generated on an interval $[ct_{min}, ct_{max}]$, we set ct_{min} to 35 and ct_{max} to 80.

To study the performance of our method, we compared the ratio between each makespan value obtained by PLS algorithm with HEFT and LS algorithm, the optimal solution obtained by *CPLEX* and the lower bound $C_{max1'}$ obtained by ($P1$). Table 2 shows the average results obtained on 10 instances given in column Inst of each application size given in the second column using *CPLEX*, HEFT, PLS and LS algorithms. We show the average time that was needed to *CPLEX* to provide the optimal solution using (Opt). We only have the result for the first two instances due to the large running time for instances with more than 60 tasks ($> 4h$). Then, we show the results obtained by HEFT, PLS and LS algorithm. GAP columns give the average ratio between the makespan obtained by each method compared to $C_{max1'}$ using the following formula: $GAP = \frac{\text{method makespan} - C_{max1'}}{C_{max1'}} \times 100$. Time columns show the average time that was needed for each method to provide a solution. Best columns present the number of instances where each algorithm provides better or the same solution than other methods. A line Average is added at the end of each table which represents the average of the values each column.

Table 2. *CPLEX*, HEFT, LS and PLS algorithms results.

Inst	Number of tasks	*CPLEX*		HEFT			LS algorithm			PLS algorithm		
		Optimal	Time	GAP	Time	Best	GAP	Time	Best	GAP	Time	Best
test_1	10	✓	0.35 s	33.23%	0.0016 s	2	20.84%	0.25 s	7	21.24%	0.008 s	7
test_2	30	✓	59.58 s	38.11%	0.0049 s	4	52.34%	0.600 s	0	43.19%	0.028 s	5
test_3	60	X	X	25.89%	0.009 s	4	24.86%	0.29 s	6	24.81%	0.081 s	7
test_4	100	X	X	15.80%	0.017 s	0	6.85%	0.198 s	8	6.46%	0.184 s	9
test_5	200	X	X	14.56%	0.044 s	0	1.34%	0.51 s	7	1.08%	0.68 s	6
test_6	400	X	X	11.80%	0.19 s	0	0.25%	1.72 s	7	0.31%	2.52 s	6
test_7	500	X	X	11.50%	0.26 s	0	0.16%	1.84 s	6	0.11%	2.33 s	9
test_8	600	X	X	11.53%	0.61 s	0	0.32%	2.06 s	6	0.237%	2.09 s	7
test_9	800	X	X	11.78%	1.40 s	0	0.14%	3.15 s	7	0.13%	4.09 s	6
test_10	1000	X	X	12.11%	1.90 s	0	0.052%	4.32 s	7	0.06%	5.26 s	7
Average	/	/	/	18.63%	0.44 s	6%	10.71%	1.49 s	61%	9.76%	1.72 s	69%

For the running time, HEFT algorithm requires less time than PLS and LS algorithms to provide a solution. PLS algorithm is the most efficient method with a gap of 9.76% and 69% of better solutions compared to other methods. Its average running time is 1.72 s, which is slightly higher than the running time of LS algorithm.

6   Conclusion and Perspectives

This paper presents an efficient algorithm to solve the problem of scheduling parallel applications on hybrid platforms with communication delays. The objective is to minimise the total execution time (makespan).

After modelling the problem, we proposed a three-phase algorithm; the first two phases consist in solving linear formulations to find the type of processor assigned to execute each task. In the third phase, we compute the start execution time of each task to generate a feasible schedule. Tests on large instances close to reality demonstrated the efficiency of our method comparing to other methods and shows the limits of solving the problem with a solver such as *CPLEX*.

A proof of the performance guarantee for PLS algorithm was initiated. In future works, we will focus on finding the value of α to have a fixed bound on the ratio between $\widehat{C}_{max}$ and $C^{\star}_{max}$. Tests on real applications and an extension to more general heterogeneous platforms is also planned.

References

1. Shen, L., Choe, T.-Y.: Posterior Task scheduling algorithms for heterogeneous computing systems. In: Daydé, M., Palma, J.M.L.M., Coutinho, Á.L.G.A., Pacitti, E., Lopes, J.C. (eds.) VECPAR 2006. LNCS, vol. 4395, pp. 172–183. Springer, Heidelberg (2007). https://doi.org/10.1007/978-3-540-71351-7_14

2. Benoit, A., Pottier, L., Robert, Y.: Resilient co-scheduling of malleable applications. Int. J. High Perform. Comput. Appl. **32**(1), 89–103 (2018)
3. Ullman, J.D.: Np-complete scheduling problems. J. Comput. Syst. Sci. **10**(3), 384–393 (1975)
4. Imreh, C.: Scheduling problems on two sets of identical machines. Computing **70**(4), 277–294 (2003)
5. Marchal, L., Canon, L.-C., Vivien, F.: Low-cost approximation algorithms for scheduling independent tasks on hybrid platforms. Ph.D. thesis, Inria-Research Centre Grenoble-Rhône-Alpes (2017)
6. Kedad-Sidhoum, S., Monna, F., Mounié, G., Trystram, D.: A family of scheduling algorithms for hybrid parallel platforms. Int. J. Found. Comput. Sci. **29**(01), 63–90 (2018)
7. Kedad-Sidhoum, S., Monna, F., Trystram, D.: Scheduling tasks with precedence constraints on hybrid multi-core machines. In: IPDPSW, pp. 27–33. IEEE (2015)
8. Amaris, M., Lucarelli, G., Mommessin, C., Trystram, D.: Generic algorithms for scheduling applications on hybrid multi-core machines. In: Rivera, F.F., Pena, T.F., Cabaleiro, J.C. (eds.) Euro-Par 2017. LNCS, vol. 10417, pp. 220–231. Springer, Cham (2017). https://doi.org/10.1007/978-3-319-64203-1_16
9. Topcuoglu, H., Hariri, S., Min-you, W.: Performance-effective and low-complexity task scheduling for heterogeneous computing. IEEE Trans. Parallel Distrib. Syst. **13**(3), 260–274 (2002)
10. Boeres, C., Rebello, V.E.F., et al.: A cluster-based strategy for scheduling task on heterogeneous processors. In: 16th Symposium on Computer Architecture and High Performance Computing, SBAC-PAD 2004, pp. 214–221. IEEE (2004)
11. Yang, T., Gerasoulis, A.: DSC: scheduling parallel tasks on an unbounded number of processors. IEEE Trans. Parallel Distrib. Syst. **5**(9), 951–967 (1994)
12. Garey, M.R., Johnson, D.S.: Complexity results for multiprocessor scheduling under resource constraints. SIAM J. Comput. **4**(4), 397–411 (1975)
13. Khan, M.A.: Scheduling for heterogeneous systems using constrained critical paths. Parallel Comput. **38**(4–5), 175–193 (2012)
14. Kushwaha, S., Kumar, S.: An investigation of list heuristic scheduling algorithms for multiprocessor system. IUP J. Comput. Sci. **11**(2) (2017)
15. Aba, M.A., Zaourar, L., Munier, A.: Approximation algorithm for scheduling applications on hybrid multi-core machines with communications delays. In: 2018 IEEE IPDPSW, pp. 36–45. IEEE (2018)
16. Zaourar, L., Aba, M.A., Briand, D., Philippe, J.-M.: Modeling of applications and hardware to explore task mapping and scheduling strategies on a heterogeneous micro-server system. In: IPDPSW, pp. 65–76. IEEE (2017)
17. IBM: Ibm ilog cplex v12.5 user's manual for cplex. http://www.ibm.com
18. Aba, M.A., Pallez, G., Munier-Kordon, A.: Scheduling on two unbounded resources with communication costs (2019)
19. Bodin, B., Lesparre, Y., Delosme, J.-M., Munier-Kordon, A.: Fast and efficient dataflow graph generation. In: Proceedings of the 17th International Workshop on Software and Compilers for Embedded Systems. ACM (2014)

Anchored Rescheduling Problems Under Generalized Precedence Constraints

Pascale Bendotti[1,2], Philippe Chrétienne[2], Pierre Fouilhoux[2],
and Adèle Pass-Lanneau[1,2(✉)]

[1] EDF R&D, 91120 Palaiseau, France
{pascale.bendotti,adele.pass-lanneau}@edf.fr
[2] Sorbonne Université, CNRS, LIP6, 75005 Paris, France
{philippe.chretienne,pierre.fouilhoux}@lip6.fr

Abstract. The anchored rescheduling problem, recently introduced in the literature, is to find a schedule under precedence constraints with a maximum number of prescribed starting times. Namely, prescribed starting times may correspond to a former schedule that must be modified while maintaining a maximum number of starting times unchanged. In the present work two extensions are investigated. First we introduce a new tolerance feature, so that starting times can be considered as unchanged when modified less than a tolerance threshold. The sensitivity of the anchored rescheduling problem to tolerance is studied. Second we consider generalized precedence constraints, which include, e.g., deadline constraints. Altogether this leads to a more realistic rescheduling problem. The main result is to show that the problem is polynomial. We discuss how to benefit from the polynomiality result in a machine scheduling environment.

Keywords: Rescheduling · Anchored jobs · Generalized precedence graph · Deadline constraints

1 Introduction

The solution of an optimization problem is often not computed from scratch. For example if the optimization problem is solved on a regular basis, then the solution computed at some point in time must take into account the solution of the previously solved instance. Importantly, practitioners may require that former decisions are maintained, that is, the solution must not change too much over time. Such a stability of solutions is needed in a variety of industrial applications. A similar situation can be observed in a decision-aiding process, when users may want to impose some decisions. In both cases it is necessary to solve the optimization problem while taking into account a subset of decisions to stick to, if possible.

In the present work, we investigate this issue for project scheduling problems under precedence constraints, where a set of jobs J must be given starting times

© Springer Nature Switzerland AG 2020
M. Baïou et al. (Eds.): ISCO 2020, LNCS 12176, pp. 156–166, 2020.
https://doi.org/10.1007/978-3-030-53262-8_13

$y \in \mathbb{R}_+^J$. Given a subset $I \subseteq J$, a partial assignment of starting times $(x_i)_{i \in I}$ is called a *baseline*. Given an instance $\mathscr{I}$ of project scheduling problem Π, a baseline $(x_i)_{i \in I}$ and a solution y of $\mathscr{I}$, the *anchorage level* introduced in [1] is $\sigma(x, y) = |\{i \in I : x_i = y_i\}|$, i.e., the number of *anchored jobs* that have the same starting time in the baseline and in the new solution y. This criterion can be integrated into a reoptimization problem as follows.

ANCHRE(Π)
Input: instance $\mathscr{I}$ of problem Π, baseline $(x_i)_{i \in I}$
Problem: find y a schedule of $\mathscr{I}$ such that $\sigma(x, y)$ is maximized.

In practice the decision maker may consider that a change of the starting time of a job is negligible within some tolerance. Given a tolerance vector $\varepsilon \in \mathbb{R}_+^J$, baseline $(x_i)_{i \in I}$ and solution y, job $i \in I$ is *ε-anchored* if $|x_i - y_i| \leq \varepsilon_i$. The *$\varepsilon$-anchorage level* is $\sigma_\varepsilon(x, y) = |\{i \in I : |x_i - y_i| \leq \varepsilon_i\}|$. The corresponding reoptimization problem is then

ε-ANCHRE(Π)
Input: instance $\mathscr{I}$ of problem Π, baseline $(x_i)_{i \in I}$, tolerance $\varepsilon \in \mathbb{R}_+^J$
Problem: find y a schedule of $\mathscr{I}$ such that $\sigma_\varepsilon(x, y)$ is maximized.

Note that the baseline $(x_i)_{i \in I}$ may be issued from the solution of a previous instance $\mathscr{I}^0$ of problem Π. However in the sequel no specific assumption is made on x.

Generalized Precedence Constraints. Let us now define the project scheduling problem under generalized precedence constraints (GenPrec). Consider a set of jobs $J = \{1, \ldots, n\}$, and a directed graph $G = (\{0, \ldots, n+1\}, \mathcal{A})$. Let $G(a)$ be the weighted digraph obtained by adding arc weights $a \in \mathbb{R}^{\mathcal{A}}$ to the digraph G. Let us denote by (i, j, a_{ij}) a weighted arc of $G(a)$. The weighted digraph $G(a)$ defines an instance of the (GenPrec) scheduling problem if it satisfies the following assumptions:

(i) there is no circuit of positive length in $G(a)$
(ii) for every job $i \in \{1, \ldots, n\}$ there exists at least a path of non-negative length from 0 to i and from i to $n+1$ in $G(a)$.

The (GenPrec) problem is to find a schedule x of jobs $\{0, \ldots, n+1\}$ so that

$$x_j - x_i \geq a_{ij} \text{ for every arc } (i, j, a_{ij}) \text{ of } G(a).$$

Note that w.l.o.g. we set $x_0 = 0$ in every schedule. Assumption (i) ensures the existence of a feasible schedule for the instance $G(a)$ (see e.g. [2]). From assumption (ii), it comes that in every feasible schedule and for every $i \in \{1, \ldots, n\}$, the inequality $0 \leq x_i \leq x_{n+1}$ holds: job 0 and job $n+1$ then represent the beginning and the end of the schedule respectively.

Various constraints can be modeled within this framework. A classical precedence constraint $x_j - x_i \geq p_i$ can be represented by an arc (i, j, p_i), where p_i is the processing time of job i. The special case of acyclic precedence graph $G(p)$ is denoted by (Prec).

For illustrative purpose, an example of a generalized precedence graph $G(a)$ with $n = 6$ jobs is represented in Fig. 1. It features circuits, and arcs with negative weights, e.g., the arc $(6, 4, -6)$ corresponds to constraint $y_6 - y_4 \leq 6$.

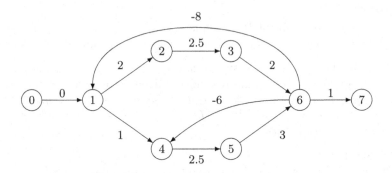

Fig. 1. Generalized precedence graph $G(a)$ with 6 jobs.

A feasible schedule y is represented in Fig. 2. Job i is represented as a rectangle of length $\max_{(i,j)\in\mathcal{A}} a_{ij}$, note e.g. that $a_{12} \neq a_{14}$. Baseline $x = (0, 2.5, 4.5, 2, 4.5, 6)$ is also represented. For baseline x and no tolerance, only jobs $\{1, 5\}$ are anchored. For the same baseline and with tolerance $\varepsilon = (0, 0.25, 0.25, 0, 0, 0)$, jobs $\{1, 2, 3, 5\}$ are ε-anchored.

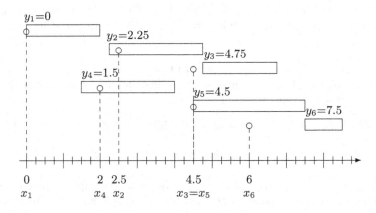

Fig. 2. Schedule $y = (0, 2.25, 4.75, 1.5, 4.5, 7.5)$ feasible for $G(a)$ with 6 jobs.

Related Work. The computation of a new schedule when disruptions occur has been studied in the literature under the name of *rescheduling* or *reactive scheduling* (see, e.g., [3] for a survey). Simple rules to restore a feasible schedule include right-shifting rules [4]. Rescheduling procedures with a stability feature, to avoid changes w.r.t. a baseline schedule, have been studied in [5–7]. In most of these procedures the objective is to minimize a continuous deviation measure between starting times in the baseline and the new schedule. The anchorage level was introduced in [1] as a combinatorial stability criterion, which is the number of unchanged starting times.

Rescheduling problems are to be contrasted with *proactive* approaches, in which a suitable baseline solution is found [8]. A procedure to find stable baseline schedule, under a stochastic continuous stability criterion, was proposed in [9]. A proactive approach involving the anchorage level was studied in [1]. A robust problem where some jobs have guaranteed starting times was studied in [10]. A related framework is *recoverable robust optimization*, that encompasses 2-stage robust problems where the second stage, or *recovery* stage, corresponds to a rescheduling problem [11,12]. The authors of [11] note that it is suitable that the rescheduling problem is computationally easy, so that the recoverable robust problem is reasonably tractable.

Finally note that reoptimization with a stability feature has recently raised attention on other combinatorial problems, e.g., spanning tree, matching, network problems. Proposed approaches include reoptimization with transition costs [13] or incremental problems [14]. This line of research tackles classical problems where decisions are naturally modelled with binary variables, in contrast with the scheduling problems considered here where decisions are continuous starting times.

Contribution and Outline. The anchored rescheduling problem was studied in [1] for project scheduling under precedence constraints only. In the present work, the main result is to extend the study to generalized precedence, thus allowing for a larger variety of constraints such as deadlines or time windows. We also tackle tolerance through ε-anchored jobs, which leads to a more realistic anchored rescheduling problem. We show how the result can be used in a machine scheduling case: while the anchored rescheduling problem is NP-hard, we show that a simpler, yet practically attractive, variant is polynomial.

In Sect. 2 the problem ε-ANCHRE(GenPrec) is proven to be solvable in polynomial time. In Sect. 3.1 we correct a flawed complexity result from [1] in the case of time-window constraints. In Sect. 3.2 we use our framework in a machine scheduling variant. In Sect. 4 we analyze the impact of tolerance ε on the optimum of the anchored rescheduling problem.

2   Polynomiality of ε-AnchRe(GenPrec)

In this section we prove

Theorem 1. ε-ANCHRE*(GenPrec) is solvable in polynomial time.*

Let $((x_i)_{i \in I}, G(a), \varepsilon)$ be an instance of ε-ANCHRE(GenPrec). By testing assumptions (i) and (ii), it can be checked in polynomial time that the weighted digraph $G(a)$ is a valid instance of the (GenPrec) problem.

Let $H \subseteq I$ be a subset of jobs. The set H is x-*compatible* if there exists a schedule x of $G(a)$ such that all jobs in H are ε-anchored with respect to x. Solving ε-ANCHRE(GenPrec) is exactly finding a set H x-compatible of maximum size and an associated solution y. Consider an *auxiliary graph* G_H defined by copying the graph $G(a)$, then adding for every job $i \in H$ two new arcs $(0, i, x_i - \varepsilon_i)$ and $(i, 0, -(x_i + \varepsilon_i))$. The auxiliary graph defines an instance of the (GenPrec) problem, whose constraints are the constraints from $G(a)$, and the new arcs constraints $y_i \geq x_i - \varepsilon_i$ and $-y_i \geq -(x_i + \varepsilon_i)$, that is exactly $|y_i - x_i| \leq \varepsilon_i$. Hence there is a one-to-one correspondence between the schedules of the auxiliary graph G_H, and the schedules of the instance $G(a)$ in which all jobs in H are ε-anchored with respect to the baseline x. Thus we obtain

Proposition 1. *The set H is x-compatible if and only if the auxiliary graph G_H has no positive circuit.*

Let us now show that the absence of positive circuit in G_H is also equivalent to $H \cup \{0\}$ being an antichain in an appropriate poset. Let us denote by $\ell(P)$ the length of a directed path P in $G(a)$. For every pair of distinct jobs $i, j \in \{0, 1, \ldots, n\}$, let $L_{G(a)}(i, j)$ be the maximum length of a directed path from i to j in $G(a)$. By convention it is equal to $-\infty$ if there is no such path. A relation $\mathcal{R}$ on the set of jobs $\{0, 1, \ldots, n\}$ is defined by:

$$i \mathcal{R} j \quad \text{if and only if} \quad i = j \quad \text{or} \quad x_i - \varepsilon_i + L_{G(a)}(i, j) > x_j + \varepsilon_j$$

where we define $\varepsilon_0 = 0$ for the simplicity of notation. In particular, if $i \neq j$ and $i \mathcal{R} j$, it implies that value $L_{G(a)}(i, j)$ is finite and there exists a path from i to j in $G(a)$.

Lemma 1. *The relation $\mathcal{R}$ is a partial order on the set of jobs $\{0, 1, \ldots, n\}$.*

Proof. Relation $\mathcal{R}$ is clearly reflexive. Relation $\mathcal{R}$ is antisymmetric: if $i \mathcal{R} j$ and $j \mathcal{R} i$ with $i \neq j$, then there exists a longest path P_{ij} from i to j and a longest path P_{ji} from j to i in $G(a)$. The circuit obtained by closing P_{ij} with P_{ji} has length $L_{G(a)}(i, j) + L_{G(a)}(j, i)$, which is non-positive by assumption (i) on $G(a)$. Furthermore $L_{G(a)}(i, j) + L_{G(a)}(j, i) > x_j + \varepsilon_j - (x_i - \varepsilon_i) + x_i + \varepsilon_i - (x_j - \varepsilon_j) = 2(\varepsilon_i + \varepsilon_j) \geq 0$, a contradiction. Finally relation $\mathcal{R}$ is transitive: consider three pairwise distinct jobs i, j, k such that $i \mathcal{R} j$ and $j \mathcal{R} k$. There exists two paths P_{ij} and P_{jk} in $G(a)$, hence their concatenation (P_{ij}, P_{jk}) forms a path from i to k. It comes $x_i - \varepsilon_i + L_{G(a)}(i, k) \geq x_i - \varepsilon_i + L_{G(a)}(i, j) + L_{G(a)}(j, k) > x_j + \varepsilon_j + L_{G(a)}(j, k) > x_k + \varepsilon_k + 2\varepsilon_j \geq x_k + \varepsilon_k$. Hence $i \mathcal{R} k$. $\square$

Proposition 2. *The auxiliary graph G_H has no positive circuit if and only if $H \cup \{0\}$ is an antichain of the poset $(\{0, 1, \ldots, n\}, \mathcal{R})$.*

Proof. Assume that $H \cup \{0\}$ is not an antichain. Then there exists two distinct jobs i and j in $H \cup \{0\}$ such that $i\mathcal{R}j$. Hence there exists a longest path P_{ij} from i to j in $G(a)$. Consider the circuit C obtained by closing the path P_{ij} with the new arc $(0, i, x_i - \varepsilon_i)$ if $i \neq 0$ and with the new arc $(j, 0, -(x_j + \varepsilon_j))$ if $j \neq 0$. Then the length of C is $L_{G(a)}(i, j) + x_i - \varepsilon_i - (x_j + \varepsilon_j)$. Note that it is valid even if $i = 0$ or $j = 0$. This length is positive from the definition of $\mathcal{R}$, hence the auxiliary graph contains a positive circuit.

Conversely, assume that there exists a positive circuit in the auxiliary graph. Then there exists a positive circuit C that contains every vertex at most once. Since $G(a)$ contains no positive circuit, the circuit C contains at least one new arc, and consequently it contains vertex 0 exactly once. It follows that C contains one or two successive new arcs. Let P be the path obtained by removing the new arcs from the circuit C. Let i and j be the first and last vertex of P respectively. The length of C can be written $x_i - \varepsilon_i + \ell(P) - (x_j + \varepsilon_j)$ (again it is valid even if $i = 0$ or $j = 0$). By assumption this length is positive, then with $\ell(P) \leq L_{G(a)}(i, j)$ it comes $x_i - \varepsilon_i + L_{G(a)}(i, j) > x_j + \varepsilon_j$ and $i\mathcal{R}j$. □

Remark that all results presented here can be extended to the case of asymmetric tolerance intervals, that is, the case where the starting time of job i is considered as unchanged if $y_i \in [x_i - \varepsilon_i^-, x_i + \varepsilon_i^+]$ with two distinct parameters $\varepsilon_i^-, \varepsilon_i^+ \geq 0$.

Proof (of Theorem 1). With Proposition 1 and Proposition 2, a set H is x-compatible if and only if $H \cup \{0\}$ is an antichain of the poset $(\{0, 1, \ldots, n\}, \mathcal{R})$. Hence solving the reactive problem is tantamount to finding a maximum size antichain $H \cup \{0\}$ of the poset. The latter problem can be solved in polynomial time [15]. Note also that given a set H^* of maximum size, a corresponding reactive solution y^* can be found in polynomial time by computing any schedule of G_{H^*}. Hence the problem ε-ANCHRE(GenPrec) is polynomial-time solvable. □

Algorithmically, a max-size antichain can be found through a combinatorial algorithm such as Dilworth's algorithm [15], or through linear programming. A non-compact characterization of the associated polytope is known, together with a polynomial separation algorithm [16]. Furthermore weights may be introduced so that the objective is to maximize the total weight of ε-anchored jobs. Then the linear programming approach allows to search for a max-weight antichain for any weight function, while Dilworth's algorithm requires integer weights.

3   Particular Cases

In this section we consider particular cases which can benefit from the polynomiality of ε-ANCHRE(GenPrec) and deal with constraints that arise in practice.

3.1   Anchored Rescheduling with Deadline Constraints

In this section we correct a flawed complexity result from [1].

Using generalized precedence constraints, time window constraints of form $x_i \in [l_i, u_i]$ can be modelled with two arcs $(0, i, l_i)$ and $(i, 0, -u_i)$ (recall that $x_0 = 0$ w.l.o.g.). In [1], the so-called Anchor-Reactive CPM-Scheduling Problem with Time Windows (ARSPTW) was defined as a variant of ANCHRE(Prec) where the baseline is a complete schedule ($I = J$), and a time window (or deadline) constraint is imposed on the new schedule: $y_{n+1} \leq B$. Deadline B is part of the instance of the ARSPTW. It follows that the ARSPTW is a special case of ε-AnchReopt(GenPrec). Indeed given an instance $(x, G(p), B)$ of the ARSPTW, a corresponding instance of ε-AnchReopt(GenPrec) is built as follows: set the baseline equal to x; form the (Prec) scheduling instance with the graph $G(p)$ and an additional arc $(n + 1, 0, -B)$; set the tolerance to zero. A consequence of Theorem 1 is then

Corollary 1. *The ARSPTW is polynomial-time solvable.*

The proof of Theorem 4.3 from [1] incorrectly stated the NP-hardness of the ARSPTW. It relied on a reduction from the so-called Maximum Complete Bipartite Subgraph problem (MCBS): given a bipartite graph $G = (L \cup R, E)$ with n non isolated nodes and an integer k, is there a complete bipartite subgraph of G with at least k nodes? The correct reference to Garey and Johnson [17] requires that the complete bipartite subgraph is *balanced*, i.e., it has the same number of nodes in L and R. Without this condition, the MCBS problem is polynomial-time solvable.

3.2 Towards Machine Rescheduling

A question is to benefit from the polynomiality result of Theorem 1 in a machine environment. Consider a set of m machines, and a set of jobs J to be scheduled under precedence constraints represented by precedence graph $G(p)$. A solution of the problem, denoted by $(m|\text{Prec})$, is then formed with a vector of starting times $x \in \mathbb{R}_+^J$, and an assignment of jobs on machines. The anchored rescheduling problem can be considered as previously, that is,

ε-ANCHRE($m|$Prec)
Input: integer m, precedence graph $G(p)$, baseline $(x_i)_{i \in I}$, tolerance $\varepsilon \in \mathbb{R}_+^J$
Problem: find y a schedule of $G(p)$ on m machines such that $\sigma_\varepsilon(x, y)$ is maximized.

We first note that

Proposition 3. ANCHRE*(m|Prec) is NP-hard, even for* $m = 1$.

Indeed, it was proven in [18] that the ANCHRE($m|$Prec) problem is NP-complete on one machine, even when there is no precedence constraints, by a reduction from 3-Partition.

Anchored Rescheduling for Fixed Sequence. Consider now the following variant. Given a baseline solution, let $\mathcal{S} = (S_1, \ldots, S_m)$ be a collection of sequences, where sequence S_k is the ordered list of jobs processed on machine k in the baseline. We consider the anchored rescheduling problem where it is required that the new solution y is consistent with the sequences from $\mathcal{S}$, i.e.,

ε-ANCHRE$(m|\text{Prec})$-fixedSeq
Input: integer m, precedence graph $G(p)$, baseline $(x_i)_{i \in I}$, sequences $\mathcal{S}$, tolerance $\varepsilon \in \mathbb{R}_+^J$
Problem: find y a schedule of $G(p)$ on m machines with sequences $\mathcal{S}$ such that $\sigma_\varepsilon(x, y)$ is maximized.

Theorem 2. ε-ANCHRE$(m|\text{Prec})$-fixedSeq is solvable in polynomial time.

Proof. Let $G^{\mathcal{S}}(p)$ denote the precedence graph formed with $G(p)$ and additional arcs (i, j) for every pair i, j of successive jobs in a sequence of $\mathcal{S}$. Then y is a schedule of $G(p)$ on m machines with sequence $\mathcal{S}$ if and only if it is a schedule of the instance $G^{\mathcal{S}}(p)$ of (Prec). Hence solving ε-ANCHRE$(m|\text{Prec})$-fixedSeq is equivalent to solving ε-ANCHRE(Prec) for the precedence graph $G^{\mathcal{S}}(p)$. From Theorem 1, it comes that ε-ANCHRE$(m|\text{Prec})$-fixedSeq can be solved in polynomial time. □

In this variant, rescheduling does not impair neither the assignment of jobs on machines nor the sequence of jobs on machines. The schedule y being much more constrained than in ε-ANCHRE$(m|\text{Prec})$, less jobs can be ε-anchored. However the sequences of jobs on machines can be regarded as decisions that are maintained during rescheduling. Rescheduling with fixed sequence thus serves a similar purpose as anchored jobs, by the stabilization of decisions.

A practical justification for rescheduling with fixed sequence is that the order of jobs on machines may be difficult or costly to revise if the instance changes. The sequence of jobs on a machine may require preparation, while it is easier to adjust only starting times of jobs.

Rescheduling while maintaining the structure of the schedule w.r.t. resources (e.g., the sequence of jobs on machines) was considered in the literature in the context of robust approaches for the resource-constrained project scheduling problem [19]. However the authors of [19] did not consider any criterion to maintain starting times.

4 Sensitivity Analysis of ε-AnchRe(GenPrec) with Respect to Tolerance

Regarding tolerance ε as a new input of the rescheduling problem, a natural question is the sensitivity of the optimal value of ε-ANCHRE(GenPrec) to ε. In this section, we study the behavior of the rescheduling optimal value in the case where tolerance is given by a single parameter $\epsilon \geq 0$, that is, $\varepsilon_i = \epsilon$ for every $i \in \{1, \ldots, n\}$. Similar results hold if every ε_i is any affine function of ϵ. Given a baseline x and an instance $G(a)$, let

$$OPT(\epsilon) = \max_{\substack{y \text{ schedule} \\ \text{of } G(a)}} \sigma_\epsilon(x, y)$$

For every pair of distinct jobs $i, j \in \{0, \ldots, n\}$, define $b_{ij} = \frac{1}{2}(L_{G(a)}(i,j) - (x_j - x_i))$ if $i \neq 0$ and $j \neq 0$, and $b_{ij} = L_{G(a)}(i,j) - (x_j - x_i)$ if $i = 0$ or $j = 0$. Let $B = \{b_{ij}, i, j \in \{0, \ldots, n\}, i \neq j\}$.

Proposition 4. *The function $OPT(\cdot)$ is piecewise constant and non-decreasing on $\mathbb{R}_+$. Moreover every breakpoint ϵ^* of function OPT belongs to set B.*

Proof. The function $OPT(\cdot)$ is integer-valued. Moreover the function $\epsilon \mapsto \sigma_\epsilon(x, y)$ is non-decreasing hence $OPT(\cdot)$ is also non-decreasing. Let $\epsilon \geq 0$. Define $b^\epsilon = \max\{b \in B, \ b \leq \epsilon\}$. The claim is that $OPT(\epsilon) = OPT(b^\epsilon)$. Consider the poset $(\{0, \ldots, n\}, \mathcal{R})$ introduced in Sect. 2. The relation $\mathcal{R}$ depends on the tolerance and it can be rewritten as follows: for two distinct jobs $i, j \in \{0, \ldots, n\}$, $i\mathcal{R}j$ if and only if $b_{ij} > \epsilon$. Moreover, from the definition of b^ϵ, for every pair i, j, the inequality $b_{ij} > \epsilon$ is equivalent to $b_{ij} > b^\epsilon$. Hence, the relation $\mathcal{R}$ is the same for tolerance ϵ and for tolerance b^ϵ. From Proposition 2, for every ϵ, the value $OPT(\epsilon)$ is the maximum size of an antichain containing job 0 in the poset $(\{0, \ldots, n\}, \mathcal{R})$. Since the poset remains the same for tolerance ϵ and tolerance b^ϵ, it comes $OPT(\epsilon) = OPT(b^\epsilon)$. From the claim, it follows that OPT is constant on every interval of form $[b, b'[$ where b and b' are two successive points of B. Hence any breakpoint of the function must be a point in B. $\square$

Consider now the problem of minimizing the tolerance while ensuring that at least k jobs are ε-anchored:

MINTOLERANCE
Input: precedence graph $G(a)$, baseline x, integer k
Problem: find ϵ such that $OPT(\epsilon) \geq k$ and ϵ is minimized.

From Proposition 4 the optimum of MINTOLERANCE can be found in set B. Values in B can be computed in polynomial time, and $|B| \leq n^2 + n$, which implies:

Corollary 2. MINTOLERANCE *can be solved in polynomial time.*

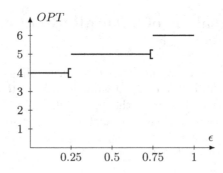

Fig. 3. Anchored rescheduling optimal value $OPT(\epsilon)$ for instance from Fig. 1 and baseline $x = (0, 2.5, 4.5, 2, 4.5, 6)$.

Consider the instance with 6 jobs from Fig. 1. Figure 3 shows its associated rescheduling optimal value OPT for the baseline $x = (0, 2.5, 4.5, 2, 4.5, 6)$. The computation of the set B for this instance leads to four possible breakpoint values $(B = \{0, 0.25, 0.5, 0.75\})$, but only three of them are actual breakpoints of OPT. Namely, changing the tolerance within the range $[0.25; 0.75[$ has no impact on the number of ε-anchored jobs.

5   Conclusion

We studied a rescheduling problem with the objective of maximizing the number of jobs whose starting times correspond to the baseline, within tolerance ε. The problem was shown to be polynomial, even under generalized precedence constraints, including deadlines or time windows constraints. It was also shown that this polynomiality result can be used in the case of machine anchored rescheduling with fixed sequence of jobs on machines.

If the anchored rescheduling problem of an NP-hard problem is bound to remain NP-hard, it is interesting to stress that a polynomial variant can be of practical interest. A perspective in the same spirit is to study anchored rescheduling problems under other resource constraints. The considered rescheduling problems should also be integrated into a recoverable robust problem, in order to find a good baseline schedule that could be marginally modified if disruptions occur.

References

1. Bendotti, P., Chrétienne, P., Fouilhoux, P., Quilliot, A.: Anchored reactive and proactive solutions to the CPM-scheduling problem. Eur. J. Oper. Res. **261**, 67–74 (2017)
2. Pinedo, M.: Scheduling: Theory, Algorithms, and Systems. Prentice Hall, Upper Saddle River (2002)
3. Herroelen, W., Leus, R.: Robust and reactive project scheduling: a review and classification of procedures, p. 42. Katholieke Universiteit Leuven, Open Access publications from Katholieke Universiteit Leuven, April 2004
4. Smith, S.F.: Reactive Scheduling Systems. In: Brown, D.E., Scherer, W.T. (eds.) Intelligent Scheduling Systems. Operations Research/Computer Science Interfaces Series, vol. 3, pp. 155–192. Springer, Boston (1995). https://doi.org/10.1007/978-1-4615-2263-8_7
5. Sakkout, H., Wallace, M.: Probe backtrack search for minimal perturbation in dynamic scheduling. Constraints **5**, 359–388 (2000)
6. Calhoun, K., Deckro, R., Moore, J., Chrissis, J., Hove, J.: Planning and re-planning in project and production scheduling. Omega **30**, 155–170 (2002)
7. Artigues, C., Roubellat, F.: A polynomial activity insertion algorithm in a multi-resource schedule with cumulative constraints and multiple modes. Eur. J. Oper. Res. **127**(2), 297–316 (2000)
8. Herroelen, W., Leus, R.: Project scheduling under uncertainty: survey and research potentials. Eur. J. Oper. Res. **165**, 289–306 (2002)
9. Herroelen, W., Leus, R.: The construction of stable project baseline schedules. Eur. J. Oper. Res. **156**(3), 550–565 (2004)

10. Bendotti, P., Chrétienne, P., Fouilhoux, P., Pass-Lanneau, A.: The anchor-robust project scheduling problem, May 2019. https://hal.archives-ouvertes.fr/hal-02144834. Working paper or preprint
11. Liebchen, C., Lübbecke, M., Möhring, R., Stiller, S.: The concept of recoverable robustness, linear programming recovery, and railway applications. Robust Online Large-Scale Optim. **5868**, 1–27 (2009)
12. D'Angelo, G., Di Stefano, G., Navarra, A., Pinotti, C.: Recoverable robust timetables: an algorithmic approach on trees. IEEE Trans. Comput. **60**, 433–446 (2011)
13. Schieber, B., Shachnai, H., Tamir, G., Tamir, T.: A theory and algorithms for combinatorial reoptimization. Algorithmica **80**(2), 576–607 (2018)
14. Şeref, O., Ahuja, R.K., Orlin, J.B.: Incremental network optimization: theory and algorithms. Oper. Res. **57**(3), 586–594 (2009)
15. Dilworth, R.P.: A decomposition theorem for partially ordered sets. Ann. Math. **51**(1), 161–166 (1950)
16. Schrijver, A.: Combinatorial Optimization - Polyhedra and Efficiency. Springer, Heidelberg (2003)
17. Garey, M.R., Johnson, D.S.: Computers and Intractability: A Guide to the Theory of NP-Completeness. W. H. Freeman and Co., New York (1979)
18. Chrétienne, P.: Reactive and proactive single-machine scheduling to maintain a maximum number of starting times. Ann. Oper. Res. 1–14 (2018). https://hal.sorbonne-universite.fr/hal-02078478
19. Bruni, M., Di Puglia Pugliese, L., Beraldi, P., Guerriero, F.: An adjustable robust optimization model for the resource-constrained project scheduling problem with uncertain activity durations. Omega **71**, 66–84 (2016)

Scheduling with Non-renewable Resources: Minimizing the Sum of Completion Times

Kristóf Bérczi[1], Tamás Király[1(✉)], and Simon Omlor[2]

[1] MTA-ELTE Egerváry Research Group, Department of Operations Research, Eötvös Loránd University, Budapest, Hungary
{berkri,tkiraly}@cs.elte.hu
[2] Institute for Algorithms and Complexity, TU Hamburg, Hamburg, Germany
simon.omlor@tuhh.de

Abstract. We consider single-machine scheduling problems with a non-renewable resource. In this setting, there are n jobs, each characterized by a processing time, a weight, and a resource requirement. At given points in time, certain amounts of the resource are made available to be consumed by the jobs. The goal is to assign the jobs non-preemptively to time slots on the machine, so that each job has the required resource amount available at the start of its processing. We consider the objective of minimizing the weighted sum of completion times.

The main contribution of the paper is a PTAS for the case of 0 processing times $(1|rm = 1, p_j = 0|\sum w_j C_j)$. In addition, we show strong NP-hardness of the case of unit resource requirements and weights $(1|rm = 1, a_j = 1|\sum C_j)$, thus answering an open question of Györgyi and Kis. We also prove that the schedule corresponding to the Shortest Processing Time First ordering provides a 3/2-approximation for the latter problem.

Keywords: Scheduling · Non-renewable resources · PTAS · Approximation algorithm

1 Introduction

Scheduling problems with non-renewable resource constraints arise naturally in various areas where resources like raw materials, energy, or funding arrive at predetermined dates. In the general setting, we are given a set of jobs and a set of machines. Each job is equipped with a requirement vector that encodes the needs of the given job for the different types of resources. There is an initial stock for each resource, and some additional resource arrival times in the future are known together with the arriving quantities. The aim is to find a schedule of the jobs on the machines such that the resource requirements are met.

Supported by DAAD with funds of the Bundesministerium für Bildung und Forschung (BMBF) and by DFG project MN 59/4-1.

© Springer Nature Switzerland AG 2020
M. Baïou et al. (Eds.): ISCO 2020, LNCS 12176, pp. 167–178, 2020.
https://doi.org/10.1007/978-3-030-53262-8_14

We will use the standard $\alpha|\beta|\gamma$ notation of Graham et al. [5]. Grigoriev et al. [6] extended this notation by adding the restriction $rm = r$ to the β field, meaning that there are r resources (raw materials). In the present paper, we concentrate on problem $1|rm = 1|\sum w_j C_j$, that is, when we have a single machine, a single resource, and the goal is to minimize the weighted sum of completion times.

Previous Work. Scheduling problems with resource restrictions (sometimes called financial constraints) were introduced by Carlier [2] and Slowinski [16]. Carlier settled the computational complexity of several variants for the single machine case [2]. In particular, it was shown that $1|rm = 1|\sum w_j C_j$ is NP-hard in the strong sense. This was also proved independently by Gafarov, Lazarev and Wener in [3]. Kis [15] showed that the problem remains weakly NP-hard even when the number of resource arrival times is 2 and gave an FPTAS for $1|rm = 1, q = 2|\sum w_j C_j$. A further variant of the problem was considered in [3]. Recently, Györgyi and Kis [13,14] gave polynomial time algorithms for several special cases, and also showed that the problem remains weakly NP-hard even under the very strong assumption that the processing time, the resource requirement and the weight are the same for each job. They also provided a 2-approximation algorithm for this variant. For a constant number of resource arrival times, they gave a PTAS when the processing time equals the weight for each job, and an FPTAS when the resource requirements and weights are 1.[1]

In comparison, much more is known about the maximum makespan and maximum lateness objectives. Slowinski [16] studied the preemptive scheduling of independent jobs on parallel unrelated machines with the use of additional renewable and non-renewable resources under financial constraints. Toker et al. [17] examined a single-machine scheduling problem under non-renewable resource constraint using the makespan as a performance criterion. Xie [18] generalized this result to the problem with multiple financial resource constraints. Grigoriev et al. [6] presented polynomial time algorithms, approximations and complexity results for single-machine scheduling problems with unit or all-equal processing times. In a series of papers [7–11], Györgyi and Kis presented approximation schemes and inapproximability results both for single and parallel machine problems with the makespan and the maximum lateness objectives. In [12], they proposed a branch-and-cut algorithm for minimizing the maximum lateness.

Our Results. The first problem that we consider is $1|rm = 1, a_j = 1|\sum C_j$. The complexity of this problem was posed as an open question in [12]. We show that the problem is NP-hard in the strong sense.

Theorem 1. $1|rm = 1, a_j = 1|\sum C_j$ *is strongly NP-hard.*

Given any scheduling problem on a single machine, the *Shortest Processing Time First* (SPT) schedule orders the jobs by processing times, i.e. $p_{\mathsf{spt}^{-1}(i)} \leq$

---

[1] Just before the submission of the present paper, Györgyi and Kis published an updated version of [14] with some new results. None of our results are implied by their paper.

$p_{\mathsf{spt}^{-1}(i+1)}$ for all i. We prove that spt provides a $3/2$-approximation. We remark that it remains open whether the problem is APX-hard.

Theorem 2. *The SPT schedule gives a $\frac{3}{2}$-approximation for $1|rm = 1, a_j = 1|\sum C_j$, and the approximation guarantee is tight.*

The second problem considered is the special case when the processing time is 0 for every job. This setting is relevant to situations where processing times are negligible compared to the gaps between resource arrival times, and the bottleneck is resource availability. Examples include financial scheduling problems where the jobs are not time consuming but the availability of funding varies in time, or production problems where products are shipped at fixed time intervals and production time is negligible compared to these intervals. Note that the number of machines is irrelevant if processing times are 0.

First we present a PTAS for constant number of resource arrival times. This procedure will be used as a subroutine in our algorithm for the general case.

Theorem 3. *There exists a $(1+\frac{q}{k})$-approximation for $1|rm = 1, p_j = 0|\sum C_j w_j$ with running time $\mathcal{O}(n^{qk+1})$.*

The main contribution of the paper is a PTAS for the same problem with an arbitrary number of resource arrival times.

Theorem 4. *There exists a PTAS for $1|rm = 1, p_j = 0|\sum C_j w_j$.*

A peculiarity of the algorithm is that the PTAS for constant number of arrival times is called repeatedly on overlapping time windows, and at each call we fix only a portion of the scheduled jobs.

Due to space constraints, several proofs and details as well as further results are deferred to the full version of this paper, which is available at http://arxiv.org/abs/1911.12138.

2   Preliminaries

Throughout the paper, we will use the following notation. We are given a set J of n *jobs*. Each job $j \in J$ has a non-negative integer processing time p_j, a non-negative weight w_j, and a resource requirement a_j. The *resources arrive at time points* $t_1, \ldots, t_q$, and the *amount of resource that arrives at* t_i is denoted by b_i. We might assume that $\sum_{i=1}^{q} b_i = \sum_{j=1}^{n} a_j$ holds. We will always assume that $t_1 = 0$, as this does not effect the approximation ratio of our algorithms.

The jobs should be processed non-preemptively on a single machine. A *schedule* is an ordering of the jobs, that is, a mapping $\sigma : J \to [n]$, where $\sigma(j) = i$ means that job j is the ith job scheduled on the machine. The *completion time* of job j in schedule σ is denoted by C_j^σ. We will drop the index σ if the schedule is clear from the context. In any reasonable schedule, there is an idle time before a job j only if there is not enough resource left to start j after finishing the last job before the idle period. Hence, the completion time of job j is determined by the ordering and by the resource arrival times, as j will be scheduled at the earliest moment when the preceding jobs are already finished and the amount of available resource is at least a_j.

3   The Problem $1|rm = 1, a_j = 1| \sum C_j$

3.1   Strong NP-completeness

Proof of Theorem 1. Recall that all a_j and w_j values are 1, and each job has an integer processing time p_j. The number of resource arrival times is part of the input. We prove NP-completeness by reduction from the 3-PARTITION problem. The input contains numbers $B \in \mathbb{N}$, $n \in \mathbb{N}$, and $x_j \in \mathbb{N}$ $(j = 1, \ldots, 3n)$ such that $B/4 < x_j < B/2$ and $\sum_{j=1}^{3n} x_j = nB$. A feasible solution is a partition $J_1, \ldots, J_n$ of $[3n]$ such that $|J_i| = 3$ and $\sum_{j \in J_i} x_j = B$ for every $i \in [n]$. In contrast to the PARTITION problem, the 3-PARTITION problem remains NP-complete even when the integers x_j are bounded above by a polynomial in n. That is, the problem remains NP-complete even when the numbers in the input are represented as unary numbers [4, pages 96–105 and 224].

Let $K = 4nB$. The reduction to $1|rm = 1, a_j = 1| \sum C_j$ involves three types of jobs. *Normal jobs* correspond to the numbers x_j in the 3-PARTITION instance, so there are $3n$ of them and the processing time p_j of the j-th normal job is x_j. We further have nK *small jobs* with processing time 1, and nK *large jobs* with processing time K. There are also three types of resource arrivals. The *Type 1* resource arrival times are $i(B + K)$ $(i = 0, \ldots, n - 1)$ with three resources arriving at each. The *Type 2* arrival times are $i(B + K) + j$ $(i = 0, \ldots, n - 1,$ $j = B, \ldots, B + K - 1)$ with one resource arriving. Finally, the *Type 3* resource arrival times are $n(B + K) + iK$ $(i = 0, \ldots, nK - 1)$ with one resource arriving.

Suppose that the 3-PARTITION instance has a feasible solution $J_1, \ldots, J_n$. We consider the following schedule S: resources of Type 1 are used by normal jobs, such that jobs in J_i are scheduled between $(i - 1)(B + K)$ and $iB + (i - 1)K$ (in spt order). Type 2 resources are used by small jobs that start immediately. Type 3 resources are used by the large jobs that also start immediately at the resource arrival times (see Fig. 1).

Instead of $\sum C_j$, we consider the equivalent shifted objective function $\sum (C_j - t_j - p_j)$, where t_j is the arrival time of the resource used by job j and p_j is its processing time – we assume without loss of generality that resources are used by jobs in order of arrival. Note that all terms of $\sum (C_j - t_j - p_j)$ are nonnegative. As small jobs and large jobs start immediately at the arrival of the corresponding resource in schedule S, their contribution to the shifted objective function is 0. The jobs in J_i have total processing time B, and their contribution to the shifted objective function is two times the processing time of the shortest job plus the processing time of the second shortest job, which is at most B. Hence the schedule S has objective value at most nB.

We claim that if the 3-PARTITION instance has no feasible solution, then the objective value of any schedule is strictly larger than nB. First, notice that if a large job is scheduled to start before time $n(B + K)$, then $\sum (C_j - t_j - p_j)$ has a term strictly larger than nB as there is a resource that arrives while the large job is processed and is not used for more than nB time units. Similarly, if the first large job starts at $n(B + K)$ but uses a resource that arrived earlier, then the resource that arrives at $n(B + K)$ is not used for more than nB time units.

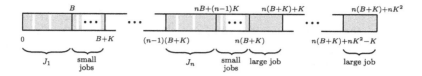

Fig. 1. The schedule corresponding to a feasible solution of 3-PARTITION.

We can conclude that the first large job uses the resource arriving at $n(B + K)$. If the first large job does not start at $n(B + K)$, then all large jobs have positive contribution to the objective value, so again, the objective value is larger than nB. We can therefore assume that the large jobs start exactly at $n(B + K) + iK$ ($i = 0, \ldots, nK - 1$) and that there is no idle time before $(B + K)n$. In particular, this means that all other jobs are already completed at time $(B + K)n$.

Consider Type 2 resources arriving at $i(B + K) + j$ ($j = B, \ldots, B + K - 1$) for some fixed i. If the first or the second resource is not used immediately, then none of the subsequent ones are, so the objective value is more than nB. Hence, the first resource must be used immediately by a small job.

Suppose that some resource in this interval is used by a normal job. If it is followed by a small job, then we may improve the objective value by exchanging the two. Thus, in this case, we can assume that the last resource of the interval is used by a normal job, and also the Type 1 resources arriving at $(i+1)(B+K)$ are used by normal jobs. But this is impossible, because normal jobs have processing time at least $B/4 + 1$, and a small job starts at time $(i + 1)(B + K) + B$.

To sum up, we can assume that all resources of Type 2 are used immediately by small jobs. This means that normal jobs have to use resources of Type 1, and must exactly fill the gaps of length B between the arrival of resources of Type 2. This is only possible if the 3-partition instance has a feasible solution, concluding the proof of Theorem 1.   □

3.2   Shortest Processing Time First for Unit Resource Requirements

We now show that scheduling the jobs according to an spt ordering provides a $3/2$-approximation for the problem with unit weights and unit resource requirements, thus proving Theorem 2.

Proof of Theorem 2. To prove the theorem consider any instance I. We denote the completion times for the spt ordering by C_j and their sum by spt. Furthermore, let $S_{\mathsf{spt}^{-1}(i)} := C_{\mathsf{spt}^{-1}(i)} - p_{\mathsf{spt}^{-1}(i)}$ denote the starting time of the ith job in the spt schedule. Let opt be the optimal schedule for I. We denote the completion times for opt by C'_j and their sum by opt. Let $S'_{\mathsf{opt}^{-1}(i)} := C'_{\mathsf{opt}^{-1}(i)} - p_{\mathsf{opt}^{-1}(i)}$ denote the starting time of the ith job in the optimal schedule.

Our strategy is to simplify the instance by revealing its structural properties while not decreasing $\frac{\mathsf{spt}}{\mathsf{opt}}$. We only state the claims needed for the proof here; their proofs can be found in the full version of the paper [1]. First we modify the resource arrival times.

Claim 1. *We may assume that the ith resource arrives at $S'_{\mathrm{opt}^{-1}(i)}$ for $i = 1,\dots,n$, and that there is no idle time in schedule opt, that is, $S'_{\mathrm{opt}^{-1}(i)} = C'_{\mathrm{opt}^{-1}(i-1)}$ for $i = 2,\dots,n$.*

Next, we modify the instance to have $0-1$ processing times.

Claim 2. *We may assume that $p_{\mathrm{opt}^{-1}(1)} > p_{\mathrm{spt}^{-1}(1)}$ and that $p_{\mathrm{spt}^{-1}(1)} = 0$. Furthermore, we may assume that $p_j \in \{0,1\}$ for all $j \in J$.*

Finally, we modify the order of the jobs in the optimal solution. If opt and spt process a job of length 0 at the same time, then we can remove the job from the instance and reduce the number of resources that arrive at this time by 1. This will reduce opt and spt by the same amount.

Let t be the time at which schedule spt first starts to process a job of length 1. On one hand, opt does not process jobs of length 0 before t by the above argument. On the other hand, there is no idle time after t in spt, because that would mean idle time in opt. Thus, if we move all jobs of length 0 and their corresponding resource arrivals in opt to time t, then spt does not change but opt decreases. We may thus assume that schedule opt processes every job of length 0 at t.

We conclude that opt first processes k_1 jobs of length 1, then k_1 jobs of length 0 and then k_2 jobs of length 1, while spt starts with the jobs of length 0 having a lot of idle time in the beginning and then consecutively processes all jobs of length 1. The weighted sums of completion times are then given by $\mathsf{opt} = \frac{k_1(k_1+1)}{2} + k_1^2 + k_2k_1 + \frac{k_2(k_2+1)}{2}$ and $\mathsf{spt} = \frac{k_1(k_1-1)}{2} + k_2k_1 + \frac{k_1(k_1+1)}{2} + (k_1 + k_2)k_1 + \frac{k_2(k_2+1)}{2}$. We get $\frac{3}{2}\mathsf{opt} - \mathsf{spt} = \frac{k_1^2}{4} + \frac{k_2^2}{4} - \frac{k_1 k_2}{2} + \frac{3k_1+k_2}{4} \geq \frac{(k_1-k_2)^2}{4} \geq 0$, showing that the approximation factor is at most $\frac{3}{2}$.

Setting $k_2 = k_1$ and letting k_1 go to infinity gives us a sequence of instances such that $\frac{\mathsf{spt}}{\mathsf{opt}}$ converges to $\frac{3}{2}$ as we have $\mathsf{spt} = \frac{9}{2}k_1^2 + \mathcal{O}(k_1)$ and $\mathsf{opt} = 3k_1^2 + \mathcal{O}(k_1)$. This concludes the proof of Theorem 2. $\qquad\square$

4   The Problem $1|rm = 1, p_j = 0| \sum C_j w_j$

In this section we consider problem $1|rm = 1, p_j = 0| \sum C_j w_j$, another special case of $1|rm = 1| \sum C_j w_j$. Since the processing times are 0, every job is processed at one of the arrival times in any optimal schedule. Thus, a schedule can be represented by a mapping $\pi : J \to [q]$, where $\pi(j)$ denotes the index of the resource arrival time when job j is processed. A schedule is feasible if the resource requirements are met, that is, if $\sum_{j:\pi(j)\leq k} a_j \leq \sum_{i\leq k} b_i$ for all $1 \leq k \leq q$. As we assume that $\sum_i b_i = \sum_j a_j$ holds, this is equivalent to $\sum_{j:\pi(j)\geq k} a_j \geq \sum_{i\geq k} b_i$ for all $1 \leq k \leq q$.

Define $B_k = \sum_{i\geq k} b_i$, and consider the set of jobs that are not processed before a given time point t_i. Then the second inequality says that if the resource requirements of these jobs add up to at least B_i, then our schedule is feasible.

4.1   PTAS for Constant q

The aim of this section is to give a PTAS for the case when the number of resource arrival times is a constant. The algorithm is a generalization of a well known PTAS for the knapsack problem, and will be used later as a subroutine in the PTAS for an arbitrary number of resource arrival times. The idea is to choose a number $k \in \mathbb{Z}_+$, guess the k heaviest jobs that are processed at each resource arrival time t_i, and then determine the remaining jobs that are scheduled at t_i in a greedy manner. Since we go over all possible sets containing at most k jobs for each resource arrival time, there is an exponential dependence on the number q of resource arrival times in the running time.

Algorithm 1. PTAS for $1|rm = 1, p_j = 0| \sum C_j w_j$ when q is a constant.

Input: Jobs J with $|J| = n$, resource requirements a_j, weights w_j, resource arrival times $t_1 \leq \ldots \leq t_q$ and resource quantities $b_1, \ldots b_q$.
Output: A feasible schedule π.
1: **for all** subpartitions $S_1 \cup \cdots \cup S_q \subseteq J$ with $|S_i| \leq k$ for $i > 1$ **do**
2:     Set $A = 0$.
3:     Set $W = 0$.
4:     **for** i from 0 to $q - 2$ **do**
5:         **for** $j \in S_{q-i}$ **do**
6:             $\pi(j) = q - i$
7:             $A \leftarrow A + a_j$
8:         **if** $|S_{q-i}| = k$ **then**
9:             $W \leftarrow \max\{W, \min\{w_j : j \in S_{q-i}\}\}$
10:             **while** $A < B_{q-i}$ **do**
11:                 **if** there exists an unassigned job j with $w_j \leq W$ **then**
12:                     Let j be an unassigned job with $w_j \leq W$ minimizing w_j/a_j.
13:                     $\pi(j) = q - i$
14:                     $A \leftarrow A + a_j$
15:                 **else**
16:                     **break**
17:     For all remaining jobs set $\pi(j) = 1$.
18: Let π be the best schedule found.
19: **return** π

Proof of Theorem 3. We claim that Algorithm 1 satisfies the requirements of the theorem. Let π^{opt} be an optimal schedule and define $J_i^{\mathrm{opt}} = \{j \in J : \pi^{\mathrm{opt}}(j) = i\}$. Let S_i^{opt} be the set of the k heaviest jobs in J_i^{opt} if $|J_i^{\mathrm{opt}}| \geq k$, otherwise let $S_i^{\mathrm{opt}} = J_i^{\mathrm{opt}}$. Let $J_i = \{j \in J : \pi(j) = i\}$ denote the set of jobs assigned to time t_i in our solution. In each iteration of the *for* loop of Step 4, let j_i be the last job added to J_i if such a job exists.

Assume that we are at the iteration of the algorithm when the subpartition $S_1^{\mathrm{opt}} \cup \cdots \cup S_q^{\mathrm{opt}}$ is considered in Step 1. Let $W_{q-\ell}$ denote the value of W at the end of the iteration of the *for* loop corresponding to $i = \ell$ in Step 4. By Steps 3 and 9, we have $W_{q-\ell} \leq \frac{1}{k} \sum_{i=\ell}^{q} \sum_{j \in J_i^{\mathrm{opt}}} w_j$. As our algorithm always picks the most inefficient job, we also have $\sum_{i=\ell}^{q} \sum_{j \in J_i \setminus \{j_i\}} w_j \leq \sum_{i=\ell}^{q} \sum_{j \in J_i^{\mathrm{opt}}} w_j$, where $J_i \setminus \{j_i\} = J_i$ if j_i is not defined for i.

Combining these two observations, for $\ell = 1, \ldots, q$ we get

$$\sum_{i=\ell}^{q} \sum_{j \in J_i} w_j = \sum_{i=\ell}^{q} \sum_{j \in J_i \setminus \{j_i\}} w_j + \sum_{i=\ell}^{q} w_{j_i}$$

$$\leq \sum_{i=\ell}^{q} \sum_{j \in J_i^{\mathrm{opt}}} w_j + (q - \ell + 1) \cdot W_\ell \leq (1 + \tfrac{q}{k}) \sum_{i=\ell}^{q} \sum_{j \in J_i^{\mathrm{opt}}} w_j,$$

where the first inequality follows from the fact that $w_{j_i} \leq W_i \leq W_\ell$ whenever $i \geq \ell$. This proves that the schedule that we get is a $(1 + \frac{q}{k})$-approximation.

We get a factor of n^{qk} in the running time for guessing the sets S_k. Assigning the remaining jobs can be done in linear time by ordering the jobs and using AVL-trees, thus we get an additional factor of n. In order to get a PTAS, we set $k = \frac{\varepsilon}{q}$, concluding the proof of the theorem. $\qquad\square$

4.2   PTAS for Arbitrary q

We turn to the proof of the main result of the paper. The first idea is to shift resource arrival times to powers of $1 + \varepsilon$, for a suitably small ε.

Let $\mathcal{I}$ be an instance of $1|rm = 1, p_j = 0| \sum_j C_j w_j$. We assume that resource arrival times are integer, and that $t_1 = 0$, $t_2 = 1$. We define a new instance $\mathcal{I}'$ of $1|rm = 1, p_j = 0| \sum_j C_j w_j$ with shifted resource arrival times as follows. Set $t_1' = 0$ and $t_i' = (1+\varepsilon)^{i-2}$ for $i = 2, \ldots, \lceil \log_{1+\varepsilon}(t_q) \rceil + 2$. Moreover, let $b_1' = b_1, b_2' = b_2$ and $b_i' = \sum [b_i : t_i \in ((1 + \varepsilon)^{i-3}, (1 + \varepsilon)^{i-2}]$ for $i = 3, \ldots, \lceil \log_{1+\varepsilon}(t_q) \rceil + 2$.

The proof of the following claim is the same as that of Claim 12 in [1].

Claim 3. *A solution to $\mathcal{I}$ with weighted sum of completion times W can be transformed into a solution of $\mathcal{I}'$ with weighted sum of completion times at most $(1 + \varepsilon)W$. Furthermore, any feasible schedule for $\mathcal{I}'$ is also feasible for $\mathcal{I}$.* $\qquad\square$

Due to the claim, we may assume that the positive arrival times are powers of $1 + \varepsilon$. For convenience of notation, we will assume in this subsection that the largest arrival time is 1, and arrival times are indexed in decreasing order, starting with $t_0 = 1$. That is, $t_i = (1 + \varepsilon)^{-i}$ $(i = 0, \ldots, q - 2)$, and $t_{q-1} = 0$. We will also assume that for a given constant r, $b_{q-r-1} = \cdots = b_{q-2} = 0$. This can be achieved by adding r dummy arrival times.

Proof of Theorem 4. Let us fix an even integer r and $\varepsilon > 0$; we will later assume that r is very large compared to ε^{-1}. We assume that resource arrival times are as described above. The algorithm is given as Algorithm 2.

In the algorithm, we fix jobs at progressively decreasing arrival times, by using the PTAS of the previous section for $r + 1$ arrival times on different instances except for the first step, when we may use the PTAS for less than $r + 1$

Algorithm 2. PTAS for $1|rm = 1, p_j = 0| \sum C_j w_j$

Input: Jobs J with $|J| = n$, resource requirements a_j, weights w_j; an even integer r; resource quantities $b_0, \ldots b_{q-1}$ such that $b_{q-r-1} = \cdots = b_{q-2} = 0$ and $\sum a_j = \sum b_i$. We assume resource arrival times are $t_i = (1 + \varepsilon)^{-i}$ $(i = 0, \ldots, q - 2)$, $t_{q-1} = 0$.
Output: A feasible schedule π.

1: **for** ℓ from 1 to $r/2$ **do**
2:    Obtain instance $\mathcal{I}'$ with $r/2 + \ell + 1$ arrival times by moving arrivals before $t_{r/2+\ell-1}$ to 0.
3:    Run Algorithm 1 on $\mathcal{I}'$ to get schedule σ.
4:    Let $A = B = 0$.
5:    **for** i from 0 to $\ell - 1$ **do**
6:        For every $j \in \sigma^{-1}(i)$, fix $\pi_\ell(j) = i$.
7:        $A \leftarrow A + \sum_{j \in \sigma^{-1}(i)} a_j$
8:        $B \leftarrow B + b_i$
9:    **for** j from 2 to $\lfloor 2(q - 1 - \ell)/r \rfloor$ **do**
10:       Let $s = (j - 2)r/2 + \ell$.
11:       Obtain instance $\mathcal{I}'$ with arrival times $t_s, t_{s+1}, \ldots, t_{s+r-1}, 0$: remove arrivals after t_s, remove $\max\{A - B, 0\}$ latest remaining resources, and move all arrivals before t_{s+r-1} to 0.
12:       Let $A = B = 0$.
13:       Run Algorithm 1 on $\mathcal{I}'$ to get schedule σ.
14:       **for** i from s to $s + r/2 - 1$ **do**
15:           For every $j \in \sigma^{-1}(i)$, fix $\pi_\ell(j) = i$.
16:           $A \leftarrow A + \sum_{j \in \sigma^{-1}(i)} a_j$
17:           $B \leftarrow B + b_i$
18:       For all unscheduled jobs j, set $\pi_\ell(j) = q - 1$.
19: Let π be the best schedule among $\pi_1, \ldots, \pi_{r/2}$.
20: **return** π

---

arrival times. We will run our algorithm $r/2$ times with slight modifications, and pick the best result. Each run is characterized by a parameter $\ell \in \{1, \ldots, r/2\}$.

In the first step, we consider arrival times $t_0, t_1, \ldots, t_{r/2+\ell-1}, 0$. We move the resources arriving before $t_{r/2+\ell-1}$ to 0, and use the PTAS for $r/2 + \ell + 1$ arrival times on this instance. We fix the jobs that are scheduled at arrival times $t_0, t_1, \ldots, t_{\ell-1}$.

Consider now the jth step for some $j \geq 2$. Define $s = (j - 2)r/2 + \ell$ and consider arrival times $t_s, t_{s+1}, \ldots, t_{s+r-1}, 0$. Move the resources arriving before t_{s+r-1} to 0, and decrease $b_s, b_{s+1}, \ldots$ in this order as needed, so that the total requirement of unfixed jobs equals the total resource. Use the PTAS for $r + 1$ arrival times on this instance. Fix the jobs that are scheduled at arrival times $t_s, t_{s+1}, \ldots, t_{s+r/2-1}$. The algorithm runs while $s + r - 1 \leq q - 2$, i.e., $jr/2 + \ell \leq q - 1$. Since the smallest r arrival times (except for 0) are dummy arrival times, the algorithm considers all resource arrivals.

The schedule given by the algorithm is clearly feasible, because when jobs at t_i are fixed, the total resource requirement of jobs starting no earlier than t_i is at least the total amount of resource arriving no earlier than t_i. To analyze the approximation ratio, we introduce the following notation: W_i is the total weight that the algorithm schedules at t_i; W_i' is the weight that the algorithm temporarily schedules at t_i when i is in the interval $[t_{s+r/2}, t_{s+r-1}]$ (or, in the first step, in the interval $[t_\ell, t_{\ell+r/2-1}]$); W_i^* is the total weight scheduled at t_i in the optimal solution.

Since we use the PTAS for $r/2 + \ell + 1$ arrival times in the first step, we have $\sum_{i=0}^{\ell-1}(1+\varepsilon)^{-i}W_i + \sum_{i=\ell}^{\ell+r/2-1}(1+\varepsilon)^{-i}W_i' \leq (1+\varepsilon)\sum_{i=0}^{\ell+r/2-1}(1+\varepsilon)^{-i}W_i^*$, as the right-hand side is $(1+\varepsilon)$ times the objective value of the feasible solution obtained from the optimal solution by moving jobs arriving before $t_{\ell+r/2-1}$ to 0.

For $s = jr/2 + \ell$, we compare the output of the PTAS with a different feasible solution: we schedule total weight W_i' at t_i for $i = s, s+1, \ldots, s+r/2-1$, total weight W_i^* at t_i for $i = s+r/2+1, \ldots, s+r-1$, and at $t_{s+r/2}$ we schedule all jobs that are no earlier than $t_{s+r/2}$ in the optimal schedule but are no later than $t_{s+r/2}$ in the PTAS schedule. We get the inequality

$$\sum_{i=jr/2+\ell}^{(j+1)r/2+\ell-1}(1+\varepsilon)^{-i}W_i + \sum_{i=(j+1)r/2+\ell}^{(j+2)r/2+\ell-1}(1+\varepsilon)^{-i}W_i' \leq (1+\varepsilon)\left(\sum_{i=jr/2+\ell}^{(j+1)r/2+\ell-1}(1+\varepsilon)^{-i}W_i'\right.$$
$$\left. + \sum_{i=(j+1)r/2+\ell}^{(j+2)r/2+\ell-1}(1+\varepsilon)^{-i}W_i^* + (1+\varepsilon)^{-(j+1)r/2-\ell}\sum_{i=0}^{(j+1)r/2+\ell-1}W_i^*\right).$$

The sum of these inequalities gives

$$\sum_{i=0}^{q-2}(1+\varepsilon)^{-i}W_i \leq \varepsilon\sum_{i=\ell}^{q-2}(1+\varepsilon)^{-i}W_i' + (1+\varepsilon)\sum_{i=0}^{q-2}(1+\varepsilon)^{-i}W_i^*$$
$$+(1+\varepsilon)\sum_{i=0}^{q-2}\left(\sum_{j:jr/2+\ell>i}(1-\varepsilon)^{-(jr/2+\ell)}\right)W_i^*. \tag{1}$$

To bound the first term on the right hand side of (1), first we observe that $\sum_{i=\ell}^{r/2+\ell-1}(1+\varepsilon)^{-i}W_i' \leq (1+\varepsilon)\sum_{i=0}^{r/2+\ell-1}(1+\varepsilon)^{-i}W_i^*$, because the left side is at most the value of the PTAS in the first step, while the right side is $(1+\varepsilon)$ times the value of a feasible solution. Similarly,

$$\sum_{i=(j+1)r/2+\ell}^{(j+2)r/2+\ell-1}(1+\varepsilon)^{-i}W_i'$$
$$\leq (1+\varepsilon)\left(\sum_{i=jr/2+\ell}^{(j+2)r/2+\ell-1}(1+\varepsilon)^{-i}W_i^* + (1+\varepsilon)^{-jr/2-\ell}\sum_{i=0}^{jr/2+\ell-1}W_i^*\right),$$

because the left side is at most the value of the PTAS in the $(j+1)$-th step, and the right side is $(1+\varepsilon)$ times the value of the following feasible solution: take the optimal solution, move jobs scheduled before $t_{(j+2)r/2+\ell-1}$ to 0, and move jobs scheduled after $t_{jr/2+\ell}$ to $t_{jr/2+\ell}$. Adding these inequalities, we get

$$\varepsilon \sum_{i=\ell}^{q-2} (1+\varepsilon)^{-i} W_i'$$

$$\leq \varepsilon(1+\varepsilon) \left(2 \sum_{i=0}^{q-2} (1+\varepsilon)^{-i} W_i^* + \sum_{i=0}^{q-2} \left(\sum_{j:jr/2+\ell>i} (1+\varepsilon)^{-jr/2-\ell} \right) W_i^* \right)$$

$$\leq \varepsilon \left(2(1+\varepsilon) + \frac{(1+\varepsilon)^{r/2}}{(1+\varepsilon)^{r/2}-1} \right) \sum_{i=0}^{q-2} (1+\varepsilon)^{-i} W_i^*.$$

The last expression is at most 4ε times the optimum value if r is large enough.

The last term of the right side of (1) is too large to get a bound that proves a PTAS. However, we can bound the *average* of these terms for different values of ℓ. The average is

$$(1+\varepsilon)\frac{2}{r} \sum_{\ell=1}^{r/2} \sum_{i=0}^{q-2} \left(\sum_{j:jr/2+\ell>i} (1-\varepsilon)^{-(jr/2+\ell)} \right) W_i^*$$

$$\leq (1+\varepsilon)\frac{2}{r} \sum_{i=0}^{q-2} \left(\sum_{j=1}^{\infty} (1+\varepsilon)^{-j} \right) (1-\varepsilon)^{-i} W_i^* = (1+\varepsilon)\frac{2}{r\varepsilon} \sum_{i=0}^{q-2} (1-\varepsilon)^{-i} W_i^*,$$

which is at most ε times the optimum if r is large enough. To summarize, we obtained that for large enough r, the average objective value of our algorithm for $\ell = 1, 2, \ldots, r/2$ is upper bounded by

$$4\varepsilon \sum_{i=0}^{q-2} (1+\varepsilon)^{-i} W_i^* + (1+\varepsilon) \sum_{i=0}^{q-2} (1+\varepsilon)^{-i} W_i^* + \varepsilon \sum_{i=0}^{q-2} (1+\varepsilon)^{-i} W_i^* = (1+6\varepsilon) \sum_{i=0}^{q-2} (1+\varepsilon)^{-i} W_i^*,$$

which is $(1+6\varepsilon)$ times the objective value of the optimal solution. This proves that the algorithm that chooses the best of the $r/2$ runs is a PTAS. $\square$

Acknowledgement. The authors are grateful to Erika Bérczi-Kovács and to Matthias Mnich for the helpful discussions. Kristóf Bérczi was supported by the János Bolyai Research Fellowship of the Hungarian Academy of Sciences and by the ÚNKP-19-4 New National Excellence Program of the Ministry for Innovation and Technology. Projects no. NKFI-128673 and ED18-1-2019-0030 have been implemented with the support provided from the National Research, Development and Innovation Fund of Hungary, financed under the FK_18 and Thematic Excellence Programme funding schemes. Tamás Király was supported by NKFIH grant number K120254.

References

1. Bérczi, K., Király, T., Omlor, S.: Scheduling with non-renewable resources: minimizing the sum of completion times (2019). https://arxiv.org/abs/1911.12138. Preprint
2. Carlier, J.: Problèmes d'ordonnancement à contraintes de ressources: algorithmes et complexité. Université Paris VI-Pierre et Marie Curie, Institut de programmation (1984)

3. Gafarov, E.R., Lazarev, A.A., Werner, F.: Single machine scheduling problems with financial resource constraints: some complexity results and properties. Math. Soc. Sci. **62**(1), 7–13 (2011)
4. Garey, M.R., Johnson, D.S.: Computers and Intractability. A Guide to the Theory of NP-Completeness (1979)
5. Graham, R.L., Lawler, E.L., Lenstra, J.K., Kan, A.R.: Optimization and approximation in deterministic sequencing and scheduling: a survey. In: Annals of discrete mathematics, vol. 5, pp. 287–326. Elsevier (1979)
6. Grigoriev, A., Holthuijsen, M., Van De Klundert, J.: Basic scheduling problems with raw material constraints. Naval Res. Logist. (NRL) **52**(6), 527–535 (2005)
7. Györgyi, P.: A ptas for a resource scheduling problem with arbitrary number of parallel machines. Oper. Res. Lett. **45**(6), 604–609 (2017)
8. Györgyi, P., Kis, T.: Approximation schemes for single machine scheduling with non-renewable resource constraints. J. Sched. **17**(2), 135–144 (2013). https://doi.org/10.1007/s10951-013-0346-9
9. Györgyi, P., Kis, T.: Approximability of scheduling problems with resource consuming jobs. Ann. Oper. Res. **235**(1), 319–336 (2015). https://doi.org/10.1007/s10479-015-1993-3
10. Györgyi, P., Kis, T.: Reductions between scheduling problems with non-renewable resources and knapsack problems. Theoret. Comput. Sci. **565**, 63–76 (2015)
11. Györgyi, P., Kis, T.: Approximation schemes for parallel machine scheduling with non-renewable resources. Eur. J. Oper. Res. **258**(1), 113–123 (2017)
12. Györgyi, P., Kis, T.: Minimizing the maximum lateness on a single machine with raw material constraints by branch-and-cut. Comput. Ind. Eng. **115**, 220–225 (2018)
13. Györgyi, P., Kis, T.: Minimizing total weighted completion time on a single machine subject to non-renewable resource constraints. J. Sched. **22**(6), 623–634 (2019). https://doi.org/10.1007/s10951-019-00601-1
14. Györgyi, P., Kis, T.: New complexity and approximability results for minimizing the total weighted completion time on a single machine subject to non-renewable resource constraints. arXiv preprint arXiv:2004.00972 (2020). Earlier version: EGRES Technical Report 2019–05, egres.elte.hu
15. Kis, T.: Approximability of total weighted completion time with resource consuming jobs. Oper. Res. Lett. **43**(6), 595–598 (2015)
16. Slowiński, R.: Preemptive scheduling of independent jobs on parallel machines subject to financial constraints. Eur. J. Oper. Res. **15**(3), 366–373 (1984)
17. Toker, A., Kondakci, S., Erkip, N.: Scheduling under a non-renewable resource constraint. J. Oper. Res. Soc. **42**(9), 811–814 (1991)
18. Xie, J.: Polynomial algorithms for single machine scheduling problems with financial constraints. Oper. Res. Lett. **21**(1), 39–42 (1997)

Arc-Flow Approach for Parallel Batch Processing Machine Scheduling with Non-identical Job Sizes

Renan Spencer Trindade[1]($\boxtimes$), Olinto C. B. de Araújo[2], and Marcia Fampa[3]

[1] LIX, CNRS, École Polytechnique, Institut Polytechnique de Paris,
Palaiseau, France
rst@lix.polytechnique.fr
[2] CTISM, Universidade Federal de Santa Maria, Santa Maria, RS, Brazil
olinto@ctism.ufsm.br
[3] IM, PESC/COPPE, Universidade Federal do Rio de Janeiro,
Rio de Janeiro, RJ, Brazil
fampa@cos.ufrj.br

Abstract. Problems of minimizing makespan in scheduling batch processing machines are widely exploited by academic literature, mainly motivated by burn-in tests in the semiconductor industry. The problem addressed in this work consists of grouping jobs into batches and scheduling them in parallel machines. The jobs have non-identical size and processing times. The total size of the batch cannot exceed the capacity of the machine. The processing time of each batch will be equal to the longest processing time among all the jobs assigned to it. This paper proposes an arc-flow based model for minimizing makespan on parallel processing machines $P_m|s_j, B|C_{max}$. The mathematical model is solved using CPLEX, and computational results show that the proposed models have a better performance than other models in the literature.

Keywords: Parallel batch processing machine · Scheduling · Makespan · Arc-flow

1 Introduction

Scheduling is a widely used decision-making process in resource allocation and allows optimization in most production systems, information processing, transport, and distribution configurations, and several other real-world environments. This paper focuses on scheduling problems in Batch Processing Machines (BPM), that have been extensively explored in the literature, motivated by a large number of applications in industries and also by the challenging solution of real world

R. S. Trindade—Partially supported by a Ph.D. scholarship from the Brazilian National Council for Scientific and Technological Development (CNPq) [grant number 142205/2014-1] and by CNPq grant 303898/2016-0.
M. Fampa—Supported in part by CNPq grants 303898/2016-0 and 434683/2018-3.

© Springer Nature Switzerland AG 2020
M. Baïou et al. (Eds.): ISCO 2020, LNCS 12176, pp. 179–190, 2020.
https://doi.org/10.1007/978-3-030-53262-8_15

problems. The main goal in these problems is to group jobs in batches and process them simultaneously in a machine, to facilitate the tasks and to reduce the time spent in handling the material. Although there are many variations of the problem involving BPM, the version addressed in this work are more suitable to model the scheduling problems that arise in reliability tests in the semiconductor industry, in operations called burn-in, presented in [22].

The burn-in operation is used to test electronic circuits and consists of designating them to industrial ovens, submitting them to thermal stress for a long period. The test of each circuit is considered here as a job and requires a minimum time inside the oven, which is referred to as the processing machine. The jobs need to be placed on a tray, respecting the capacity of the machine. The burn-in tests are a bottleneck in final testing operations, and the efficient scheduling of these operations aims to maximize productivity. The processing time to test an electronic circuit can reach up to 120 hours in a constant temperature around 120°C, as presented in [13]. On tests reported in [19] and [7], a liquid crystal display usually takes 6 hours to complete the reliability test, which reinforces the importance of an efficient scheduling.

The research on BPM is recent, compared to the history of the semiconductor manufacturing, and consists of grouping the jobs into batches. The publication [18] reviews the research done on scheduling models considering batch processing machines. A survey related to BPM problems research found in [15], analyzing publications between 1986 and 2004 (part of 2004 only). Another survey that focus on BPM problems published in [16] and reveals that p-batching is much more important in semiconductor manufacturing comparing with s-batching.

This paper considers $P_m|s_j, B|C_{\max}$ problem. In the literature, the works that address it are mostly extensions of the works published for the single machine version of the problem. In [3], the simulated annealing meta-heuristic is applied, and an Mixed Integer Linear Programming (MILP) formulation is presented for the problem. This work also proves the NP-hard complexity of the problem, and shows results for instances with up to 50 jobs. In [11], a hybrid genetic algorithm is used to compute solutions for instances with up to 100 jobs, considering 2 and 4 parallel machines. In [8] a new application of the genetic algorithm is proposed, which solves instances with up to 100 jobs, also on 2 and 4 parallel machines. In [6] an approximation algorithm is presented for the problem, with the approximation factor of 2. Finally, two other works that apply meta-heuristics ([5] and [10]), use the ant colony method and a meta-heuristic based on a max-min ant system for this problem. In [5], results for instances with up to 500 jobs on 4 and 8 parallel machines are shown, whereas, in [10], instances are solved with up to 100 jobs, on 2, 3, and 4 parallel machines. In [21] and [20], the authors propose a new formulation focused on symmetry breaking constraints.

We propose an arc-flow formulation for problem $P_m|s_j, B|C_{\max}$. The paper is organized as follows: In Sect. 2, we introduce problem $P_m|s_j, B|C_{\max}$ and present two formulations from the literature. In Sect. 3, we present an arc-flow based formulation for the problem. In Sect. 4, we discuss our numerical experiments comparing the arc-flow formulation to formulations from the literature. In Sect. 5, we present some concluding remarks and discuss future work.

2   Problem Definition

The problem can be formally defined as follows. Given a set $J := \{1, \ldots, n_J\}$ of jobs, each job $j \in J$ has a processing time p_j and a size s_j. Each of them must be assigned to a batch $k \in K := \{1, \ldots, n_K\}$, not exceeding a given capacity limit B of the processing machine, i.e., the sum of the sizes of the jobs assigned to a single batch cannot exceed B. We assume that $s_j \leq B$, for all $j \in J$. The batches must be assigned to a specific machine $M := \{1, \ldots, n_M\}$. All machines are identical, and each one has its own processing time, defined by the time of the last batch processed on the machine. The processing time P_k of each batch $k \in K$ is defined as longest processing time among all jobs assigned to it, i.e., $P_k := \max\{p_j : j \text{ is assigned to } k\}$. Jobs cannot be split between batches. It is also not possible to add or remove jobs from the machine while the batches are being processed. The goal is to design and schedule the batches so that the makespan ($C_{\max}$) is minimized, where the design of a batch is defined as the set of jobs assigned to it, to schedule the batches means to define the ordering in which they are processed in the machine, and the makespan is defined as the time required to finish processing the last machine.

Consider the following decision variables, for all $j \in J$, $k \in K$, and $m \in M$:

$$x_{jkm} = \begin{cases} 1, \text{ if job } j \text{ is assigned to batch } k \text{ processed in machine } m; \\ 0, \text{ otherwise.} \end{cases} \tag{1}$$

$$P_{km} : \text{time to process batch } k \text{ in machine } m. \tag{2}$$

$$C_{max} : \text{the makespan.} \tag{3}$$

In [3] the following MILP formulation is proposed for $P_m|s_j, B|C_{\max}$:

$$(\text{MILP}) \quad \min C_{\max}, \tag{4}$$

$$\sum_{k \in K} \sum_{m \in M} x_{jkm} = 1, \qquad\qquad \forall j \in J, \tag{5}$$

$$\sum_{j \in J} \sum_{m \in M} s_j x_{jkm} \leq B, \qquad\qquad \forall k \in K, \tag{6}$$

$$P_{km} \geq p_j x_{jkm}, \qquad\qquad \forall j \in J, \forall k \in K, \forall m \in M, \tag{7}$$

$$C_{\max} \geq \sum_{k \in K} P_{km}, \qquad\qquad \forall m \in M, \tag{8}$$

$$x_{jkm} \in \{0,1\}, \qquad\qquad \forall j \in J, \forall k \in K, \forall m \in M. \tag{9}$$

The objective function (4) minimizes the makespan. Constraints (5) and (6) ensure that each job is assigned to a single batch and a single machine, respecting the capacity of the machine. Constraints (7) determine the processing time of batch k in machine m. Constraints (8) determine the makespan, which is given by the longest sum of the processing times of all batches, among all machines. Note that formulation (MILP) takes into account that $n_K = n_J$, and therefore, all batches assigned to all machines on a given solution can be indexed by distinct indexes. Note that constraints (6) take into account the fact that, although we have batches indexed by a given k, corresponding to all machines, a job can only be assigned to one of them, because of constraints (5). Therefore, a job j is only assigned to a unique pair (k, m).

(MILP) can be considered highly symmetrical concerning the order in which the batches are scheduled in each one of the parallel machines. This is because the same solution can be represented in different ways, just by changing the sequence order of the batches. In [21], and [20] the symmetry mentioned above is considered and a symmetry breaking procedure is used. At first, the variables x_{jkm} are replaced by two binary variables x_{jk}, which determine only the design of the batches, and the binary variables y_{km}, which determine whether or not batch k is processed in machine m. This replacement significantly reduces the number of binary variables. Furthermore, [21] presents a new formulation for the problem, where symmetric solutions are eliminated from the feasible set of (MILP), with the following approach. Firstly, the indexes of the jobs are defined by ordering them by their processing times. More specifically, it is considered that $p_1 \leq p_2 \leq \ldots \leq p_{n_J}$. Secondly, it is determined that batch k can only be used if job k is assigned to it, for all $k \in K$. Thirdly, it is determined that job j can only be assigned to batch k if $j \leq k$. Considering the above, the following formulation for $1|s_j, B|C_{\max}$ is proposed in [21]:

$$(\text{MILP}^+) \quad \min \ C_{\max}, \tag{10}$$

$$\sum_{k \in K : k \geq j} x_{jk} = 1, \qquad\qquad \forall j \in J, \tag{11}$$

$$\sum_{j \in J : j \leq k} s_j x_{jk} \leq B x_{kk}, \qquad\qquad \forall k \in K, \tag{12}$$

$$x_{jk} \leq x_{kk}, \qquad\qquad \forall j \in J, \forall k \in K, \tag{13}$$

$$x_{kk} \leq \sum_{m \in M} y_{km}, \qquad\qquad \forall k \in K, \tag{14}$$

$$C_m \geq \sum_{k \in K} p_k y_{km}. \qquad\qquad \forall m \in M, \tag{15}$$

$$C_{\max} \geq C_m \qquad\qquad \forall m \in M, \tag{16}$$

$$x_{jk} \in \{0, 1\} \qquad\qquad \forall j \in J, \forall k \in K : j \leq k. \tag{17}$$

The objective function (10) minimizes the makespan given by the latest time to finish processing all batches in all machines. Constraints (11) determine that each job j is assigned to a single batch k, such that $k \geq j$. Constraints (12) determine that the batches do not exceed the capacity of the machine. They also ensure that each batch k is used if and only if job k is assigned to it. Constraints

(13) are redundant together with (12), but are included to strengthen the linear relaxation of the model. Constraints (14) ensure that each used batch is assigned to a machine. Constraints (15) and (16) determine the makespan.

3   Arc Flow Approach

The arc flow approach has been used recently in classical optimization problems and allows modeling with a pseudo-polynomial number of variables and constraints. For a cutting-stock problem, [23] proposes a branch-and-price approach for an arc-flow formulation. Next, it was extended for the bin-packing problem in [24]. An alternative arc-flow formulation for the cutting-stock problem is proposed in [1] and [2], which uses a graph compression technique. These formulations were recently tested and compared in [9] against several other models and problem-specific algorithms on one-dimensional bin packing and cutting stock problems. The results show that the arc-flow formulation outperforms all other models. In [14] the arc-flow model and the one-cut model are compared for the one-dimensional cutting-stock problem, and reduction techniques for both approaches are presented.

For the scheduling area, we are only aware of two works that consider the arc-flow approach. In [12] the problem of scheduling a set of jobs on a set of identical parallel machines, with the aim of minimizing the total weighted completion time, $P||\sum W_j C_j$ is considered. In [17] the makespan minimization problem on identical parallel machines, $P||C_{max}$ is considered. It is important to note that these works do not consider more complex features in scheduling problems, such as batching machines, non-identical job sizes, and machine capacity.

The idea in this section is to formulate problem $P_m|s_j, B|C_{\mathrm{max}}$ as a problem of determining flows in graphs. With this goal, we initially define a directed graph $G = (V, A)$, in which each physical space of the batch with capacity B is represented by a node, i.e., $V = \{0, \dots, B\}$. The set of directed arcs A is divided into three subsets: the set of *job arcs* A^J, the set of *loss arcs* A^L, and the set with a *feedback arc* A^F. Therefore, $A = A^J \cup A^L \cup A^F$. Each arc (i, j) of the subset A^J represents the existence of at least one job k of size s_k, such that $s_k = j - i$. The subset A^J is more specifically defined as $A^J := \{(i, j) : \exists k \in J, s_k = j - i \wedge i, j \in V \wedge i < j\}$. To compose valid paths and represent all possible solutions, it is necessary to include the *loss arcs* in G, which represent empty spaces at the end of a batch. The subset of arcs A^L is more specifically defined as $A^L := \{(i, B) : i \in V \wedge 0 < i < B\}$. Finally, the *feedback arc* is used to connect the last node to the first one, defined as $A^F := \{(B, 0)\}$.

The graph G is then replicated for each different processing time of the problem in our modeling approach. Each replicated graph will be referred to as an arc-flow structure for our problem. We consider $P := \{P_1, \dots, P_\delta\}$ as the set with all the different processing times among all jobs, and $T := \{1, \dots, \delta\}$ as the set of indexes corresponding to the arc-flow structures in the problem formulation.

A variable $w_{t,m}$ is created to determine the number of batches with processing time P_t that will be allocated on the machine m. Considering $NT_{\ell,t}$ ($NT_{\ell,t}^+$) as the number of jobs of size S_ℓ and processing time $= P_t$ ($\leq P_t$), and NJ_t as the number of jobs with processing time P_t, our new formulation is presented below.

$f_{i,j,t}$: flow on *job arc* $(i,j) \in A^J$ in arc-flow structure t. The variable indicates the quantity of batches created with position i occupied by jobs with size $j-i$.
$y_{i,j,t}$: flow on the *loss arc* $(i,B) \in A^L$ in arc-flow structure t.
v_t: flow on the *feedback arc* in arc-flow structure t. The variable indicates the number of batches required with processing time P_t.
$z_{c,t}$: number of jobs with size c, not allocated in the batches with processing time smaller than or equal to P_t. Theses jobs are allowed to be allocated in the batches with processing time P_{t+1}.
$w_{t,m}$: number of batches with processing time P_t, allocated to machine m.

(FLOW$_2$) min C_{max} $\hspace{6cm}$ (18)

$$\left(\sum_{(i,j)\in A^J} f_{i,j,t} + \sum_{(i,j)\in A^L} y_{i,j,t} \right)$$

$$-\left(\sum_{(j,i)\in A^J} f_{j,i,t} + \sum_{(j,i)\in A^L} y_{j,i,t} \right) = \begin{cases} -v_t & \text{if } j = 0; \\ v_t & \text{if } j = B; \\ 0 & \text{if } 0 < j < B. \end{cases} \qquad t \in T \quad (19)$$

$$NT_{c,t} - \sum_{\substack{(i,j)\in A^J: \\ j-i=c}} f_{i,j,t} = \begin{cases} z_{c,t} & \text{if } t = 1; \\ -z_{c,t-1} & \text{if } t = \delta; \\ z_{c,t} - z_{c,t-1} & \text{if } 1 < t < \delta. \end{cases} \qquad c \in \{1..B\} \quad (20)$$

$$\sum_{m\in M} w_{t,m} \geq v_t \qquad\qquad\qquad t \in T \quad (21)$$

$$\sum_{t\in T} P_t w_{t,m} \leq C_{max} \qquad\qquad\qquad m \in M \quad (22)$$

$$f_{i,j,t} \leq min(NJ_t, NT^+_{j-i,t}), \; f_{i,j,t} \in \mathbb{Z} \qquad t \in T, (i,j) \in A^J \quad (23)$$

$$v_t \leq NJ_t, \; v_t \in \mathbb{Z} \qquad\qquad\qquad t \in T \quad (24)$$

$$y_{i,j,t} \leq NJ_t, \; y_{i,j,t} \in \mathbb{Z} \qquad\qquad t \in T, (i,j) \in A^L \quad (25)$$

$$z_{c,t} \leq NT^+_{c,t}, \; z_{c,t} \in \mathbb{Z} \qquad\qquad t \in T : t < \delta, c \in \{1..B\} \quad (26)$$

$$w_{t,m} \in \mathbb{Z} \qquad\qquad\qquad t \in T, m \in M \quad (27)$$

The objective function (18) minimizes the makespan. The set of flow conservation constraints are defined by constraints (19). Constraints (20) ensure that all jobs are assigned and also control the number of jobs to be assigned to each arc-flow structure. Constraints (21) ensure that all batches used are assigned to a machine. Constraints (22) determine the makespan as the time required to finish processing the last batch on all machines. Constraints (23–27) define the domains of the variables and their respective upper bounds. We emphasize that (21) and (22) are the constraints that make it possible for the arc-flow model to handle batch allocation on parallel machines.

4   Computational Results

The models presented in this chapter were compared through computational tests performed. The set was created by the authors of [4], who kindly sent them to us, to use in our work. Wee use the CPLEX version 12.7.1.0, configured to run in only one thread to not benefit from the processor parallelism. We used a computer with a 2.70 GHz Intel Quad-Core Xeon E5-2697 v2 processor and 64 GB of RAM. The computational time to solve each instance was limited in 1800 s.

The set of test instances for problem $P_m|s_j, B|C_{max}$ is the same considered in [4] for the $1|s_j, B|C_{max}$ problem. For each job j, an integer processing time p_j and an integer job size s_j were generated from the respective uniform distribution depicted in Table 1. In total, 4200 instances were generated, 100 for each of the 42 different combinations of number and size of the jobs. We test each instance with three different numbers of parallel machines.

Table 1. Parameter settings.

Number of jobs (n_J)	Processing time (p_J)	Jobs size	Machine capacity (B)	Parallel machines (n_M)
10, 20, 50, 100	p_1: $[1, 10]$	s_1: $[1, 10]$	$B = 10$	2, 4, 8
200, 300, 500	p_2: $[1, 20]$	s_2: $[2, 4]$		
		s_3: $[4, 8]$		

We present in Table 2, 3 and 4 comparison results among the arc flow formulation proposed in this work and another two from the literature. All values presented are the average results computed over the instances of the same configuration, as described in Table 1.

The comparative tests clearly show that formulation (FLOW) is superior to (MILP) and (MILP$^+$), especially when the number of jobs increases. Model (FLOW) did not prove the optimality of only one instance from the set of test problems. For instances with 20 jobs or less, (MILP$^+$) can solve some instances in less computational time than (FLOW), but the difference between times is always a fraction of a second. Additionally, the duality gaps shown for (MILP) reveal the difficulty in obtaining good lower bounds.

Unlike what we have with models (MILP) and (MILP$^+$), the number of variables in (FLOW) does not grow when the number of jobs increases. Moreover, the flow graph does not change in this case. Only the bounds on the variables change. The flow graphs of two distinct instances will be the same if the settings in the parameters Processing Time, Job Size and Machine Capacity are the same. In fact, this is a very important characteristic of the flow approach. We finally note that the computational time to construct the graphs for the flow formulation was not considered in these times. However, the maximum time to construct a graph for any instance in our experiments was 0.008 s.

Table 2. Computational results for $P_m|s_j.B|C_{\max}$ - 2 parallel machines.

Instance		(MILP)			(MILP$^+$)			(FLOW)		
Jobs	Type	$C_{\max}$	$T(s)$	Gap	$C_{\max}$	$T(s)$	Gap	$C_{\max}$	$T(s)$	Gap
2 parallel machines										
10	p_1s_1	18.76	0.13	0.00	18.76	0.01	0.00	18.76	0.02	0.00
10	p_1s_2	11.03	0.05	0.00	11.03	0.02	0.00	11.03	0.02	0.00
10	p_1s_3	22.13	0.19	0.00	22.13	0.01	0.00	22.13	0.00	0.00
10	p_2s_1	34.50	0.12	0.00	34.50	0.01	0.00	34.50	0.03	0.00
10	p_2s_2	21.71	0.05	0.00	21.71	0.02	0.00	21.71	0.03	0.00
10	p_2s_3	40.87	0.17	0.00	40.87	0.01	0.00	40.87	0.01	0.00
20	p_1s_1	34.27	1308.41	5.54	34.27	0.03	0.00	34.27	0.04	0.00
20	p_1s_2	18.83	884.08	8.16	18.83	0.11	0.00	18.83	0.04	0.00
20	p_1s_3	42.13	1412.74	6.27	42.13	0.02	0.00	42.13	0.01	0.00
20	p_2s_1	66.79	1287.70	4.35	66.79	0.03	0.00	66.79	0.09	0.00
20	p_2s_2	36.87	651.70	7.05	36.87	0.15	0.00	36.87	0.09	0.00
20	p_2s_3	79.82	1395.83	5.60	79.82	0.02	0.00	79.82	0.01	0.00
50	p_1s_1	83.07	–	58.36	82.30	2.48	0.00	82.30	0.08	0.00
50	p_1s_2	46.56	–	59.68	43.94	529.33	0.52	43.94	0.07	0.00
50	p_1s_3	101.74	–	60.69	101.30	0.02	0.00	101.30	0.01	0.00
50	p_2s_1	159.08	–	61.30	157.52	5.12	0.00	157.52	0.33	0.00
50	p_2s_2	88.96	–	62.44	84.32	478.37	0.19	84.32	0.55	0.00
50	p_2s_3	192.95	–	64.02	192.34	0.03	0.00	192.34	0.02	0.00
100	p_1s_1	171.60	–	87.71	159.78	192.10	0.07	159.78	0.11	0.00
100	p_1s_2	98.19	–	86.49	85.56	1743.59	1.73	85.56	0.10	0.00
100	p_1s_3	206.66	–	86.52	198.75	0.15	0.00	198.75	0.01	0.00
100	p_2s_1	328.38	–	89.47	305.58	84.36	0.02	305.58	0.42	0.00
100	p_2s_2	188.69	–	88.60	163.39	1770.79	1.21	163.31	1.58	0.00
100	p_2s_3	398.94	–	89.28	383.73	0.20	0.00	383.73	0.03	0.00
200	p_1s_1	Unperformed			314.93	332.53	0.05	314.92	0.07	0.00
200	p_1s_2				167.44	–	1.58	166.97	0.15	0.00
200	p_1s_3				393.36	79.14	0.01	393.36	0.02	0.00
200	p_2s_1				599.00	495.03	0.05	598.96	0.67	0.00
200	p_2s_2				320.16	–	1.52	318.85	3.43	0.00
200	p_2s_3				752.78	42.39	0.00	752.78	0.05	0.00
300	p_1s_1				464.59	639.71	0.08	464.54	0.10	0.00
300	p_1s_2				250.62	–	1.89	248.06	0.14	0.00
300	p_1s_3				587.49	241.24	0.02	587.49	0.02	0.00
300	p_2s_1				897.09	764.46	0.05	897.00	0.57	0.00
300	p_2s_2				487.55	–	2.09	481.61	3.25	0.00
300	p_2s_3				1123.96	274.67	0.02	1123.96	0.09	0.00
500	p_1s_1				772.54	1084.33	0.11	772.38	0.11	0.00
500	p_1s_2				421.92	–	1.98	415.76	0.17	0.00
500	p_1s_3				975.15	382.95	0.02	975.15	0.01	0.00
500	p_2s_1				1483.02	1365.87	0.09	1482.58	0.59	0.00
500	p_2s_2				806.24	–	2.10	794.00	2.78	0.00
500	p_2s_3				1851.16	488.38	0.01	1851.16	0.06	0.00

Table 3. Computational results for $P_m|s_j.B|C_{\max}$ - 4 parallel machines.

Instance		(MILP)			(MILP$^+$)			(FLOW)		
Jobs	Type	$C_{\max}$	$T(s)$	Gap	$C_{\max}$	$T(s)$	Gap	$C_{\max}$	$T(s)$	Gap
4 parallel machines										
10	$p_1 s_1$	10.87	0.16	0.00	10.87	0.02	0.00	10.87	0.02	0.00
10	$p_1 s_2$	9.49	0.10	0.00	9.49	0.01	0.00	9.49	0.01	0.00
10	$p_1 s_3$	12.18	0.25	0.00	12.18	0.02	0.00	12.18	0.01	0.00
10	$p_2 s_1$	20.26	0.16	0.00	20.26	0.02	0.00	20.26	0.03	0.00
10	$p_2 s_2$	18.68	0.11	0.00	18.68	0.01	0.00	18.68	0.02	0.00
10	$p_2 s_3$	22.67	0.23	0.00	22.67	0.02	0.00	22.67	0.01	0.00
20	$p_1 s_1$	17.47	1316.19	8.11	17.47	0.05	0.00	17.47	0.06	0.00
20	$p_1 s_2$	10.43	56.14	0.49	10.43	0.32	0.00	10.43	0.05	0.00
20	$p_1 s_3$	21.29	1629.93	11.78	21.29	0.03	0.00	21.29	0.01	0.00
20	$p_2 s_1$	33.95	1122.49	5.29	33.95	0.07	0.00	33.95	0.14	0.00
20	$p_2 s_2$	20.51	92.62	0.64	20.51	0.35	0.00	20.51	0.14	0.00
20	$p_2 s_3$	40.21	1731.24	12.27	40.21	0.05	0.00	40.21	0.02	0.00
50	$p_1 s_1$	42.69	–	70.03	41.43	2.54	0.00	41.43	0.11	0.00
50	$p_1 s_2$	23.76	–	58.83	22.18	269.31	0.56	22.18	0.26	0.00
50	$p_1 s_3$	51.85	–	71.91	50.90	0.05	0.00	50.90	0.01	0.00
50	$p_2 s_1$	81.23	–	71.19	78.97	0.90	0.00	78.97	0.80	0.00
50	$p_2 s_2$	45.55	–	57.37	42.38	283.70	0.19	42.38	1.41	0.00
50	$p_2 s_3$	97.89	–	73.39	96.40	0.07	0.00	96.40	0.04	0.00
100	$p_1 s_1$	93.06	–	93.44	80.09	82.33	0.05	80.09	0.18	0.00
100	$p_1 s_2$	50.26	–	81.80	43.06	1409.33	1.67	43.04	0.26	0.00
100	$p_1 s_3$	110.60	–	92.94	99.64	0.55	0.00	99.64	0.02	0.00
100	$p_2 s_1$	177.17	–	93.54	153.03	51.98	0.02	153.03	1.83	0.00
100	$p_2 s_2$	96.18	–	86.63	82.03	1679.11	1.32	81.88	3.12	0.00
100	$p_2 s_3$	213.47	–	93.38	192.11	0.52	0.00	192.11	0.05	0.00
200	$p_1 s_1$	Unperformed			157.71	209.19	0.06	157.70	0.16	0.00
200	$p_1 s_2$				84.05	1788.54	1.69	83.67	0.53	0.00
200	$p_1 s_3$				196.93	38.03	0.01	196.93	0.02	0.00
200	$p_2 s_1$				299.78	396.85	0.06	299.75	21.23	0.00
200	$p_2 s_2$				160.95	–	1.94	159.68	31.02	0.01
200	$p_2 s_3$				376.64	59.36	0.01	376.64	0.07	0.00
300	$p_1 s_1$				232.56	422.79	0.09	232.52	0.17	0.00
300	$p_1 s_2$				126.22	–	2.40	124.28	0.33	0.00
300	$p_1 s_3$				293.99	146.32	0.03	293.99	0.02	0.00
300	$p_2 s_1$				448.84	568.62	0.06	448.79	1.19	0.00
300	$p_2 s_2$				244.96	–	2.49	241.07	21.59	0.00
300	$p_2 s_3$				562.24	230.89	0.02	562.24	0.11	0.00
500	$p_1 s_1$				386.62	1009.64	0.15	386.47	0.17	0.00
500	$p_1 s_2$				211.91	–	2.31	208.12	0.40	0.00
500	$p_1 s_3$				487.84	266.44	0.03	487.84	0.02	0.00
500	$p_2 s_1$				741.93	1306.98	0.12	741.56	1.91	0.00
500	$p_2 s_2$				405.30	–	2.58	397.30	134.68	0.01
500	$p_2 s_3$				925.84	345.92	0.02	925.84	0.07	0.00

Table 4. Computational results for $P_m|s_j.B|C_{max}$ - 8 parallel machines.

Instance		(MILP)			(MILP$^+$)			(FLOW)		
Jobs	Type	C_{max}	$T(s)$	Gap	C_{max}	$T(s)$	Gap	C_{max}	$T(s)$	Gap
8 parallel machines										
10	p_1s_1	9.54	0.23	0.00	9.54	0.01	0.00	9.54	0.02	0.00
10	p_1s_2	9.49	0.25	0.00	9.49	0.01	0.00	9.49	0.02	0.00
10	p_1s_3	9.42	0.33	0.00	9.42	0.01	0.00	9.42	0.01	0.00
10	p_2s_1	18.55	0.21	0.00	18.55	0.01	0.00	18.55	0.02	0.00
10	p_2s_2	18.68	0.24	0.00	18.68	0.01	0.00	18.68	0.02	0.00
10	p_2s_3	18.27	0.34	0.00	18.27	0.01	0.00	18.27	0.01	0.00
20	p_1s_1	10.51	276.62	2.44	10.51	0.09	0.00	10.51	0.06	0.00
20	p_1s_2	9.81	2.76	0.00	9.81	0.07	0.00	9.81	0.03	0.00
20	p_1s_3	11.61	760.27	7.24	11.60	0.15	0.00	11.60	0.01	0.00
20	p_2s_1	20.76	328.01	3.34	20.76	0.13	0.00	20.76	0.18	0.00
20	p_2s_2	19.52	2.80	0.00	19.52	0.08	0.00	19.52	0.04	0.00
20	p_2s_3	22.31	958.29	8.28	22.30	0.26	0.00	22.30	0.04	0.00
50	p_1s_1	22.30	–	55.90	20.96	2.99	0.00	20.96	0.25	0.00
50	p_1s_2	12.83	1783.20	27.02	11.77	850.12	4.12	11.77	0.39	0.00
50	p_1s_3	26.78	–	62.67	25.71	0.10	0.00	25.71	0.02	0.00
50	p_2s_1	42.41	–	55.68	39.72	1.25	0.00	39.72	2.47	0.00
50	p_2s_2	24.41	1775.39	28.90	22.46	1198.77	3.91	22.45	11.60	0.00
50	p_2s_3	50.33	–	61.21	48.45	0.17	0.00	48.45	0.08	0.00
100	p_1s_1	59.57	–	96.84	40.34	51.78	0.05	40.34	0.19	0.00
100	p_1s_2	28.80	–	84.72	21.82	872.63	2.04	21.75	1.49	0.00
100	p_1s_3	69.72	–	98.03	50.07	0.22	0.00	50.07	0.03	0.00
100	p_2s_1	123.38	–	97.70	76.81	59.45	0.01	76.81	10.88	0.00
100	p_2s_2	57.82	–	94.95	41.34	1251.21	1.45	41.23	37.99	0.03
100	p_2s_3	139.99	–	98.19	96.34	0.81	0.00	96.34	0.11	0.00
200	p_1s_1	Unperformed			79.13	213.02	0.12	79.10	0.15	0.00
200	p_1s_2				42.41	1599.90	1.97	42.10	0.95	0.00
200	p_1s_3				98.74	2.72	0.00	98.74	0.02	0.00
200	p_2s_1				150.18	341.94	0.08	150.14	54.67	0.01
200	p_2s_2				81.27	1796.74	2.64	80.08	141.71	0.04
200	p_2s_3				188.53	21.83	0.01	188.53	0.14	0.00
300	p_1s_1				116.58	480.57	0.16	116.51	18.20	0.01
300	p_1s_2				63.40	1779.89	2.44	62.39	0.57	0.00
300	p_1s_3				147.25	94.40	0.03	147.25	0.02	0.00
300	p_2s_1				224.76	641.81	0.12	224.61	2.84	0.00
300	p_2s_2				123.61	–	3.20	120.78	358.94	0.13
300	p_2s_3				281.31	115.95	0.02	281.31	0.31	0.00
500	p_1s_1				193.74	1067.41	0.29	193.53	0.24	0.00
500	p_1s_2				106.63	1787.69	2.72	104.38	1.67	0.00
500	p_1s_3				244.11	134.96	0.03	244.11	0.04	0.00
500	p_2s_1				371.50	1136.26	0.20	371.03	2.08	0.00
500	p_2s_2				203.42	–	2.83	198.96	376.70	0.09
500	p_2s_3				463.16	189.74	0.02	463.16	0.24	0.00

The results show that instances of configuration s_2 require more computational time and are more difficult compared to the other instances for all formulations. The reason for this is the small sizes of the jobs when compared to the machine capacity, which allows more combinations of assignment to a batch.

5 Conclusion

In this paper we propose a new arc-flow formulation for minimizing makespan on parallel batch machines, considering non-identical job sizes. The computational results reveal that this new approach is much more efficient than those previously published in the literature. It is able to solve instances up to 500 jobs, which have never been solved before, with low computational times. Even for the most difficult instances, for which the model failed to prove optimality, the results are very close to the optimal with gaps between 0.13–0.01%. One of the best advantages of the arc-flow model is that the number of variables does not increase if the number of jobs of the instance increases.

As future work, it is interesting to investigate whether this approach can be applied to other variants of scheduling problems, such as considering incompatible families or jobs with non-identical release times.

References

1. Brandão, F., Pedroso, J.P.: Bin packing and related problems: general arc-flow formulation with graph compression. Comput. Oper. Res. **69**, 56–67 (2016). https://doi.org/10.1016/j.cor.2015.11.009
2. Brandão, F.D.A.: Cutting & packing problems: general arc-flow formulation with graph compression. Ph.D. thesis, Universidade do Porto (2017)
3. Chang, P.Y., Damodaran, P., Melouk, S.: Minimizing makespan on parallel batch processing machines. Int. J. Prod. Res. **42**(19), 4211–4220 (2004). https://doi.org/10.1080/00207540410001711863
4. Chen, H., Du, B., Huang, G.Q.: Scheduling a batch processing machine with non-identical job sizes: a clustering perspective. Int. J. Prod. Res. **49**(19), 5755–5778 (2011). https://doi.org/10.1080/00207543.2010.512620
5. Cheng, B., Wang, Q., Yang, S., Hu, X.: An improved ant colony optimization for scheduling identical parallel batching machines with arbitrary job sizes. Appl. Soft Comput. **13**(2), 765–772 (2013). https://doi.org/10.1016/j.asoc.2012.10.021
6. Cheng, B., Yang, S., Hu, X., Chen, B.: Minimizing makespan and total completion time for parallel batch processing machines with non-identical job sizes. Appl. Math. Model. **36**(7), 3161–3167 (2012). https://doi.org/10.1016/j.apm.2011.09.061
7. Chung, S., Tai, Y., Pearn, W.: Minimizing makespan on parallel batch processing machines with non-identical ready time and arbitrary job sizes. Int. J. Prod. Res. **47**(18), 5109–5128 (2009). https://doi.org/10.1080/00207540802010807
8. Damodaran, P., Hirani, N.S., Gallego, M.C.V.: Scheduling identical parallel batch processing machines to minimise makespan using genetic algorithms. Eur. J. Ind. Eng. **3**(2), 187 (2009). https://doi.org/10.1504/EJIE.2009.023605
9. Delorme, M., Iori, M., Martello, S.: Bin packing and cutting stock problems: mathematical models and exact algorithms. Eur. J. Oper. Res. **255**(1), 1–20 (2016). https://doi.org/10.1016/j.ejor.2016.04.030

10. Jia, Z.H., Leung, J.Y.T.: A meta-heuristic to minimize makespan for parallel batch machines with arbitrary job sizes. Eur. J. Oper. Res. **240**(3), 649–665 (2015). https://doi.org/10.1016/j.ejor.2014.07.039

11. Kashan, A.H., Karimi, B., Jenabi, M.: A hybrid genetic heuristic for scheduling parallel batch processing machines with arbitrary job sizes. Comput. Oper. Res. **35**(4), 1084–1098 (2008). https://doi.org/10.1016/j.cor.2006.07.005

12. Kramer, A., Dell'Amico, M., Iori, M.: Enhanced arc-flow formulations to minimize weighted completion time on identical parallel machines. Eur. J. Oper. Res. **275**(1), 67–79 (2019). https://doi.org/10.1016/j.ejor.2018.11.039

13. Lee, C.Y., Uzsoy, R., Martin-Vega, L.A.: Efficient algorithms for scheduling semi-conductor burn-in operations. Oper. Res. **40**(4), 764–775 (1992). https://doi.org/10.1287/opre.40.4.764

14. Martinovic, J., Scheithauer, G., de Carvalho, J.M.V.: A comparative study of the arcflow model and the one-cut model for one-dimensional cutting stock problems. Eur. J. Oper. Res. **266**(2), 458–471 (2018). https://doi.org/10.1016/j.ejor.2017.10.008

15. Mathirajan, M., Sivakumar, A.: A literature review, classification and simple meta-analysis on scheduling of batch processors in semiconductor. Int. J. Adv. Manuf. Technol. **29**(9–10), 990–1001 (2006). https://doi.org/10.1007/s00170-005-2585-1

16. Mönch, L., Fowler, J.W., Dauzère-Pérès, S., Mason, S.J., Rose, O.: A survey of problems, solution techniques, and future challenges in scheduling semiconductor manufacturing operations. J. Sched. **14**(6), 583–599 (2011). https://doi.org/10.1007/s10951-010-0222-9

17. Mrad, M., Souayah, N.: An arc-flow model for the makespan minimization problem on identical parallel machines. IEEE Access **6**, 5300–5307 (2018). https://doi.org/10.1109/ACCESS.2018.2789678

18. Potts, C.N., Kovalyov, M.Y.: Scheduling with batching: a review. Eur. J. Oper. Res. **120**(2), 228–249 (2000). https://doi.org/10.1016/S0377-2217(99)00153-8

19. Tai, Y.: The study on the production scheduling problems for liquid crystal display module assembly factories. Ph.D. thesis, National Chiao Tung University (2008). https://ir.nctu.edu.tw/bitstream/11536/57924/1/381201.pdf

20. Trindade, R.S.: Modelling batch processing machines problems with symmetry breaking and arc flow formulation. Ph.D. thesis, Universidade Federal do Rio de Janeiro (2019). https://www.cos.ufrj.br/index.php/pt-BR/publicacoes-pesquisa/details/15/2902

21. Trindade, R.S., de Araújo, O.C.B., Fampa, M.H.C., Müller, F.M.: Modelling and symmetry breaking in scheduling problems on batch processing machines. Int. J. Prod. Res. **56**(22), 7031–7048 (2018). https://doi.org/10.1080/00207543.2018.1424371

22. Uzsoy, R.: Scheduling a single batch processing machine with non-identical job sizes. Int. J. Prod. Res. **32**(7), 1615–1635 (1994). https://doi.org/10.1080/00207549408957026

23. de Carvalho, J.V.: Exact solution of cutting stock problems using column generation and branch-and-bound. Int. Trans. Oper. Res. **5**(1), 35–44 (1998). https://doi.org/10.1111/j.1475-3995.1998.tb00100.x

24. de Carvalho, J.V.: Exact solution of bin-packing problems using column generation and branch-and-bound. Ann. Oper. Res. **86**, 629–659 (1999). https://doi.org/10.1023/A:1018952112615

Matching

Dynamic and Stochastic Rematching for Ridesharing Systems: Formulations and Reductions

Gabriel Homsi[1(✉)], Bernard Gendron[1], and Sanjay Dominik Jena[2]

[1] Department of Computer Science and Operations Research,
Université de Montréal and CIRRELT, Montreal, Canada
{gabriel.homsi,bernard.gendron}@cirrelt.ca
[2] Department of Analytics, Operations and Information Technology,
École des Sciences de la Gestion, Université du Québec à Montréal and CIRRELT,
Montreal, Canada
sanjay.jena@cirrelt.ca

Abstract. We introduce a dynamic and stochastic rematching problem with applications in request matching for ridesharing systems. We propose three mathematical programming formulations that can be used in a rolling horizon framework to solve this problem. We show how these models can be simplified provided that specific conditions that are typically found in practice are met.

Keywords: Ridesharing · Request matching · Stochastic programming

1 Introduction

Ridesharing systems are matching agencies for drivers and riders that are interested in sharing commute expenses. Over time, these systems receive requests from their customers corresponding to the intent of engaging in ridesharing as a driver or as a rider for a certain itinerary. In this context, an itinerary is composed of an origin, a destination, the desired departure time, and the desired arrival time. A common goal of ridesharing systems is to create matches that generate profit and that promote customer engagement. We use the term *match* to refer to the pairing of two requests, meaning that the customers behind these requests are assigned to travel together to fulfill their corresponding itineraries. A driver request may be matched to a rider request if their itineraries are compatible and if the corresponding ridesharing trip generates value for the participants. The value of a ridesharing trip is often assumed to be the amount of travel distance savings generated by the trip when compared to the individual trips for each participant [1,2].

In this work, we study a ridesharing system that matches requests that arrive dynamically and may unmatch requests whose corresponding rides have not yet started. We assume to have access to forecasts on the probability that future requests exist. To avoid compromising customer engagement, we allow for defining a penalty on unmatch operations, which may correspond to a discount on future trips offered to the unmatched customers. These penalties may

© Springer Nature Switzerland AG 2020
M. Baïou et al. (Eds.): ISCO 2020, LNCS 12176, pp. 193–201, 2020.
https://doi.org/10.1007/978-3-030-53262-8_16

depend, for example, on the amount of time since the requests have been created. Unmatching a request that was released much earlier would therefore lead to a large penalty. Nevertheless, unmatching requests may be desirable if a new request becomes available such that it allows for a highly profitable match with a currently-matched request.

Research on optimization for ridesharing started gaining traction with the work of [1], where the authors studied a dynamic driver-rider matching problem. Several further studies explored specific attributes of ridesharing systems. To name a few, [3] studied the impact of meeting points in ridesharing systems, [4] studied the impact of participant time flexibility in ridesharing, [5] studied the integration of a ridesharing system with public transit, and [2] investigated the impact of matching stability on a dynamic ride-matching system. For surveys on ridesharing and related shared mobility systems, we refer the reader to [6,7] and [8]. Our work extends the ridesharing request matching literature by addressing the stochasticity in requests, the unmatching of requests, and the time-dependency of matching profits and unmatching penalties. In summary, our contributions are threefold: 1) we introduce a new dynamic and stochastic rematching problem 2) we propose three mathematical programming formulations to solve this problem, and 3) we show how these formulations can be simplified under specific but realistic conditions.

2   Problem Definition

Requests in a ridesharing system arrive continuously over a planning horizon $T = \{1, 2, \ldots, h\}$. Let $G = (V, E)$ be a bipartite graph where V is the set of requests and E is the set of edges between compatible requests. The set of requests is partitioned into a set of driver requests D and a set of rider requests R, such that $V = D \cup R$, $D \cap R = \emptyset$, and $E \subseteq D \times R$. A pair of requests $(ij) \in D \times R$ does not belong to E if the itineraries of i and j are incompatible or if the corresponding ridesharing trip does not generate value for its participants.

At each time period $t \in T$, a pair of requests $(ij) \in E$ can be matched for a profit of c_{ij}^t and unmatched for a cost of d_{ij}^t. We assume that $c_{ij}^t \leq d_{ij}^t$, i.e., it is never profitable to unmatch a pair of requests and then match it in the same time period. A pair of requests (ij) is said to be active at the beginning of the time period t if it was not unmatched since the last period it has been matched. For each request $i \in V$, let r_i be its release time and b_i be its latest possible match time. The latest possible match time may correspond to the desired departure time or to the latest time period that a customer is willing to wait for a match. The pair of requests $(ij) \in E$ can only be matched or unmatched at time period $t \in T$ if both requests have already been released and if they are still available for matching. Hence, matching and unmatching is only possible if $t \in W_{ij}$, where

$$W_{ij} = \{t \in T \mid \max(r_i, r_j) \leq t \leq \min(b_i, b_j)\}.$$

If the condition above is not met, we interdict matches and unmatches by assigning $c_{ij}^t = -M$ and $d_{ij}^t = M$, where M is a sufficiently big constant such that an optimal solution never has the pair of requests (ij) matched or unmatched outside W_{ij}.

The objective of the ridesharing system is to dynamically match and unmatch driver and rider requests such that the net profit over the planning horizon is maximized. In the following, we present three mathematical programming formulations that can be used in a rolling horizon framework to provide matches and rematches at specific time periods throughout the planning horizon. We first present a myopic formulation that can be used when no forecasts on future demand are available. Then, we present a static formulation that can be used when the information for all time periods is known in advance, or when a sufficiently accurate forecast is available. Finally, we present a stochastic formulation that can be used when sufficient historical information is known to accurately generate multiple scenarios that are representative of future demand.

3   The Myopic Problem

When no forecasts on future demand are available, a myopic optimization problem can be defined. For each pair $(ij) \in E$, let x_{ij}^t be a binary variable equal to 1 if and only if the pair (ij) is matched at time period t, y_{ij}^t be a binary variable equal to 1 if and only if the pair (ij) is unmatched at time period t, and a_{ij}^t be a binary constant equal to 1 if and only if (ij) is active at the beginning of t. The myopic problem of matching and unmatching requests such that the net profit at time period $t \in T$ is maximized can be formulated as below

$$f_{\text{MYO}}(t, a^t) := \max \sum_{(ij) \in E} (c_{ij}^t x_{ij}^t - d_{ij}^t y_{ij}^t) \tag{1}$$

$$\text{s.t.} \sum_{(ij) \in \delta(v)} (x_{ij}^t - y_{ij}^t) \le 1 - \sum_{(ij) \in \delta(v)} a_{ij}^t \qquad \forall v \in V \tag{2}$$

$$y_{ij}^t \le a_{ij}^t \qquad\qquad \forall (ij) \in E \tag{3}$$

$$x_{ij}^t, y_{ij}^t \in \{0, 1\} \qquad\qquad \forall (ij) \in E. \tag{4}$$

The objective function (1) maximizes the net profit of matching and unmatching requests at time period t. Constraints (2) ensure that requests can only be matched if they are inactive. Constraints (3) ensure that requests can only be unmatched if they are active.

The formulation above can be rewritten as a maximum-weight bipartite matching problem. Let $E_0 = \{(ij) \in E \mid a_{ij}^t = 0\}$ and $E_1 = E \setminus E_0$. For each pair $(ij) \in E_0$, let x_{ij}^t be a binary variable equal to 1 if and only if the inactive pair (ij) is matched at time period t. For each pair $(ij) \in E_1$, let z_{ij}^t be a binary variable equal to 1 if and only if the active pair (ij) is not unmatched at time period t. The bipartite matching reformulation is defined below

$$f'_{\text{MYO}}(t) := \max \sum_{(ij) \in E_0} c_{ij}^t x_{ij}^t + \sum_{(ij) \in E_1} d_{ij}^t (z_{ij}^t - 1) \tag{5}$$

$$\text{s.t.} \sum_{(ij)\in\delta(v)\cap E_0} x^t_{ij} + \sum_{(ij)\in\delta(v)\cap E_1} z^t_{ij} \leq 1 \qquad \forall v \in V. \tag{6}$$

The objective function (5) maximizes the net profit of matching and unmatching requests at time period t. If $z^t_{ij} = 0$, then (ij) is unmatched, which yields a penalty of $-d^t_{ij}$ in the objective function. Constraints (6) ensure that requests can be matched at most once, either in E_0 or in E_1. Despite having two sets of variables, the formulation above is equivalent to a classic bipartite matching formulation. Nevertheless, we have decided to use distinct variable names to highlight the differences between matches in E_0 and matches in E_1.

4   The Static Problem

When the requests released throughout all time periods are known in advance, a multiperiod static problem that provides matches and unmatches for the whole planning horizon can be formulated. This model can be used as a benchmark for other models evaluated in the rolling horizon framework, and is defined below

$$\max \sum_{t\in T} \sum_{(ij)\in E} (c^t_{ij} x^t_{ij} - d^t_{ij} y^t_{ij}) \tag{7}$$

$$\text{s.t.} \sum_{\ell=1}^{t} \sum_{(ij)\in\delta(v)} (x^\ell_{ij} - y^\ell_{ij}) \leq 1 \qquad \forall t \in T, v \in V \tag{8}$$

$$y^t_{ij} \leq \sum_{\ell=1}^{t-1} (x^\ell_{ij} - y^\ell_{ij}) \qquad \forall t \in T, (ij) \in E \tag{9}$$

$$x^t_{ij}, y^t_{ij} \in \{0,1\} \qquad \forall t \in T, (ij) \in E. \tag{10}$$

The objective function (7) maximizes the net profit over the full planning horizon. Constraints (8) ensure that requests can only be matched if they are inactive. Constraints (9) ensure that requests can only be unmatched if they are active.

Assumption 1. *The profit of matching a pair of requests is never bigger than the cost of unmatching it, i.e.,*

$$c^t_{ij} \leq d^k_{ij} \qquad \forall t \in T, k \in T, t \leq k.$$

Proposition 1. *If Assumption 1 holds for all pairs of requests, then it is never necessary to unmatch in an optimal solution for (7)–(10).*

Proof. Let $(\bar{x},\bar{y})$ be an optimal solution for (7)–(10) and $z(\bar{x},\bar{y})$ its objective function value. Assume that there exists a pair $(ij) \in E$ and time periods $t \in T$ and $k \in T$ with $t \leq k$ such that $\bar{x}^t_{ij} = 1$ and $\bar{y}^k_{ij} = 1$. As $c^t_{ij} - d^k_{ij} \leq 0$, there exists a solution $(\hat{x},\hat{y})$ similar to $(\bar{x},\bar{y})$, except for $\hat{x}^t_{ij} = 0$ and $\hat{y}^k_{ij} = 0$. Consequently, $z(\hat{x},\hat{y}) \geq z(\bar{x},\bar{y})$, which either contradicts the optimality of $(\bar{x},\bar{y})$, or shows that both solutions have the same objective function value.   □

If Assumption 1 holds, by Proposition 1, it follows that the static problem can be reduced by removing all unmatching variables, which gives the formulation below

$$\max \sum_{t \in T} \sum_{(ij) \in E} c_{ij}^t x_{ij}^t \tag{11}$$

$$\text{s.t.} \sum_{t \in T} \sum_{(ij) \in \delta(v)} x_{ij}^t \leq 1 \qquad \forall v \in V \tag{12}$$

$$x_{ij}^t \in \{0, 1\} \qquad \forall t \in T, (ij) \in E. \tag{13}$$

As each pair $(ij) \in E$ can be matched at most once, it is more profitable to match (ij) at the time period that maximizes $c_{ij}^t, \forall t \in T$. Thus, the multiple time periods can be represented implicitly and the formulation above can be reduced to a maximum-weight bipartite matching problem, as below

$$\max \sum_{(ij) \in E} \hat{c}_{ij} x_{ij} \tag{14}$$

$$\text{s.t.} \sum_{(ij) \in \delta(v)} x_{ij} \leq 1 \qquad \forall v \in V, \tag{15}$$

where $\hat{c}_{ij} = \max\{c_{ij}^t \mid t \in T\}$. Let $\bar{x}$ be an optimal solution for Eqs. (14)–(15). If $\bar{x}_{ij} = 1$, then (ij) is matched in time period $t = \arg\max_{t \in T} c_{ij}^t$.

The static formulation can be adapted to be used in a rolling horizon framework if a forecast such as the expected future demand is available. The static model that matches and unmatches requests for a time period $k \in T$ while taking into consideration a forecast for periods $(k + 1), \ldots, h$ is defined below

$$f_{\text{STAT}}(k, a^k) := \max \sum_{t=k}^{h} \sum_{(ij) \in E} (c_{ij}^t x_{ij}^t - d_{ij}^t y_{ij}^t) \tag{16}$$

$$\text{s.t.} \sum_{\ell=k}^{t} \sum_{(ij) \in \delta(v)} (x_{ij}^\ell - y_{ij}^\ell) \leq 1 - \sum_{(ij) \in \delta(v)} a_{ij}^k \quad \forall t = k, \ldots, h, v \in V \tag{17}$$

$$y_{ij}^t \leq a_{ij}^k + \sum_{\ell=1}^{t-1} (x_{ij}^\ell - y_{ij}^\ell) \qquad \forall t = k, \ldots, h, (ij) \in E \tag{18}$$

$$x_{ij}^t, y_{ij}^t \in \{0, 1\} \qquad \forall t = k, \ldots, h, (ij) \in E. \tag{19}$$

The formulation above is similar to the single-scenario case of the two-stage stochastic programming formulation defined next.

5   The Stochastic Problem

To address the uncertainty on future demand, we introduce a two-stage stochastic programming formulation. In the first stage, matching and unmatching decisions are given for requests available at the current time period $t \in T$. In the second stage, these decisions are made for the remainder of the planning horizon, given the first-stage decisions and a sample realization of future requests.

Let S be the set of second-stage scenarios and p_s the probability associated with each scenario $s \in S$. We assume that each scenario $s \in S$ contains a full realization of the planning horizon since the time period $t + 1$. For each request $i \in V$ and scenario $s \in S$, let ξ_i^s be a random variable defined below

$$\xi_i^s = \begin{cases} 1 \text{ if request } i \text{ is available for matching in scenario } s, \\ 0 \text{ otherwise.} \end{cases}$$

For each pair $(ij) \in E$ and time period $k = (t + 1), \ldots, h$, if $\xi_i^s = 0$ or $\xi_j^s = 0$, then $c_{ij}^{ks} = -M$ and $d_{ij}^{ks} = M$. Otherwise, $c_{ij}^{ks} = c_{ij}^k$ and $d_{ij}^{ks} = d_{ij}^k$. Moreover, if a request $i \in V$ is released before the second stage, i.e., $r_i \leq t$, then we assume that $\xi_i^s = 1, \forall s \in S$. The two-stage stochastic programming formulation is defined as below

$$f_{\mathrm{STO}}(t) := \max \sum_{(ij) \in E_0} c_{ij}^t x_{ij}^t + \sum_{(ij) \in E_1} d_{ij}^t (z_{ij}^t - 1) + \sum_{s \in S} p_s Q(t + 1, s, a^{t+1}) \quad (20)$$

$$\text{s.t.} \quad \sum_{(ij) \in E_0 \cap \delta(v)} x_{ij}^t + \sum_{(ij) \in E_1 \cap \delta(v)} z_{ij}^t \leq 1 \qquad \forall v \in V \quad (21)$$

$$a_{ij}^{t+1} = \begin{cases} x_{ij}^t \text{ if } (ij) \in E_0, \\ z_{ij}^t \text{ otherwise.} \end{cases} \qquad \forall (ij) \in E \quad (22)$$

$$x_{ij}^t \in \{0, 1\} \qquad \forall (ij) \in E_0 \quad (23)$$

$$z_{ij}^t \in \{0, 1\} \qquad \forall (ij) \in E_1. \quad (24)$$

The objective function (20) maximizes the net profit at time period t plus the expect net profit for time periods $(t + 1), \ldots, h$. Constraints (21) match and unmatch requests for period t. Equations (22) define the values of $a_{ij}^{t+1}, \forall (ij) \in E$. The second-stage problem is a multiperiod problem over the time periods $(t + 1), \ldots, h$, and is defined for each scenario $s \in S$ as below

$$Q(k, s, a^k) := \max \sum_{t=k}^{h} \sum_{(ij) \in E} (c_{ij}^{ts} x_{ij}^{ts} - d_{ij}^{ts} y_{ij}^{ts}) \quad (25)$$

$$\text{s.t.} \sum_{\ell=k}^{t} \sum_{(ij)\in\delta(v)} (x_{ij}^{\ell s} - y_{ij}^{\ell s}) \leq 1 - \sum_{(ij)\in\delta(v)} a_{ij}^{k} \qquad \forall t = k, \ldots, h, v \in V \qquad (26)$$

$$y_{ij}^{ts} \leq a_{ij}^{k} + \sum_{\ell=k}^{t-1}(x_{ij}^{\ell s} - y_{ij}^{\ell s}) \qquad \forall t = k, \ldots, h, (ij) \in E \qquad (27)$$

$$x_{ij}^{ts}, y_{ij}^{ts} \in \{0,1\} \qquad \forall t = k, \ldots, h, (ij) \in E. \qquad (28)$$

We now show how the two-stage formulation can be reduced if some realistic conditions are met.

5.1 Reductions Based on the System Environment

Motivated by the fact that unmatching close to the departure time is typically not user-friendly, we study how to reduce the stochastic model if the penalty of unmatching does not become less expensive over time.

Assumption 2. *For each $(ij) \in E$, the unmatching costs are non-decreasing in W_{ij}, i.e.,*

$$d_{ij}^{t} \leq d_{ij}^{k} \qquad \forall t \in W_{ij}, k \in W_{ij}, t < k$$

Proposition 2. *If Assumption 2 holds, then for each $(ij) \in E$, d_{ij}^{t} is non-decreasing in $\{t \in T \mid \min(W_{ij}) \leq t \leq h\}$.*

Proof. For each $t \in T$, if $t \in W_{ij}$, then $d_{ij}^{t} \leq M$. Otherwise, if $\max(W_{ij}) < t \leq h$, then $d_{ij}^{t} = M$. □

Proposition 3. *If Assumptions 1 and 2 hold, then the two-stage problem can be reduced such that the second stage has a single period.*

Proof. Let $(\bar{x}, \bar{a}, \bar{y})$ be an optimal solution for the two-stage problem. For each $(ij) \in E$ where $\bar{a}_{ij}^{k} = 1$, the corresponding time period t for when (ij) was last matched must be in W_{ij}. Moreover, by Proposition 2, it follows that the second-stage penalties for unmatching (ij) are non-decreasing. Thus, k is the best period to unmatch (ij) in the second stage. Together with Proposition 1, it follows that an optimal solution will never unmatch the same pair more than once in the second stage. Consequently, the second-stage unmatching variables $y_{ij}^{ts}, \forall t = (k+1), \ldots, h, (ij) \in E$ can be set to 0. As a result, the time periods for the second-stage matching variables can be represented implicitly, which gives us the single-period second-stage problem below

$$Q'(k, s, a^{k}) := \max \sum_{(ij)\in E} (\hat{c}_{ij}^{s} x_{ij}^{s} - d_{ij}^{ks} y_{ij}^{ks}) \qquad (29)$$

$$\text{s.t.} \sum_{(ij)\in\delta(v)} (x_{ij}^s - y_{ij}^{ks}) \leq 1 - \sum_{(ij)\in\delta(v)} a_{ij}^k \qquad \forall v \in V \qquad (30)$$

$$y_{ij}^{ks} \leq a_{ij}^k \qquad \forall(ij) \in E \qquad (31)$$

$$x_{ij}^s, y_{ij}^{ks} \in \{0,1\} \qquad \forall(ij) \in E, \qquad (32)$$

where $\hat{c}_e^s = \max\{c_{ij}^{ts} \mid t = k, \dots, h\}$. Together, Eqs. (20)–(24) and (29)–(32) form a reduced two-stage model. $\qquad\qquad\qquad\qquad\qquad\qquad\qquad\qquad\qquad\qquad\qquad\square$

5.2   Reductions Based on the First-Stage Solution Structure

In certain situations, it is important to efficiently solve the second-stage problem given a first-stage solution. For example, when the two-stage problem is solved independently within a mathematical decomposition method. We provide two reductions for the second-stage problem, based on the first-stage solution.

No Matches Before the Second Stage. If $a_{ij}^k = 0, \forall(ij) \in E$, then the second-stage problem is independent of the first stage, and is defined as follows

$$Q''(k,s,\mathbf{0}) := \max \sum_{t=k}^{h} \sum_{(ij)\in E} (c_{ij}^{ts} x_{ij}^{ts} - d_{ij}^{ts} y_{ij}^{ts}) \qquad (33)$$

$$\text{s.t.} \sum_{\ell=k}^{t} \sum_{(ij)\in\delta(v)} (x_{ij}^{\ell s} - y_{ij}^{\ell s}) \leq 1 \qquad \forall t = k, \dots, h, v \in V \qquad (34)$$

$$y_{ij}^{ts} \leq \sum_{\ell=k}^{t-1} (x_{ij}^{\ell s} - y_{ij}^{\ell s}) \qquad \forall t = k, \dots, h, (ij) \in E \qquad (35)$$

$$x_{ij}^{ts}, y_{ij}^{ts} \in \{0,1\} \qquad \forall t = k, \dots, h, (ij) \in E, \qquad (36)$$

which has the same structure as the static problem defined in Eqs. (7)–(10). It follows that if Assumption 1 holds, then the formulation above can be rewritten as a maximum-weight bipartite matching problem.

Matches Before the Second Stage. If Assumption 1 holds, then the second-stage unmatching of pairs in $\{(ij) \in E \mid a_{ij}^k = 0\}$ is never profitable, even if $\{(ij) \in E \mid a_{ij}^k = 1\} \neq \emptyset$. Thus, the variables $y_{ij}^{ts}, \forall t = k, \dots, h, (ij) \in E, a_{ij}^k = 0$ can be set to 0.

Although the second-stage problem can be reduced in such cases, these reductions do not apply to the full two-stage problem defined in Eqs. (20)–(28).

6 Conclusions and Future Work

We have introduced a matching and rematching problem with applications in request matching for ridesharing systems. We have presented three mathematical formulations that can be used in a rolling horizon framework. We discussed how to reduce these formulations provided that specific conditions that are typically found in practice are met. In some cases, these formulations can be reduced to a simple maximum-weight bipartite matching formulation. Some opportunities for future work are the evaluation of the proposed models on a rolling horizon framework and the development of efficient decomposition methods that exploit the proposed model reduction techniques.

References

1. Agatz, N.A.H., Erera, A.L., Savelsbergh, M.W.P., Wang, X.: Dynamic ride-sharing: a simulation study in metro Atlanta. Transp. Res. Part B Methodol. **45**(9), 1450–1464 (2011)
2. Wang, X., Agatz, N., Erera, A.: Stable matching for dynamic ride-sharing systems. Transp. Sci. **52**(4), 850–867 (2018)
3. Stiglic, M., Agatz, N., Savelsbergh, M., Gradisar, M.: The benefits of meeting points in ride-sharing systems. Transp. Res. Part B Methodol. **82**, 36–53 (2015)
4. Stiglic, M., Agatz, N., Savelsbergh, M., Gradisar, M.: Making dynamic ride-sharing work: the impact of driver and rider flexibility. Transp. Res. Part E Logist. Transp. Rev. **91**, 190–207 (2016)
5. Stiglic, M., Agatz, N., Savelsbergh, M., Gradisar, M.: Enhancing urban mobility: integrating ride-sharing and public transit. Comput. Oper. Res. **90**, 12–21 (2018)
6. Agatz, N., Erera, A., Savelsbergh, M., Wang, X.: Optimization for dynamic ride-sharing: a review. Eur. J. Oper. Res. **223**(2), 295–303 (2012)
7. Furuhata, M., Dessouky, M., Ordóñez, F., Brunet, M.E., Wang, X., Koenig, S.: Ridesharing: the state-of-the-art and future directions. Transp. Res. Part B Methodol. **57**, 28–46 (2013)
8. Mourad, A., Puchinger, J., Chu, C.: A survey of models and algorithms for optimizing shared mobility. Transp. Res. Part B Methodol. **123**, 323–346 (2019)

The Distance Matching Problem

Péter Madarasi[✉]

Department of Operations Research, Eötvös Loránd University, Budapest, Hungary
madarasi@cs.elte.hu

Abstract. This paper introduces the *d-distance matching problem*, in which we are given a bipartite graph $G = (S, T; E)$ with $S = \{s_1, \ldots, s_n\}$, a weight function on the edges and an integer $d \in \mathbb{Z}_+$. The goal is to find a maximum weight subset $M \subseteq E$ of the edges satisfying the following two conditions: i) the degree of every node of S is at most one in M, ii) if $s_i t, s_j t \in M$, then $|j - i| \geq d$. The question arises naturally, for example, in various scheduling problems.

We show that the problem is NP-complete in general and give an FPT algorithm parameterized by d. We also settle the case when the size of T is constant. From an approximability point of view, we consider a local search algorithm that achieves an approximation ratio of $3/2 + \epsilon$ for any constant $\epsilon > 0$ in the unweighted case. We show that the integrality gap of the natural integer programming model is at most $2 - \frac{1}{2d-1}$, and give an LP-based approximation algorithm for the weighted case with the same guarantee. We also present a combinatorial $(2 - \frac{1}{d})$-approximation algorithm. The novel approaches used in the analysis of the integrality gap and the approximation ratio of locally optimal solutions might be of independent combinatorial interest.

Keywords: Distance matching · Parameterized algorithms · Approximation algorithms · Integrality gap · Shift scheduling · Restricted matching

1 Introduction

In the *perfect d-distance matching problem*, given are a bipartite graph $G = (S, T; E)$ with $S = \{s_1, \ldots, s_n\}$, $T = \{t_1, \ldots, t_k\}$, a weight function on the edges $w : E \to \mathbb{R}_+$ and an integer $d \in \mathbb{Z}_+$. The goal is to find a maximum weight subset $M \subseteq E$ of the edges such that the degree of every node of S is one in M and if $s_i t, s_j t \in M$, then $|j - i| \geq d$. In the (non-perfect) *d-distance matching problem*, some of the nodes of S might remain uncovered. Note that the order of nodes in $S = \{s_1, \ldots, s_n\}$ affects the set of feasible d-distance matchings, but the order of $T = \{t_1, \ldots, t_k\}$ is indifferent.

An application of this problem for $w \equiv 1$ is as follows. Imagine n consecutive all-day events $s_1, \ldots, s_n$ each of which must be assigned one of k watchmen

Supported by the ÚNKP-19-3 New National Excellence Program of the Ministry for Innovation and Technology.

$t_1, \ldots, t_k$. For each event s_i, a set of possible watchmen is given – those who are qualified to be on guard at event s_i. Appoint exactly one watchman to each of the events such that no watchman is assigned to more than one of any d consecutive events, where $d \in \mathbb{Z}_+$ is given. In the weighted version of the problem, let $w_{s_i t_j}$ denote the level of safety of event s_i if watchman t_j is on watch, and the objective is to maximize the level of overall safety.

As another application of the above question, consider n items $s_1, \ldots, s_n$ one after another on a conveyor belt, and k machines $t_1, \ldots, t_k$. Each item s_i is to be processed on the conveyor belt by one of the qualified machines $N(s_i) \subseteq \{t_1, \ldots, t_n\}$ such that if a machine processes item s_i, then it can not process the next $d - 1$ items—because the conveyor belt is running.

Previous Work. Observe that in the special case $d = |S|$, one gets the classic (perfect) bipartite matching problem. For $d = 1$, the problem reduces to the b-matching problem, and one can show that it is a special case of the circulation problem for $d = 2$. The perfect d-distance matching problem is a special case of the list coloring problem on interval graphs [5] and the frequency assignment problem [6].

Our Results. This paper settles the complexity of the distance matching problem, and gives an FPT algorithm [4] parameterized by d. An efficient algorithm for constant T is also given. We present an LP-based $(2 - \frac{1}{2d-1})$-approximation algorithm for the weighted distance matching problem, which implies that the integrality gap of the natural IP model is at most $2 - \frac{1}{2d-1}$. A combinatorial $(2 - \frac{1}{d})$-approximation algorithm is also described for the weighted case. One of the main contributions of the paper is a $(3/2 + \epsilon)$-approximation algorithm for the unweighted case for any constant $\epsilon > 0$ for the unweighted case. The proof is based on revealing the structure of locally optimal solutions recursively. Motivated by the second application above, we give a polynomial time algorithm to find a permutation of S (i.e. the items on the conveyor belt) such that the weight of the optimal d-distance matching becomes as large as possible. The full version of the paper with detailed proofs and further results is available on arXiv [7].

Notation. Throughout the paper, assume that $G = (S, T; E)$ contains no loops or parallel edges, unless stated otherwise. Let $\Delta(v)$ and $N(v)$ denote the set of incident edges to node v and the neighbors of v, respectively. For a subset $X \subseteq E$ of the edges, $N_X(v)$ denotes the neighbors of v for edge set X. $\deg(v)$ is the degree of node v. Let $L_d(s_i) = \{s_{\max(i-d+1,1)}, \ldots, s_i\}$ and $R_d(s_i) = \{s_i, \ldots, s_{\min(i+d-1,|S|)}\}$. The maximum of the empty set is $-\infty$ by definition. Given a function $f : A \to B$, both $f(a)$ and f_a denote the value f assigns to $a \in A$, and let $f(X) = \sum_{a \in X} f(a)$ for $X \subseteq A$. Let χ_Z denote the characteristic vector of set Z, i.e. $\chi_Z(y) = 1$ if $y \in Z$, and 0 otherwise. Occasionally, the braces around sets consisting of a single element are abandoned, e.g. $\chi_e = \chi_{\{e\}}$ for $e \in E$.

2   Complexity and Tractable Cases

Theorem 1. *It is NP-complete to decide if a graph has a perfect d-distance matching, even if the maximum degree of the graph is at most 4.*

Sketch of the Proof. The proof of the theorem is more involved, hence we only present the main idea here. One of Karp's 21 NP-complete problems [1], the 3-dimensional matching can be reduced to the following problem. Given a bipartite graph $G = (S, T; E)$ and $S_1, S_2 \subseteq S$ s.t. $S_1 \cup S_2 = S$, decide whether there exists $M \subseteq E$ for which $|M| = |S|$ and both $M \cap E_1$ and $M \cap E_2$ are matchings, where E_i denotes the edges induced by T and S_i for $i = 1, 2$. The latter problem can be reduced to the d-distance matching problem in a non-trivial way, hence the claim follows. The 3-dimensional matching problem is NP-complete even if the maximum degree is 4 [2, p. 221]. As the reductions do not increase the maximum degree, the hardness of the bounded degree case follows.   □

2.1   FPT Algorithm Parameterized by d

In what follows, an FPT algorithm parameterized by d is presented for the weighted (perfect) d-distance matching problem. First observe that the weighted d-distance matching problem easily reduces to the perfect case by adding a new node t_s to T and a new edge st_s of weight zero for each $s \in S$, therefore the algorithm is given only for the weighted perfect d-distance matching problem. The next claim gives a way to reduce the problem so that it admits an efficient dynamic programming solution.

Lemma 1. *If* $\deg(s) \geq 2d$ *for* $s \in S$, *then one of the incident edges can be removed without changing the weight of the optimal perfect d-distance matching.*

Proof. Let st be a minimum weight edge incident to node s. In order to prove that st can be removed, it suffices to show that there is a maximum weight d-distance matching that does not use edge st. Given a d-distance matching M that contains edge st, let $Z \subseteq T$ denote the nodes that M assigns to $L_d(s) \cup R_d(s)$. Since $|Z| \leq 2d - 1$, there exists a node $t' \in N(s) \setminus Z$ for which $w_{st} \leq w_{st'}$. To complete the proof, observe that $M' = (M \cup \{st'\}) \setminus \{st\}$ is a d-distance matching of weight at least $w(M)$, which does not contain edge st.   □

Based on Lemma 1, any instance of the weighted d-distance matching problem can be reduced so that the degree of each node $s \in S$ is at most $2d - 1$. Note that the reduction can be performed in linear time if the edges are already sorted by their weights at each node $s \in S$. In what follows, a dynamic programming approach is presented to solve the reduced problem. For $i \geq d$, let $f(s_i, z_1, \ldots, z_d)$ denote the weight of the maximum weight d-distance matching if the problem is restricted to the first i nodes of S and s_{i-j+1} is assigned to its neighbor z_j for $j = 1, \ldots, d$. Formally, $f(s_i, z_1, \ldots, z_d) = -\infty$ if $z_1, \ldots, z_d$ are not distinct, otherwise, it is defined by the following recursive formula.

$$f(s_i, z_1, \ldots, z_d) = \begin{cases} w_{s_i z_1} + \displaystyle\max_{t \in \Delta(s_{i-d})} f(s_{i-1}, z_2, \ldots, z_d, t) & \text{if } i > d \\ \displaystyle\sum_{j=1}^{d} w_{s_j z_{d-j+1}} & \text{if } i = d \end{cases} \quad (1)$$

where $i \geq d$, $s_i \in S$ and $z_j \in N(s_{i-j+1})$ for $j = 1, \ldots, d$.

The weight of the optimal d-distance matching is $\max\{f(s_n, z_1, \ldots, z_d) : z_j \in N(s_{n-j+1})$ for $j = 1, \ldots, d\}$. Observe that the number of subproblems is $\mathcal{O}(n(2d-1)^d)$, since the degree of each $s \in S$ is at most $2d-1$ by Lemma 1. Recursion (1) gives a way to compute $f(s_i, z_1, \ldots, z_d)$ in $\mathcal{O}(d)$ steps. Therefore the overall running time of the algorithm is $\mathcal{O}(dn(2d-1)^d + poly(n + |T|))$.

2.2 Efficient Algorithm for Constant $|T|$

To obtain an algorithm running in $\mathcal{O}(n|T|d^{|T|})$ steps, consider the following subproblems. Let $f(s_i, d_1, \ldots, d_{|T|})$ denote the weight of the optimal perfect d-distance matching if the problem is restricted to $s_1, \ldots, s_i$, and t_j can not be matched to nodes $s_{i-d_j+1}, \ldots, s_i$. The algorithm is similar to the previous one, hence the details are left to the full version of the paper.

3 LP-Based Approach

This section proves that the integrality gap of the natural integer programming model for the weighted d-distance matching problem is at most $2 - \frac{1}{2d-1}$, and presents an LP-based $(2 - \frac{1}{2d-1})$-approximation algorithm. We also investigate the integrality of *LP1* and *LP2* in special cases. First consider the relaxation of the natural 0–1 integer programming formulation of the weighted d-distance matching problem.

$$\max \sum_{st \in E} w_{st} x_{st} \qquad (LP1)$$

s.t.

$$x \in \mathbb{R}_+^E \qquad (2a)$$

$$\sum_{st \in \Delta(s)} x_{st} \le 1 \qquad \forall s \in S \qquad (2b)$$

$$\sum_{s't \in E : s' \in R_d(s)} x_{s't} \le 1 \qquad \forall s \in S, t \in T \qquad (2c)$$

One gets the relaxation of the 0–1 integer programming formulation (*LP2*) of the weighted perfect d-distance matching problem by tightening (2b) to equality.

The following definition and lemma play a central role in the LP-based approximation algorithm presented at the end of this section.

Definition 1. *Given a feasible solution x to LP1, an order of the edges $e_1 = s^1 t^1, \ldots, e_m = s^m t^m$ is θ-flat with respect to x if*

$$\xi_i + \bar{\xi}_i \le \theta - x_{e_i} \qquad (3)$$

holds for each $i = 1, \ldots, m$, where $\xi_i = \sum\{x_{e_j} : j > i, e_j \in \Delta(s^i)\}$ and $\bar{\xi}_i = \sum\{x_{e_j} : j > i, e_j \in \Delta(t^i), s^j \in L_d(s^i) \cup R_d(s^i)\}$.

Algorithm 1. The ordering procedure for Lemma 2

Let x be a given fractional solution to *LP1*, and $G = (S, T; E)$ a copy of the graph.
j:=1
for $i = 1, \ldots, n$ **do**
    while $\deg(s_i) \neq 0$ **do**
        Choose an edge $s_i t \in \Delta(s_i)$ for which $x_{s_i t}$ is as large as possible.
        $e_j := s_i t$
        $j := j + 1$
        $E := E \setminus \{s_i t\}$
output $e_1, \ldots, e_m$

Lemma 2. *There exists an optimal solution $x \in \mathbb{Q}^m$ of LP1 and an order $e_1 = s^1 t^1, \ldots, e_m = s^m t^m$ of the edges that is $(2 - \frac{1}{2d-1})$-flat with respect to x.*

Proof. Let $E_s \subseteq \Delta(s)$ denote the first $\min(2d - 1, \deg(s))$ largest weight edges incident to node s for each $s \in S$. Let x be an optimal solution to *LP1* for which $\gamma(x) = \sum\{x_e : e \in E \setminus \bigcup_{s \in S} E_s\}$ is minimal. By contradiction, suppose that $\gamma(x) > 0$. By definition, $\gamma(x) > 0$ implies that there exists an edge $st \in E \setminus \bigcup_{k=1}^n E_k$ for which $x_{st} > 0$. There exists edge $st' \in E_s$ s.t. $x' = x - \epsilon \chi_{st} + \epsilon \chi_{st'}$ is feasible for sufficiently small $\epsilon > 0$, otherwise $x(\bigcup\{\Delta(s') : s' \in L_d(s) \cup R_d(s)\}) \geq 2d - 1 + \epsilon$ would hold, which is not possible. Observe that $wx \leq wx'$ and $\gamma(x') < \gamma(x)$, contradicting the minimality of $\gamma(x)$. Therefore $\gamma(x) = 0$ follows, meaning that $x_e = 0$ holds for each $e \in E \setminus \bigcup_{s \in S} E_s$. Hence one can restrict the edge set of the graph to $\bigcup_{s \in S} E_s$ without change in the optimal objective value, which implies that there exists a rational optimal solution $x \in \mathbb{Q}^m$ of *LP1* with $\gamma(x) = 0$.

Let x be as above, and let $e_1 = s^1 t^1, \ldots, e_m = s^m t^m$ be the order of the edges given by Algorithm 1 for input x. To prove that this order is $(2 - \frac{1}{2d-1})$-flat with respect to x, let ξ_i and $\bar{\xi}_i$ $(i = 1, \ldots, n)$ be as in Definition 1. First observe that $\bar{\xi}_i \leq 1 - x_i$ holds for each $i = 1, \ldots, n$, because the algorithm places each edge $\bigcup_{j=1}^{i-1} \Delta(s^j)$ before e_i. Hence, to obtain (3), it suffices to prove that $\xi_i \leq 1 - \frac{1}{2d-1}$. For any node $s \in S$, if there exists an edge $st \in \Delta(s)$ for which $x_{st} \geq \frac{1}{2d-1}$, then $\xi_j \leq 1 - \frac{1}{2d-1}$ follows for each $e_j \in \Delta(s)$, since $x_e \geq \frac{1}{2d-1}$ holds for the first edge $e \in \Delta(s)$ selected by Algorithm 1. Otherwise, if there exists no edge $st \in \Delta(s)$ for which $x_{st} \geq \frac{1}{2d-1}$, then $x(\Delta(s)) < |E_s| \frac{1}{2d-1} \leq 1$, but then $x' = x + \epsilon \chi_{st'}$ is feasible for some $st' \in E_s$ and sufficiently small $\epsilon > 0$ (because $x(\bigcup\{\Delta(s') : s' \in L_d(s) \cup R_d(s)\}) < 2d - 1$)—contradicting the optimality of x. Therefore $\xi_i \leq 1 - \frac{1}{2d-1}$ follows for $i = 1, \ldots, n$, which means that the order of the edges is $(2 - \frac{1}{2d-1})$-flat. $\square$

The rest of this section presents an LP-based $(2 - \frac{1}{2d-1})$-approximation algorithm and proves that the integrality gap is at most $\theta := 2 - \frac{1}{2d-1}$.

Theorem 2. *Algorithm 2 is a θ-approximation algorithm for the weighted d-distance matching problem if a θ-flat order of the edges is given in the first step of the algorithm.*

Algorithm 2. θ-approximation algorithm for the weighted distance matching problem

1: Let $e_1, \ldots, e_m$ be a θ-flat order with respect to a solution x of *LP1* (see Lemma 2).
2: **procedure** WDMLPAPX(E,w)
3:      $E := E \setminus \{e \in E : w_e \leq 0\}$
4:      **if** $E = \emptyset$ **then**
5:          **return** $\emptyset$
6:      Let st be the first edge according to the above order that appears in E.
7:      $M' := $ WDMLPAPX$(E \setminus \{st\}, w')$, where $w' := w - w_{st}\chi_{\Delta(s) \cup \{s't \in \Delta(t):s' \in R_d(s)\}}$
8:      **if** $M' \cup \{st\}$ is a feasible d-distance matching **then**
9:          **return** $M' \cup \{st\}$
10:     **else**
11:         **return** M'

Proof. The proof is by induction on the number of edges. Let M denote the distance matching found by WDMLPAPX(E,w), and let x be as defined in Algorithm 2. In the base case, if $E = \emptyset$, then $\theta w(M) \geq wx$ holds. Let $st \in E$ be the first edge with respect to the order of the edges used by Algorithm 2. By induction, $\theta w'(M') \geq w'x$ holds for $M' = $ WDMLPAPX$(E \setminus \{st\}, w')$, where $w' = w - w_{st}\chi_{\Delta(s) \cup \{s't \in \Delta(t):s' \in R_d(s)\}}$. The key observation is that

$$\theta(w - w')(M) \geq \theta w_{st} \geq (w - w')x \tag{4}$$

follows by the definition of w' and the order of the edges. Hence, one gets that

$$\theta w(M) = \theta(w - w')(M) + \theta w'(M) \geq (w - w')x + w'x = wx, \tag{5}$$

where $w'(M) = w'(M')$ because $w'_{st} = 0$. Therefore M is indeed a θ-approximate solution, which completes the proof.   □

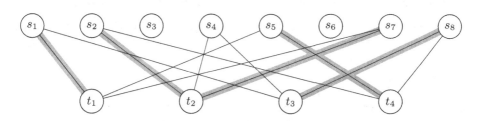

Fig. 1. For $w \equiv 1$ and $d = 5$, $x \equiv 1/2$ is an optimal solution to *LP1*, and the highlighted edges form an optimal 5-distance matching, hence the integrality gap is $6/5$.

Theorem 2 also implies that the integrality gap of *LP1* is at most θ. Note that if we have a θ'-flat order of the edges in the first step of Algorithm 2, then it outputs a θ'-approximate solution. We believe that there always exist a θ'-flat order of the edges for some $\theta' < \theta$ (i.e. it is possible to improve Lemma 2), which would automatically improve both the integrality gap and the approximation guarantee of the algorithm to θ'.

The largest known lower bound of the integrality gap of *LP1* is $6/5$ (see Fig. 1), thus it remains open whether the analysis is tight. On the other hand, one might easily show that the integrality gap of *LP2* is unbounded, as it was expected due to the complexity of the problem.

Note that both *LP1* and *LP2* are integral if $d = 2$, because the matrix of the linear programs is a network matrix [3]. If $d = |T|$, then *LP2* is integral (but the matrix is not totally unimodular – in fact, one can show that the matrix is not totally unimodular in general for $d \geq 3$).

4   A Combinatorial $(2 - \frac{1}{d})$-Approximation Algorithm

This section presents a $(2 - \frac{1}{d})$-approximation algorithm for the weighted distance matching problem. Let $k \in \{d-1, \ldots, 3d-3\}$ be such that $2d-1$ divides $|S|+k$, and add k new dummy nodes $s_{n+1}, \ldots, s_{n+k}$ to the end of S in this order. Let us consider the extended node set in cyclic order. Observe that the new cyclic problem is equivalent to the original one. Let H_j denote the graph induced by $R_d(s_j) \cup T$, where $R_d(s_j)$ is the set consisting of node s_j and the next $d - 1$ nodes on its right in the new cyclic problem. Let

$$G_i = (S_i, T; E_i) = \bigcup_{j=0}^{\frac{n+k}{2d-1}-1} H_{i+j(2d-1)}$$

for $i = 1, \ldots, 2d - 1$, where $S_i \subseteq S$. For each $i = 1, \ldots, 2d - 1$, compute a maximum weigh matching M_i of G_i and let $i^* = \arg\max\{w(M_i) : i = 1, \ldots, 2d-1\}$. For example, consider the graph on Fig. 2 with $d = 3$. The nodes of G_4 are highlighted on the figure and the edges of M_4 are the wavy ones (note that $s_6, \ldots, s_{10}$ are the five dummy nodes).

Theorem 3. M_{i^*} *is a feasible d-distance matching and it is* $(2 - \frac{1}{d})$-*approximate.*

Proof. Each node of S is covered by at most one edge of M_{i^*}, as M_{i^*} is the union of matchings no two of which cover the same node of S. If $s_i t, s_j t \in M_{i^*}$, then $s_i t$ and $s_j t$ belong to two distinct $\tilde{M}_k, \tilde{M}_l$ for some k, l, hence $|i - j| \geq d$ and the feasibility of M_{i^*} follows.

To show the approximation guarantee, let M^* be an optimal d-distance matching. For each node $s \in S$, let $\mu_s \in \mathbb{R}_+$ denote the weigh of the edge covering s in M^*, and zero if M^* does not cover s. Note that $\sum_{s \in S} \mu_s = w(M^*)$ by definition, and

$$\sum_{s \in S_i} \mu_s \leq w(M_i) \tag{6}$$

follows because $\sum_{s \in S_i} \mu_s$ is the weight of a matching in G_i. Observe that

$$dw(M^*) = d \sum_{s \in S} \mu_s = \sum_{i=1}^{2d-1} \sum_{s \in S_i} \mu_s \leq \sum_{i=1}^{2d-1} w(M_i) \leq (2d - 1)w(M_{i^*}), \tag{7}$$

where the first equation holds because μ_s occurs exactly d times as a summand in $\sum_{i=1}^{2d-1} \sum_{s \in S_i} \mu_s$ for all $s \in S$, the first inequality follows from (6) and the last one holds because M_{i^*} is a largest weight matching among $M_1, \ldots, M_{2d-1}$. One gets by (7) that $w(M^*) \leq (2 - \frac{1}{d}) w(M_{i^*})$, which completes the proof. $\quad\square$

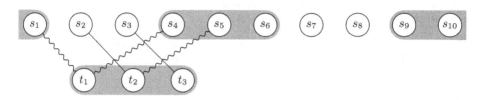

Fig. 2. Tight example for Theorem 3 in the case $d = 3$. The wavy edges form a possible output of the algorithm. (Recall that the nodes of S are in cyclic order.)

The analysis is tight in the sense that, for every $d \in \mathbb{Z}_+$, there exists a graph G for which the algorithm returns a d-distance matching M for which $w(M^*) = (2 - \frac{1}{d}) w(M)$, where M^* is an optimal d-distance matching. Let S and T consist of $2d-1$ and d nodes, respectively. Add edge $s_i t_i$ for $i = 1, \ldots, d$, and edge $s_{i+d} t_i$ for $i = 1, \ldots, d-1$. Note that the edge set is a feasible d-distance matching itself, and the above algorithm returns a matching that covers exactly d nodes of S. Hence the approximation ratio of the found solution is $\frac{2d-1}{d}$. Figure 2 shows the construction for $d = 3$, where $s_6, \ldots, s_{10}$ are the dummy nodes.

5 $(3/2 + \epsilon)$-Approximation Algorithm for the Unweighted d-distance Matching

This section investigates the approximation ratio of the so-called locally optimal solutions, and presents a $3/2 + \epsilon$ approximation algorithm for constant $\epsilon > 0$. The reader is referred to the full version of this paper for a detailed description of locally optimal solutions. First of all, consider the following notion.

Definition 2. *Given an edge $e^* \in E$, let $\mathcal{H}(e^*, M) \subseteq M$ denote the inclusion-wise minimal subset of M for which $M \backslash \mathcal{H}(e^*, M) \cup \{e^*\}$ is a d-distance matching. An edge e^* hits $e \in M$ if $e \in \mathcal{H}(e^*, M)$. Given an edge set $X \subseteq E$, let $\mathcal{H}(X, M) = \bigcup_{e^* \in X} \mathcal{H}(e^*, M)$.*

Definition 3. *A d-distance matching M is l-locally optimal if there exists no $X \subseteq E \setminus M$ s.t. $l \geq |X| > |\mathcal{H}(X, M)|$. Similarly, M is l-locally optimal with respect to M^* if M is l-locally optimal in $G' = (S, T; M \cup M^*)$, where $M^* \subseteq E$.*

In what follows, an upper bound ϱ_l is shown on the approximation ratio of l-locally optimal solutions for each $l \geq 1$, where ϱ_l is defined by the following recursion.

$$\varrho_l = \begin{cases} 3, & \text{if } l = 1 \\ 2, & \text{if } l = 2 \\ \frac{4\varrho_{l-2}-3}{2\varrho_{l-2}-1}, & \text{if } l \geq 3. \end{cases} \tag{8}$$

For $l = 1, 2, 3, 4$, the statement can be proved by a simple argument, given below. However, this approach does not seem to work in the general case. The proof of the general case, which is much more involved and quite esoteric, is deferred to the full version of the paper.

Theorem 4. *If M, M^* are d-distance matchings s.t. M is l-locally optimal with respect to M^*, then the approximation ratio $|M^*|/|M|$ is at most ϱ_l, where $l = 1, \ldots, 4$ and ϱ_l is as defined above.*

Proof. Let $M_i^* = \{e^* \in M^* : |\mathcal{H}(e^*, M)| = i\}$ for $i = 0, \ldots, 3$. Note that M_0, M_1, M_2, M_3 is a partition of M^*, and $M_0^* = \emptyset$ since each edge of M^* hits at least one edges of M if $l \geq 1$. Since each edge $e \in M$ can be hit by at most three edges of M^*, one gets that

$$3|M| \geq \sum_{e^* \in M^*} |\mathcal{H}_+(e^*, M)| = |M_1^*| + 2|M_2^*| + 3|M_3^*|. \tag{9}$$

Case $l = 1$. It easily follows from (9) that

$$|M^*| = |M_1^*| + |M_2^*| + |M_3^*| \leq |M_1^*| + 2|M_2^*| + 3|M_3^*| \leq 3|M|. \tag{10}$$

Case $l = 2$.

$$2|M^*| = 2(|M_1^*| + |M_2^*| + |M_3^*|) \leq |M_1^*| + |M_1^*| + 2|M_2^*| + 3|M_3^*|$$
$$\leq |M_1^*| + 3|M| \leq 4|M|, \tag{11}$$

where the second inequality follows from (9) and the third one holds because M is 2-locally optimal with respect to M^*.
Case $l = 3$.

$$5|M^*| = 5(|M_1^*| + |M_2^*| + |M_3^*|) = 2(|M_1^*| + 2|M_2^*| + 3|M_3^*|) + 3|M_1^*| + |M_2^*| - |M_3^*|$$
$$\leq 6|M| + 3|M_1^*| + |M_2^*| - |M_3^*| \leq 6|M| + 3|M_1^*| + |M_2^*| \leq 9|M|, \tag{12}$$

where the last inequality holds by the following claim.

Claim. $|M_2^*| \leq 3(|M| - |M_1|)$ if M is 3-locally optimal with respect to M^*.

Proof. It suffices to show that there exist d-distance matchings $\tilde{M}, \tilde{M}^*$ s.t. 1) $|\tilde{M}| = |M| - |M_1^*|$, 2) $|\tilde{M}^*| = |M_2^*|$, and 3) $\tilde{M}$ is 1-locally optimal with respect to $\tilde{M}^*$. Indeed, $|\tilde{M}^*| \leq 3|\tilde{M}|$ holds, from which one obtains the inequality to be proved by substituting 1) and 2). Let $\tilde{M} = M \setminus \mathcal{H}(M_1^*, M)$ and $\tilde{M}^* = M_2^*$. Clearly, both 1) and 2) hold. By contradiction, suppose that 3) does not hold, that is, there exists $e_1^* \in \tilde{M}^*$ s.t. $\tilde{M} \cup \{e_1^*\}$ is feasible d-distance matching. By definition, $e_1^* \in M_2^*$, therefore e_1^* hits exactly two edges e_1, e_2 in M. Neither e_1, nor e_2 are in $\tilde{M}$, thus $e_1, e_2 \in \mathcal{H}(M_1^*, M)$, that is e_j is hit by an edge $e_{j+1}^* \in M_1^*$ for j=1,2. Note that $e_1^* \neq e_2^*$, hence e_1^*, e_2^*, e_3^* are pairwise distinct edges, and $\mathcal{H}(\{e_1^*, e_2^*, e_3^*\}, M) = \{e_1, e_2\}$, contradicting that M is 3-locally optimal. □

Case $l = 4$.

$$6|M^*| = 6(|M_1^*| + |M_2^*| + |M_3^*|) = 2(|M_1^*| + 2|M_2^*| + 3|M_3^*|)$$
$$+4|M_1^*| + 2|M_2^*| \leq 6|M| + 4|M_1^*| + 2|M_2^*| \leq 10|M|, \quad (13)$$

where the first inequality holds by (9), the last one by the following claim.

Claim. $2|M_2^*| \leq 4(|M| - |M_1|)$ if M is 4-locally optimal with respect to M^*.

Proof. The proof is similar to that of the previous claim.   □

This concludes the proof of the theorem.   □

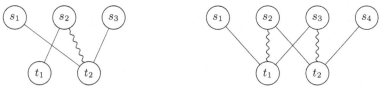

(a) The wavy edges form a 1-locally op-   (b) The wavy edges form a 2-locally optimal 2-distance matching M, and $M^* =$ timal 2-distance matching M, and $M^* =$
$E \setminus M$ is the optimal 2-distance matching, $E \setminus M$ is the optimal 2-distance matching,
hence $|M^*|/|M| = 3 = \varrho_1$.   hence $|M^*|/|M| = 2 = \varrho_2$.

Fig. 3. Tight examples for Theorem 5.

Figure 3a and 3b show that the bound given by Theorem 5 is tight for $l = 1$ and 2, respectively. A tight example for $l = 3$ is presented in the full version of the paper. It remains open whether the analysis is tight for $l \geq 4$.

It is worth noting that the proof of Theorem 4 for $l = 3, 4$ refers inductively to the case $l - 2$, which is quite unexpected. The same idea does not seem to work for $l = 5$. Based on cases $l = 1, 2, 3, 4$, one gains the following analogous computation.

$$13|M^*| = 13(|M_1^*| + |M_2^*| + |M_3^*|) = 4(|M_1^*| + 2|M_2^*| + 3|M_3^*|)$$
$$+9|M_1^*| + 5|M_2^*| + |M_3^*| \leq 12|M| + 9|M_1^*| + 5|M_2^*| + |M_3^*| \leq 21|M|, (14)$$

where the last inequality requires that $5|M_2^*| + |M_3^*| \leq 9(|M| - |M_1^*|)$. However, the latter inequality does not admit a constructive argument similarly to the cases $l = 3, 4$. To overcome this complication, consider the following extended problem setting, which surprisingly does admit a constructive argument.

Definition 4. *Let R be a set of (parallel) loops on the nodes of S. A subset $M \subseteq E \cup R$ is **(R,d)-distance matching** if it is the union of a d-distance matching and R.*

Definition 5. *Given an (R, d)-distance matching M and $sv \in (S \times T) \cup R$, let*

$$\mathcal{H}_+(sv, M) = \begin{cases} \mathcal{H}(sv, M \setminus R) \cup \{e \in R : e \text{ is incident to } s\}, & \text{if } sv \in S \times T \\ sv, & \text{if } sv \in R. \end{cases}$$

In other words, each $st \in E$ hits the edges of $\mathcal{H}(st, M)$ and all the loops incident to node s, while each loop hits only itself. Given an edge set $X \subseteq E$, let $\mathcal{H}_+(X, M) = \bigcup_{e \in X} \mathcal{H}(e, M)$.

Definition 6. *An (R, d)-distance matching M is l-**locally optimal** if there exists no $X \subseteq E \setminus M$ s.t. $l \geq |X| > |\mathcal{H}_+(X, M)|$. Similarly, M is l-**locally optimal with respect to** M^* if there exists no $X \subseteq M^* \setminus M$ s.t. $l \geq |X| > |\mathcal{H}_+(X, M)|$, where M^* in an (R, d)-distance matching.*

Note that each definition reduces to its original counterpart if $R = \emptyset$. Therefore, it suffices to show that ϱ_l is an upper bound on the approximation ratio of (R, l)-locally optimal solutions.

Theorem 5. *If M, M^* are (R, d)-distance matchings s.t. M is l-locally optimal with respect to M^*, then the approximation ratio $|M^*|/|M|$ is at most ϱ_l, where $l \geq 1$ and ϱ_l is as defined above.* □

Corollary 1. *For any constant $\epsilon > 0$, there exist a polynomial algorithm for the unweighted d-distance matching problem that achieves an approximation guarantee of $3/2 + \epsilon$.*

Proof. By Theorem 5, the approximation ratio of l-locally optimal d-distance matchings is at most ϱ_l, where ϱ_l is as defined above. It is easy to show that $\lim_{l \to \infty} \varrho_l = 3/2$. Hence for any $\epsilon > 0$, there exists $l_0 \in \mathbb{Z}_+$ s.t. $\varrho_l \leq 3/2 + \epsilon$. To complete the proof, observe that l_0 is independent from the problem size, therefore one can compute an l_0-locally optimal solution in polynomial time. □

6   Optimal Permutation of S

This section investigates a slightly different problem, motivated by the second application presented in Sect. 1. It is natural to ask whether we can find a permutation of S (i.e. the items on the conveyor belt, see Sect. 1) that maximizes the weight of the maximum weight d-distance matching. The proof of the next theorem provides a polynomial time algorithm to solve this problem. We say that a triple $y \in \mathbb{N}^{S \cup T}, z \in \mathbb{N}^E, v \in \mathbb{N}^T$ is a u-**cover** of $G = (S, T; E)$ for $u \in \mathbb{N}$, if $y_s + y_t + z_{st} \geq w_{st}$ for all $st \in E$ and $v_t + y_t \geq u$ for all $t \in T$.

Theorem 6. *The maximum weight of a d-distance matching under all permutations is equal to the minimum of $\{yb + \mathbb{1}z + \mathbb{1}v - d\lfloor \frac{n}{d} \rfloor u : \text{where } y \in \mathbb{N}^{S \cup T}, z \in \mathbb{N}^E, v \in \mathbb{N}^T \text{ is a } u\text{-cover of } G\}$, where $b \in \mathbb{N}^{S \cup T}$ is such that $b_s = 1$ for $s \in S$ and $b_t = \lceil n/d \rceil$ for $t \in T$.*

Proof. Let $n = |S|$ and let $M \subseteq E$ be a maximum weight edge set s.t. $\deg_M(s) \leq 1$ for all $s \in S$, $\deg_M(t) \leq \lceil n/d \rceil$ and the number of nodes $t \in T$ for which $\deg_M(t) = \lceil n/d \rceil$ is at most $n - \lfloor n/d \rfloor d$. Such an edge set M can be found in polynomial time by a reduction to the maximum cost circulation problem.

It is easy to see that $w(M) \geq W$. To show that $w(M) = W$, it suffices to construct a permutation of S under which M is a d-distance matching. Let $S_1, \ldots, S_{k+1}$ be a partition of S s.t. $k = \lfloor n/d \rfloor$, $|S_i| = d$ for $i = 1, \ldots, k$ and M induces a (not necessarily perfect) matching between T and S_{k+1} covering each node $t \in T$ that has degree $\lfloor n/d \rfloor + 1$. Note that $|S_{k+1}| = n - \lfloor n/d \rfloor d < d$.

Let α denote the number of edge pairs $st, s't \in M$ s.t. $s, s' \in S_i$ for some $i = 1, \ldots, k + 1$. If $\alpha = 0$, then M is a d-distance matching with respect to the order given by the concatenation of $S_1, \ldots, S_{k+1}$ if the nodes of each S_i are in appropriate order. Otherwise, let i be an index for which there exists $st, s't \in M$ s.t. $s, s' \in S_i$ and $|N_M(S_i)|$ is as small as possible ($\alpha > 0$ implies that at least one such index exists). There exists index $j \in 1, \ldots, k$ s.t. $N_M(t) \cap S_j = \emptyset$, and one can easily show that there is a node $s'' \in S_j$ for which $N_M(s'') \not\subseteq N_M(S_i)$ or $N_M(s'') = \emptyset$. By setting $S_i = S_i + s'' - s$ and $S_j = S_j + s - s''$, α decreases by one, hence the algorithm terminates in polynomial time.

One can easily derive the min-max formula using LP-duality or the max-flow min-cut theorem. □

Remark. A similar approach solves the analogue problem for the perfect d-distance matching problem. In this case, one should look for an edge set M for which $\deg_M(s) = 1$ (instead of $\deg_M(s) \leq 1$) and repeat the proof of Theorem 6.

Acknowledgement. The author is grateful to Kristóf Bérczi and Alpár Jüttner for their valuable comments that greatly improved the manuscript. The author is indebted to an anonymous reviewer of an earlier version of the manuscript for suggesting an approach which led to the algorithm described in Sect. 4.

References

1. Karp, R.M.: Reducibility among combinatorial problems. In: Miller, R.E., Thatcher, J.W., Bohlinger, J.D. (eds.) Complexity of Computer Computations, pp. 85–103. Springer, Boston (1972). https://doi.org/10.1007/978-1-4684-2001-2_9
2. Garey, M.R., Johnson, D.S.: Computers and Intractability: A Guide to the Theory of NP-Completeness. Freeman, San Francisco (1979)
3. Frank, A.: Connections in Combinatorial Optimization. Oxford University Press, Oxford (2011)
4. Downey, R.G., Fellows, M.R.: Parameterized Complexity. Springer, Heidelberg (2012)
5. Zeitlhofer, T., Wess, B.: List-coloring of interval graphs with application to register assignment for heterogeneous register-set architectures. Sig. Process. **83**(7), 1411–1425 (2003)
6. Aardal, K.I., van Hoesel, S.P.M., Koster, A.M.C.A., et al.: Models and solution techniques for frequency assignment problems. Ann. Oper. Res. **153**, 79–129 (2007)
7. Madarasi, P.: The distance matching problem (2019). https://arxiv.org/abs/1911.12432

Notes on Equitable Partitions
into Matching Forests in Mixed Graphs
and b-branchings in Digraphs

Kenjiro Takazawa[✉]

Department of Industrial and Systems Engineering,
Faculty of Science and Engineering, Hosei University,
3-7-2, Kajino-cho, Koganei-shi, Tokyo 184-8584, Japan
takazawa@hosei.ac.jp

Abstract. An equitable partition into branchings in a digraph is a partition of the arc set into branchings such that the sizes of any two branchings differ at most by one. For a digraph whose arc set can be partitioned into k branchings, there always exists an equitable partition into k branchings. In this paper, we present two extensions of equitable partitions into branchings in digraphs: those into matching forests in mixed graphs; and into b-branchings in digraphs. For matching forests, Király and Yokoi (2018) considered a tricriteria equitability based on the sizes of the matching forest, and the matching and branching therein. In contrast to this, we introduce a single-criterion equitability based on the number of the covered vertices. For b-branchings, we define an equitability based on the size of the b-branching and the indegrees of all vertices. For both matching forests and b-branchings, we prove that equitable partitions always exist.

1 Introduction

Partitioning a finite set into its subsets with certain combinatorial structure is a fundamental topic in the fields of combinatorial optimization, discrete mathematics, and graph theory. The most typical partitioning problem is graph coloring, which amounts to partitioning the vertex set of a graph into stable sets. In particular, *equitable coloring*, in which the numbers of vertices in any two stable sets differ at most by one, has attracted researchers' interest since the famous conjecture of Erdős [13] on the existence of an equitable coloring with $\Delta + 1$ colors in a graph with maximum degree Δ, which was later proved by Hajnal and Szemerédi [20].

Equitable edge-coloring has been mainly considered in bipartite graphs: a bipartite graph with maximum degree Δ admits an equitable edge-coloring with k colors for every $k \geq \Delta$ [7,8,10,14,24]. Equitable edge-coloring in bipartite graphs can be generalized to equitable partition of the common ground set of two matroids into common independent sets, which has been a challenging topic in the literature [6,15]. One successful example is equitable partition into

© Springer Nature Switzerland AG 2020
M. Baïou et al. (Eds.): ISCO 2020, LNCS 12176, pp. 214–224, 2020.
https://doi.org/10.1007/978-3-030-53262-8_18

branchings in digraphs. The following theorem is derived from Edmonds' disjoint branchings theorem [11]. For a real number x, let $\lfloor x \rfloor$ and $\lceil x \rceil$ denote the maximum integer that is not greater than x and the minimum integer that is not less than x, respectively.

Theorem 1 (see Schrijver [27, Theorem 53.3]**).** *In a digraph $D = (V, A)$, if A can be partitioned into k branchings, then A can be partitioned into k branchings each of which has size $\lfloor |A|/k \rfloor$ or $\lceil |A|/k \rceil$.*

The aim of this paper is to extend of Theorem 1 to equitable partition into two generalizations of branchings: *matching forests* [17–19] and *b-branchings* [21]. An important feature is that, due to their structures, defining the equitability of matching forests and b-branchings is not a trivial task, as explained below.

1.1 Matching Forests

The concept of matching forests was introduced by Giles [17–19]. A mixed graph $G = (V, E, A)$ consists of the set V of vertices, the set E of undirected edges, and the set A of directed edges (arcs). We say that an undirected edge in E *covers* a vertex $v \in V$ if v is one of the endpoints of the undirected edge, and a directed edge in A *covers* v if v is the head of the directed edge. A subset of edges $F \subseteq E \cup A$ is a *matching forest* if the underlying edge set of F is a forest and each vertex is covered by at most one edge in F. It is straightforward to see that matching forests offer a common generalization of matchings in undirected graphs and branchings in digraphs: if $F \subseteq E \cup A$ is a matching forest, then $F \cap E$ is a matching and $F \cap A$ is a branching.

An equivalent definition of matching forests can be given in the following way. For a subset of undirected edges $M \subseteq E$, let $\partial M \subseteq V$ denote the set of vertices covered by at least one edge in M. For a subset of directed edges $B \subseteq A$, let $\partial B \subseteq V$ denote the set of vertices covered by at least one arc in B^1. For $F \subseteq E \cup A$, define $\partial F = \partial(F \cap E) \cup \partial(F \cap A)$. A vertex v in ∂F is said to be covered by F. Now $F \subseteq E \cup A$ is a matching forest if $M := F \cap E$ is a matching, $B := F \cap A$ is a branching, and $\partial M \cap \partial B = \emptyset$.

Previous work on matching forests includes polynomial algorithms with polyhedral description [17–19], total dual integrality of the description [26], a Vizing-type theorem on the number of matching forests partitioning the edge set $E \cup A$ [22], and reduction to linear matroid parity [27]. More recently, Takazawa [28] showed that the sets of vertices covered by the matching forests form a *delta-matroid* [3,4,9], which provides some explanation of the tractability of matching forests as well as the above polyhedral results. For two sets X and Y, let $X \triangle Y$ denote their symmetric difference, i.e., $X \triangle Y = (X \setminus Y) \cup (Y \setminus X)$. For a finite set V and its subset family $\mathcal{F} \subseteq 2^V$, the set system $(V, \mathcal{F})$ is a *delta-matroid* if it satisfies the following exchange property:

---

[1] We believe that this notation causes no confusion on the direction of the arcs, since we never refer to the set of the tails of the arcs in this paper.

For each $U_1, U_2 \in \mathcal{F}$ and $u \in U_1 \triangle U_2$, there exists $u' \in U_1 \triangle U_2$ such that $U_1 \triangle \{u, u'\} \in \mathcal{F}$.

Theorem 2 ([28]). *For a mixed graph $G = (V, E, A)$, define $\mathcal{F}_G \subseteq 2^V$ by*

$$\mathcal{F}_G = \{\partial F \mid F \subseteq E \cup A \text{ is a matching forest in } G\}.$$

Then, the set system $(V, \mathcal{F}_G)$ is a delta-matroid.

The most recent work on matching forests is due to Király and Yokoi [23], which discusses equitable partition into matching forests. They considered equitability based on the sizes of F, $F \cap E$, and $F \cap A$, and proved the following two theorems. Let $\mathbb{Z}_{++}$ denote the set of the positive integers. For $k \in \mathbb{Z}_{++}$, let $[k]$ denote the set of positive integers not greater than k, i.e., $[k] = \{1, \ldots, k\}$.

Theorem 3 (Király and Yokoi [23]). *Let $G = (V, E, A)$ be a mixed graph and $k \in \mathbb{Z}_{++}$. If $E \cup A$ can be partitioned into k matching forests, then $E \cup A$ can be partitioned into k matching forests $F_1, \ldots, F_k$ such that*

$$||F_i| - |F_j|| \le 1, \quad ||M_i| - |M_j|| \le 2, \quad and \quad ||B_i| - |B_j|| \le 2 \qquad (1)$$

for every $i, j \in [k]$, where $M_i = F_i \cap E$ and $B_i = F_i \cap A$ for each $i \in [k]$.

Theorem 4 (Király and Yokoi [23]). *Let $G = (V, E, A)$ be a mixed graph and $k \in \mathbb{Z}_{++}$. If $E \cup A$ can be partitioned into k matching forests, then $E \cup A$ can be partitioned into k matching forests $F_1, \ldots, F_k$ such that*

$$||F_i| - |F_j|| \le 2, \quad ||M_i| - |M_j|| \le 1, \quad and \quad ||B_i| - |B_j|| \le 2 \qquad (2)$$

for every $i, j \in [k]$, where $M_i = F_i \cap E$ and $B_i = F_i \cap A$ for each $i \in [k]$.

Moreover, Király and Yokoi [23] showed that both of Theorems 3 and 4 are best possible with this tricriteria equitability, by presenting examples in which (1) and (2) cannot be improved. That is, while attaining the best possible results, Theorems 3 and 4 mean that these three criteria cannot be optimized at the same time, which demonstrates a sort of obscurity of matching forests.

Now the first contribution of this paper is a new theorem on equitable partitions into matching forests. For this theorem, we introduce a new equitability, which builds upon a single criterion defined by the size of the set ∂F of the covered vertices. Namely, our equitable-partition theorem is described as follows.

Theorem 5. *Let $G = (V, E, A)$ be a mixed graph and $k \in \mathbb{Z}_{++}$. If $E \cup A$ can be partitioned into k matching forests, then $E \cup A$ can be partitioned into k matching forests $F_1, \ldots, F_k$ such that*

$$||\partial F_i| - |\partial F_j|| \le 2 \qquad (3)$$

for every $i, j \in [k]$.

We remark that this criterion of equitability is plausible in the light of the delta-matroid structure of matching forests (Theorem 2). Theorem 5 contrasts with Theorems 3 and 4 in that the value two in the right-hand side of (3) is tight: consider the case where $G = (V, E, A)$ consists of an odd number of undirected edges forming a path and no directed edges, and $k = 2$.

1.2 b-branchings

We next address equitable partition of a digraph into b-*branchings*, introduced by Kakimura, Kamiyama, and Takazawa [21]. Let $D = (V, A)$ be a digraph and let $b \in \mathbb{Z}_{++}^V$. For $X \subseteq V$, we denote $b(X) = \sum_{v \in X} b(v)$. For $F \subseteq A$ and $X \subseteq V$, let $F[X]$ denote the set of arcs in F induced by X. For $F \subseteq A$ and $v \in V$, let $d_F^-(v)$ denote the indegree of v in the subgraph (V, F), i.e., the number of arcs in F whose head is v. Now an arc set $B \subseteq A$ is a b-*branching* if

$$d_B^-(v) \leq b(v) \qquad \text{for each } v \in V, \text{ and} \tag{4}$$

$$|B[X]| \leq b(X) - 1 \qquad \text{for each nonempty subset } X \subseteq V. \tag{5}$$

Note that the branchings is a special case of b-branchings where $b(v) = 1$ for every $v \in V$. That is, b-branchings provide a generalization of branchings in which the indegree bound of each vertex $v \in V$ can be an arbitrary positive integer $b(v)$ (Condition (4)). Together with Condition (5), it yields a reasonable generalization of branchings admitting extensions of several fundamental results on branchings, such as a multi-phase greedy algorithm [2,5,12,16], a packing theorem [11], and the integer decomposition property of the corresponding polytope [1].

The packing theorem on b-branchings leads to a necessary and sufficient condition for the arc set A to be partitioned into k b-branchings [21]. In this paper, we prove that equitable partition into k b-branchings is possible provided that any partition into k b-branchings exists.

Theorem 6. *Let $D = (V, A)$ be a digraph, $b \in \mathbb{Z}_{++}^V$, and $k \in \mathbb{Z}_{++}$. If A can be partitioned into k b-branchings, then A can be partitioned into k b-branchings $B_1, \ldots, B_k$ satisfying the following:*

1. *for each $i = 1, \ldots, k$, the size $|B_i|$ is $\lfloor |A|/k \rfloor$ or $\lceil |A|/k \rceil$; and*
2. *for each $i = 1, \ldots, k$, the indegree $d_{B_i}^-(v)$ of each vertex $v \in V$ is $\lfloor d_A^-(v)/k \rfloor$ or $\lceil d_A^-(v)/k \rceil$.*

When $b(v) = 1$ for every $v \in V$, Theorem 6 exactly coincides with Theorem 1. A new feature is that our definition of equitability of b-branchings is twofold: the number of arcs in any two b-branchings differ at most one (Condition 1); and the indegrees of each vertex with respect to any two b-branchings differ at most one (Condition 2). Theorem 6 means that the optimality of these $|V| + 1$ criteria can be attained at the same time, which suggests some good structure of b-branchings.

One consequence of Theorem 6 is the integer decomposition property of the convex hull of b-branchings of fixed size. For a polytope P and a positive integer κ, define $\kappa P = \{x \mid \exists x' \in P, \ x = \kappa x'\}$. A polytope P has the *integer decomposition property* if, for every $\kappa \in \mathbb{Z}_{++}$ and every integer vector $x \in \kappa P$, there exist κ integer vectors $x_1, \ldots, x_\kappa$ such that $x = x_1 + \cdots + x_\kappa$.

For branchings, Baum and Trotter [1] showed that the branching polytope has the integer decomposition property. Moreover, McDiarmid [25] proved the integer decomposition property of the convex hull of branchings of fixed size ℓ. For b-branchings, the integer decomposition property of the b-branching polytope is proved in [21]:

Theorem 7 (Kakimura, Kamiyama and Takazawa [21]). *Let $D = (V, A)$ be a digraph and $b \in \mathbb{Z}_{++}^V$. Then, the b-branching polytope has the integer decomposition property.*

In this paper, we derive the integer decomposition property of the convex hull of b-branchings of fixed size ℓ from Theorems 6 and 7.

Theorem 8. *Let $D = (V, A)$ be a digraph, $b \in \mathbb{Z}_{++}^V$, and $\ell \in \mathbb{Z}_+$. Then, the convex hull of the incidence vectors of the b-branchings of size ℓ has the integer decomposition property.*

1.3   Organization of the Paper

The remainder of the paper is organized as follows. Section 2 is devoted to a proof for Theorem 5 on equitable partition into machining forests. In Sect. 3, we prove Theorem 6 on equitable partition into b-branchings, and then derive Theorem 8 on the integer decomposition property of the related polytope.

2   Equitable Partition into Matching Forests

The aim of this section is to prove Theorem 5. In fact, Theorem 5 is derived by extending the argument in the proof for the exchangeability of matching forests by Schrijver [26, Theorem 2]. For the sake of completeness, however, below we describe a full proof. We remark that Schrijver used this property to prove the total dual integrality of the linear system describing the matching forest polytope presented by Giles [18]. The delta-matroid structure of matching forests [28] is also derived from an in-depth analysis of the proof for the exchangeability.

Let $G = (V, E, A)$ be a mixed graph. For a branching $B \subseteq A$, let $R(B) = V \setminus \partial B$, which represents the set of root vertices of B. Similarly, for a matching $M \subseteq E$, define $R(M) = V \setminus \partial M$. Note that, for a matching M and a branching B, their union $M \cup B$ is a matching forest if and only if $R(M) \cup R(B) = V$.

A *source component* X in a digraph $D = (V, A)$ is a strong component in D such that no arc in A enters X. In what follows, a source component is often denoted by its vertex set. Observe that, for a vertex subset $V' \subseteq V$, there exists a branching B satisfying $R(B) = V'$ if and only if $|V' \cap X| \geq 1$ for every source component X in D. This fact is extended to the following lemma on the partition of the arc set into to two branchings, which can be derived from Edmonds' disjoint branchings theorem [11].

Lemma 1 (Schrijver [26]). *Let $D = (V, A)$ be a digraph, and B_1 and B_2 are branchings in D partitioning A. Then, for two vertex sets $R_1', R_2' \subseteq V$ such that $R_1' \cup R_2' = R(B_1) \cup R(B_2)$ and $R_1' \cap R_2' = R(B_1) \cap R(B_2)$, the arc set A can be partitioned into two branchings B_1' and B_2' such that $R(B_1') = R_1'$ and $R(B_2') = R_2'$ if and only if*

$$|R_1' \cap X| \geq 1 \quad and \quad |R_2' \cap X| \geq 1 \quad for \ each \ source \ component \ X \ in \ D.$$

We remark that Schrijver [27] derived Theorem 1 from Lemma 1. Here we prove that Lemma 1 further leads to Theorem 5.

Proof (Proof of Theorem 5). The case $k = 1$ is trivial, and thus let $k \geq 2$. Let $F_1, \ldots, F_k$ be matching forests minimizing

$$\sum_{1 \leq i < j \leq k} ||\partial F_i| - |\partial F_j|| \tag{6}$$

among those partitioning $E \cup A$. We prove that every pair of F_i and F_j $(i, j \in [k])$ attains (3).

Suppose to the contrary that (3) does not hold for some $i, j \in [k]$. Without loss of generality, assume

$$|\partial F_1| - |\partial F_2| \geq 3. \tag{7}$$

Let $A' = B_1 \cup B_2$. Denote the family of source components in (V, A') by $\mathcal{X}'$. If a vertex $v \in V$ belongs to $R(B_1) \cap R(B_2)$, then v has no incoming arc in A', and hence v itself forms a source component in (V, A'). Thus, for $X \in \mathcal{X}'$ with $|X| \geq 2$, it follows that $X \cap R(B_1)$ and $X \cap R(B_2)$ are not empty and disjoint with each other. Denote the family of such $X \in \mathcal{X}'$ by $\mathcal{X}''$, i.e., $\mathcal{X}'' = \{X \in \mathcal{X}' \mid |X| \geq 2\}$. For each $X \in \mathcal{X}''$, take a pair e_X of vertices of which one vertex is in $R(B_1)$ and the other in $R(B_2)$. Denote $N = \{e_X \mid X \in \mathcal{X}''\}$. Note that N is a matching.

Consider an undirected graph $H = (V, M_1 \cup M_2 \cup N)$. Observe that each vertex $v \in V$ has degree at most two: if a vertex $v \in V$ is covered by both M_1 and M_2, then it follows that $v \in R(B_1) \cap R(B_2)$, implying that v is not covered by N. Thus, H consists of a disjoint collection of paths, some of which are possibly isolated vertices, and cycles.

Proposition 1. *For a vertex v in H with degree exactly two, it holds that $v \in \partial F_1 \cap \partial F_2$.*

Proof (Proof of Proposition 1). It is clear if $v \in \partial M_1 \cap \partial M_2$, and suppose not. Without loss of generality, assume $v \in \partial M_1 \cap \partial N$. It then follows from $v \in \partial M_1$ that $v \in R(B_1)$. Since $v \in \partial N$, this implies that $v \notin R(B_2)$. We thus conclude $v \in \partial B_2 \subseteq \partial F_2$. □

By Proposition 1, each vertex $u \in \partial F_1 \triangle \partial F_2$ is an endpoint of a path in H. It then follows from (7) that there must exist a path P such that

- one endpoint u of P belongs to $\partial F_1 \setminus \partial F_2$, and
- the other endpoint u' of P belongs to ∂F_1.

We remark that u' may or may not belong to ∂F_2. It may also be the case that u' is null, i.e., u is an isolated vertex which by itself forms P.

Denote the set of vertices in P by $V(P)$, and the set of edges in P belonging to $M_1 \cup M_2$ by $E(P)$. Define $M_1' = M_1 \triangle E(P)$ and $M_2' = M_2 \triangle E(P)$. It is not difficult to see that M_1' and M_2' are matchings satisfying

$$R(M_1') = (R(M_1) \setminus V(P)) \cup (R(M_2) \cap V(P)),$$
$$R(M_2') = (R(M_2) \setminus V(P)) \cup (R(M_1) \cap V(P)).$$

Also define

$$R_1' = (R(B_1) \setminus V(P)) \cup (R(B_2) \cap V(P)),$$
$$R_2' = (R(B_2) \setminus V(P)) \cup (R(B_1) \cap V(P)).$$

It then follows that $R_1' \cup R_2' = R(B_1) \cup R(B_2)$ and $R_1' \cap R_2' = R(B_1) \cap R(B_2)$. It also follows from the construction of H that $R_i' \cap X \neq \emptyset$ for every $X \in \mathcal{X}'$ and for $i = 1, 2$. Thus, by Lemma 1, the arc set A' can be partitioned into two branchings B_1' and B_2' such that $R(B_1') = R_1'$ and $R(B_2') = R_2'$. Then it holds that

$$
\begin{aligned}
R(M_1') \cup R(B_1') &= R(M_1') \cup R_1' \\
&= ((R(M_1) \cup R(B_1)) \setminus V(P)) \cup ((R(M_2) \cup R(B_2)) \cap V(P)) \\
&= (V \setminus V(P)) \cup (V \cap V(P)) \\
&= V,
\end{aligned}
$$

and hence $F_1' := M_1' \cup B_1'$ is a matching forest in G. This is also the case with $F_2' := M_2' \cup B_2'$.

Now we have two disjoint matching forests F_1' and F_2' such that $F_1' \cup F_2' = F_1 \cup F_2$. Moreover, by the definition of P, we have that

$$||\partial F_1'| - |\partial F_2'|| = ||\partial F_1| - |\partial F_2|| - 2 \quad \text{or} \quad ||\partial F_1'| - |\partial F_2'|| = ||\partial F_1| - |\partial F_2|| - 4,$$

and in particular, by (7),

$$||\partial F_1'| - |\partial F_2'|| < ||\partial F_1| - |\partial F_2||.$$

It also follows that

$$\sum_{i \in [k] \setminus \{1,2\}} (||\partial F_1'| - |\partial F_i|| + ||\partial F_2'| - |\partial F_i||)$$
$$\leq \sum_{i \in [k] \setminus \{1,2\}} (||\partial F_1| - |\partial F_i|| + ||\partial F_2| - |\partial F_i||).$$

This contradicts the fact that $F_1, \dots, F_k$ minimize (6), and thus completes the proof of the theorem. $\square$

We remark that, if an arbitrary partition of $E \cup A$ into k matching forests is given, a partition of $E \cup A$ into k matching forests satisfying (3) can be found in polynomial time. This can be done by repeatedly applying the update of two matching forests described in the above proof. The time complexity follows from the fact that each update can be done in polynomial time and decreases the value (6) by at least two.

3   Equitable Partition into b-branchings

In this section we first prove Theorem 6, and then derive Theorems 8. In proving Theorem 6, we make use of the following lemma, which is an extension of Lemma 1 to b-branchings.

Lemma 2 ([29]). *Let $D = (V, A)$ be a digraph and $b \in \mathbb{Z}_{++}^V$. Suppose that A can be partitioned into two b-branchings $B_1, B_2 \subseteq A$. Then, for two vectors $b_1', b_2' \in \mathbb{Z}_{++}^V$ satisfying $b_1' \leq b$, $b_2' \leq b$, and $b_1' + b_2' = d_A^-$, the arc set A can be partitioned into two b-branchings B_1' and B_2' such that $d_{B_1'}^- = b_1'$ and $d_{B_2'}^- = b_2'$ if and only if*

$$b_1'(X) < b(X) \quad and \quad b_2'(X) < b(X) \quad for \; each \; source \; component \; X \; in \; D.$$

We now prove Theorem 6.

Proof (Proof of Theorem 6). The case $k = 1$ is trivial, and thus let $k \geq 2$. Let $B_1, \ldots, B_k$ be k b-branchings minimizing

$$\sum_{i \in [k]} \left(\min \left\{ \left| |B_i| - \left\lfloor \frac{|A|}{k} \right\rfloor \right|, \left| |B_i| - \left\lceil \frac{|A|}{k} \right\rceil \right| \right\} \right. \tag{8}$$

$$\left. + \sum_{v \in V} \min \left\{ \left| |d_{B_i}^-(v)| - \left\lfloor \frac{d_A^-(v)}{k} \right\rfloor \right|, \left| |d_{B_i}^-(v)| - \left\lceil \frac{d_A^-(v)}{k} \right\rceil \right| \right\} \right)$$

among those partitioning A.

Suppose to the contrary that Condition 1 or 2 does not hold for some $i \in [k]$. Then, it is straightforward to see that there exists $j \in [k]$ such that

$$\min \{ |B_i|, |B_j| \} < \frac{|A|}{k} < \max \{ |B_i|, |B_j| \}, \quad ||B_i| - |B_j|| \geq 2, \tag{9}$$

or there exist $j \in [k]$ and $v \in V$ such that

$$\min \left\{ d_{B_i}^-(v), d_{B_j}^-(v) \right\} < \frac{d_A^-(v)}{k} < \max \left\{ d_{B_i}^-(v), d_{B_j}^-(v) \right\}, \quad \left| d_{B_i}^-(v) - d_{B_j}^-(v) \right| \geq 2. \tag{10}$$

Without loss of generality, let $i = 1$ and $j = 2$, and denote $b_1 = d_{B_1}^-$ and $b_2 = d_{B_2}^-$. Let $D' = (V, B_1 \cup B_2)$. Since B_1 and B_2 are b-branchings, it directly follows the definition of b-branchings that

$$b_1(v) \leq b(v) \quad \text{for each } v \in V, \tag{11}$$

$$b_2(v) \leq b(v) \quad \text{for each } v \in V, \tag{12}$$

$$b_1(X) \leq b(X) - 1 \quad \text{for each source component } X \text{ in } D', \tag{13}$$

$$b_2(X) \leq b(X) - 1 \quad \text{for each source component } X \text{ in } D'. \tag{14}$$

Let $\mathcal{X}$ be the set of source components X in D' such that $b_1(X) + b_2(X)$ is even, and let $\mathcal{Y}$ be the set of source components Y in D' such that $b_1(Y) + b_2(Y)$ is odd. Then, define $b_1', b_2' \in \mathbb{Z}_+^V$ satisfying $b_1' + b_2' = b_1 + b_2$ in the following manner.

– For all $X \in \mathcal{X}$, take $b_1'(v), b_2'(v) \in \mathbb{Z}_+$ for all $v \in X$ so that

$$b_1'(v) = b_2'(v) = \frac{b_1(v) + b_2(v)}{2} \quad \text{if } b_1(v) + b_2(v) \text{ is even;}$$

$$|b_1'(v) - b_2'(v)| = 1 \quad \text{if } b_1(v) + b_2(v) \text{ is odd; and}$$

$$b_1'(X) = b_2'(X).$$

– For all $Y \in \mathcal{Y}$, take $b_1'(v), b_2'(v) \in \mathbb{Z}_+$ for all $v \in Y$ so that

$$b_1'(v) = b_2'(v) = \frac{b_1(v) + b_2(v)}{2} \quad \text{if } b_1(v) + b_2(v) \text{ is even;}$$

$$|b_1'(v) - b_2'(v)| = 1 \quad \text{if } b_1(v) + b_2(v) \text{ is odd;}$$

$$|b_1'(Y) - b_2'(Y)| = 1 \quad \text{for every } Y \in \mathcal{Y}; \text{ and}$$

$$\left| \sum_{Y \in \mathcal{Y}} b_1'(Y) - \sum_{Y \in \mathcal{Y}} b_2'(Y) \right| \leq 1.$$

– For $v \in V \setminus (\bigcup_{X \in \mathcal{X} \cup \mathcal{Y}} X)$, take $b_1'(v), b_2'(v) \in \mathbb{Z}_+$ so that

$$|b_1'(v) - b_2'(v)| \leq 1 \quad \text{for every } v \in V \setminus \bigcup_{X \in \mathcal{X} \cup \mathcal{Y}} X; \text{ and}$$

$$|b_1'(V) - b_2'(V)| \leq 1.$$

Now it directly follows from (11)–(14) that $b_1' \leq b$, $b_2' \leq b$, and

$$b_1'(X) \leq b(X) - 1, \quad b_2'(X) \leq b(X) - 1 \quad (X \in \mathcal{X} \cup \mathcal{Y}).$$

It then follows from Lemma 2 that there exist b-branchings B_1' and B_2' such that $B_1' \cup B_2' = B_1 \cup B_2$, $d_{B_1'}^- = b_1'$, and $d_{B_2'}^- = b_2'$. For these two b-branchings B_1' and B_2', we have that

$$\left| |B_1'| - |B_2'| \right| \leq 1 \quad \text{and} \quad \left| d_{B_1'}^-(v) - d_{B_2'}^-(v) \right| \leq 1 \quad \text{for every } v \in V. \tag{15}$$

Therefore, we can strictly decreases the value (8) by replacing B_1 and B_2, which satisfy (9) or (10), with B_1' and B_2', which satisfy (15). This contradicts the fact that $B_1 \ldots, B_k$ minimize (8). $\square$

We remark that a partition of A into k b-branchings satisfying Conditions 1 and 2 in Theorem 6 can be found in polynomial time. First, we can check if there exists a partition of A into k b-branchings and find one if exists in polynomial time [21]. If this partition does not satisfy Conditions 1 and 2, then we repeatedly apply the update of two b-branchings as shown in the above proof, which can be done in polynomial time and strictly decreases the value (8).

We conclude this paper by deriving Theorems 8 from Theorem 6.

Proof (Proof of Theorem 8). Denote the convex hull of the b-branchings in D by P, and that of the b-branchings in D of size ℓ by Q. Take a positive integer $\kappa \in \mathbb{Z}_{++}$, and let x be an integer vector in κQ. Note that $x(A) = \kappa \cdot \ell$.

Let $D' = (V, A')$ be a digraph obtained from D by replacing each arc $a \in A$ by x_a parallel arcs. Then, since $x \in \kappa Q \subseteq \kappa P$, it follows from the integer decomposition property of the b-branching polytope (Theorem 7) that x is the sum of the incidence vectors of κ b-branchings, i.e., A' can be partitioned into κ b-branchings. Here we have that $|A'| = |x(A)| = \kappa \cdot \ell$, and thus it follows from Theorem 6 that A' can be partitioned into κ b-branchings $B_1, \ldots, B_\kappa$ such that $|B_i| = \ell$ for each $i \in [\kappa]$. Therefore, x can be represented as the sum of the incidence vectors of κ b-branchings of size ℓ, i.e., integer vectors in Q, which completes the proof. $\qquad\square$

Acknowledgements. This work is partially supported by JSPS KAKENHI Grant Numbers JP16K16012, JP20K11699, Japan.

References

1. Baum, S., Trotter Jr., L.E.: Integer rounding for polymatroid and branching optimization problems. SIAM J. Algebraic Discrete Methods **2**, 416–425 (1981)
2. Bock, F.: An algorithm to construct a minimum directed spanning tree in a directed network. In: Developments in Operations Research, pp. 29–44. Gordon and Breach (1971)
3. Bouchet, A.: Greedy algorithm and symmetric matroids. Math. Program. **38**, 147–159 (1987)
4. Chandrasekaran, R., Kabadi, S.N.: Pseudomatroids. Discrete Math. **71**, 205–217 (1988)
5. Chu, Y.J., Liu, T.H.: On the shortest arborescence of a directed graph. Sci. Sinica **14**, 1396–1400 (1965)
6. Davies, J., McDiarmid, C.: Disjoint common transversals and exchange structures. J. London Math. Soc. **14**, 55–62 (1976)
7. de Werra, D.: On some combinatorial problems arising in scheduling. Can. Oper. Res. J. **8**, 165–175 (1970)
8. de Werra, D.: Decomposition of bipartite multigraphs into matchings. Zeitschrift für Oper. Res. **16**, 85–90 (1972)
9. Dress, A.W.M., Havel, T.: Some combinatorial properties of discriminants in metric vector spaces. Adv. Math. **62**, 285–312 (1986)
10. Dulmage, A.L., Mendelsohn, N.S.: Some graphical properties of matrices with non-negative entries. Aequationes Math. **2**, 150–162 (1969)

11. Edmonds, J.: Edge-disjoint branchings. In: Rustin, R. (ed.) Combinatorial Algorithms, pp. 91–96. Algorithmics Press (1973)
12. Edmonds, J.: Optimum branchings. J. Res. Natl. Bureau Stand. Sect. B **71**, 233–240 (1967)
13. Erdős, P.: Problem 9. In: Fiedler, M. (ed.) Theory of Graphs and its Applications, p. 159. Czech Academy of Sciences (1964)
14. Folkman, J., Fulkerson, D.R.: Edge colorings in bipartite graphs. In: Bose, R.C., Dowling, T.A. (eds.) Combinatorial Mathematics and Its Applications (Proceedings Conference Chapel Hill, North Carolina, 1967), pp. 561–577. The University of North Carolina Press, Chapel Hill (1969)
15. Fujishige, S., Takazawa, K., Yokoi, Y.: A note on a nearly uniform partition into common independent sets of two matroids. J. Oper. Res. Soc. Japan (to appear)
16. Fulkerson, D.R.: Packing rooted directed cuts in a weighted directed graph. Math. Program. **6**(1), 1–13 (1974)
17. Giles, R.: Optimum matching forests I: special weights. Math. Program. **22**, 1–11 (1982)
18. Giles, R.: Optimum matching forests II: general weights. Math. Program. **22**, 12–38 (1982)
19. Giles, R.: Optimum matching forests III: facets of matching forest polyhedra. Math. Program. **22**, 39–51 (1982)
20. Hajnal, A., Szemerédi, E.: Proof of a conjecture of P. Erdős. In: Erdős, P., Rényi, A., Sós, V. (eds.) Combinatorial Theory and its Applications, II (Proceedings Colloquium on Combinatorial Theory and its Applications, Balatonfüred, Hungary, 1969), North-Holland, Amsterdam, pp. 601–623 (1970)
21. Kakimura, N., Kamiyama, N., Takazawa, K.: The b-branching problem in digraphs. Discrete Appl. Math. (to appear)
22. Keijsper, J.: A Vizing-type theorem for matching forests. Discrete Math. **260**, 211–216 (2003)
23. Király, T., Yokoi, Y.: Equitable partitions into matchings and coverings in mixed graphs. CoRR abs/1811.07856 (2018)
24. McDiarmid, C.J.H.: The solution of a timetabling problem. J. Inst. Math. Appl. **9**, 23–34 (1972)
25. McDiarmid, C.: Integral decomposition in polyhedra. Math. Program. **25**(2), 183–198 (1983)
26. Schrijver, A.: Total dual integrality of matching forest constraints. Combinatorica **20**, 575–588 (2000)
27. Schrijver, A.: Combinatorial Optimization: Polyhedra and Efficiency. Springer, Heidelberg (2003)
28. Takazawa, K.: Optimal matching forests and valuated delta-matroids. SIAM J. Discrete Math. **28**, 445–467 (2014)
29. Takazawa, K.: The b-bibranching problem: TDI system, packing, and discrete convexity. CoRR abs/1802.03235 (2018)

Network Design

Quasi-Separable Dantzig-Wolfe Reformulations for Network Design

Antonio Frangioni[1]([✉]), Bernard Gendron[2,3], and Enrico Gorgone[4]

[1] Dipartimento di Informatica, Università di Pisa, Pisa, Italy
frangio@di.unipi.it
[2] Department of Computer Science and Operations Research,
Université de Montréal, Montréal, Canada
[3] Interuniversity Research Centre on Enterprise Networks,
Logistics and Transportation, CIRRELT, Montréal, Canada
bernard.gendron@cirrelt.ca
[4] Dipartimento di Matematica ed Informatica, Università di Cagliari, Cagliari, Italy
egorgone@unica.it

Abstract. Under mild assumptions that are satisfied for many network design models, we show that the Lagrangian dual obtained by relaxing the flow constraints is what we call "quasi-separable". This property implies that the Dantzig-Wolfe (DW) reformulation of the Lagrangian dual exhibits two sets of convex combination constraints, one in terms of the design variables and the other in terms of the flow variables, the latter being linked to the design variables. We compare the quasi-separable DW reformulation to the standard disaggregated DW reformulation. We illustrate the concepts on a particular case, the budget-constrained multicommodity capacitated unsplittable network design problem.

Keywords: Lagrangian relaxation · Dantzig-Wolfe reformulations · Network design

1 Introduction

We consider a large class of network design models that can be represented by the following generic mixed-integer linear program (MILP), denoted (ND) [1]:

$$v(ND) = \min cx + fy \tag{1}$$
$$Ax = b \tag{2}$$
$$Dx + Ey \geq g \tag{3}$$
$$Hy \geq p \tag{4}$$
$$x \in \mathcal{X} \subset \mathbb{R}^n_+ \tag{5}$$
$$y \in \mathcal{Y} \subset \mathbb{Z}^m_+ \tag{6}$$

where $v(M)$ denotes the optimal value of any model (M) and the rational vectors b, c, f, g, p and rational matrices A, D, E, H have appropriate dimensions. The

© Springer Nature Switzerland AG 2020
M. Baïou et al. (Eds.): ISCO 2020, LNCS 12176, pp. 227–236, 2020.
https://doi.org/10.1007/978-3-030-53262-8_19

sets $\mathcal{X}$ and $\mathcal{Y}$ are the domains of the *flow variables* x and the *design variables* y, respectively. We assume the two sets are bounded and defined by three types of constraints: integer-valued bounds on individual variables; simplex constraints on subsets of the variables, which may be included in the definition of $\mathcal{Y}$, but not in that of $\mathcal{X}$; integrality constraints, which are included in the definition of $\mathcal{Y}$, but not necessarily in that of $\mathcal{X}$. When integrality constraints are relaxed, we denote the corresponding domains of variables $\overline{\mathcal{X}}$ and $\overline{\mathcal{Y}}$.

We call (2) the *flow constraints*, (3) the *linking constraints* and (4) the *design constraints*. To analyse these constraints, we introduce the corresponding sets $\mathcal{Q}_F = \{(x,y) \in \mathcal{X} \times \mathcal{Y} \mid Ax = b\}$, $\mathcal{Q}_L = \{(x,y) \in \mathcal{X} \times \mathcal{Y} \mid Dx + Ey \geq g\}$, and $\mathcal{Q}_D = \{(x,y) \in \mathcal{X} \times \mathcal{Y} \mid Hy \geq p\}$. The associated linear programming (LP) relaxation polyhedra, obtained by substituting $\mathcal{X}$ with $\overline{\mathcal{X}}$ and $\mathcal{Y}$ with $\overline{\mathcal{Y}}$, will be denoted respectively with $\mathcal{P}_F$, $\mathcal{P}_L$, and $\mathcal{P}_D$. Note that sets $\mathcal{Q}_F$ and $\mathcal{Q}_D$ ($\mathcal{P}_F$ and $\mathcal{P}_D$) are defined on all the space, but in fact only concern a subset of the variables; we will therefore denote by $\mathcal{Q}_F^x \subset \mathbb{R}_+^n$ and $\mathcal{Q}_D^y \subset \mathbb{Z}_+^m$ (and similarly for the continuous relaxation) their projection on the set of relevant variables.

Many solution methods for network design models that can be cast as special cases of (ND) rely on *Lagrangian relaxation* strategies. These consist in relaxing, in a Lagrangian way, either the linking constraints (3) [2,3] or the flow constraints (2) [2,4–7]. These strategies give rise to two Lagrangian dual programs: the *linking relaxation dual*, noted (LD_L), and the *flow relaxation dual*, noted (LD_F), respectively.

In (LD_L), relaxing the linking constraints (3) yields a Lagrangian relaxation that can be decomposed into two independent subproblems: one in the x variables, and one in the y ones. Hence, the bound strength of (LD_L) can be estimated using the primal interpretation of Lagrangian duality: the general result of [8] reads

$$\begin{aligned} v(LD_L) &= \min\{\, cx + fy \mid (x,y) \in \mathcal{P}_L \cap conv(\mathcal{Q}_F \cap \mathcal{Q}_D)\,\} \\ &= \min\{\, cx + fy \mid (x,y) \in \mathcal{P}_L \cap conv(\mathcal{Q}_F) \cap conv(\mathcal{Q}_D)\,\} \\ &= \min\{\, cx + fy \mid (x,y) \in \mathcal{P}_L\,,\ x \in conv(\mathcal{Q}_F^x)\,,\ y \in conv(\mathcal{Q}_D^y)\,\}, \end{aligned}$$

where $conv(\mathcal{C})$ denotes the convex hull of the set $\mathcal{C}$. Since the Lagrangian relaxation is separable in two independent problems, we can write down its *Dantzig-Wolfe (DW) reformulation* using two sets of convex combination constraints. That is, being $\{x^s\}_{s \in \mathcal{S}}$ and $\{y^t\}_{t \in \mathcal{T}}$ the sets of extreme points of $conv(\mathcal{Q}_F^x)$ and $conv(\mathcal{Q}_D^y)$, respectively, one has the following explicit form

$$v(LD_L) = \min\ cx + fy \tag{1}$$
$$Dx + Ey \geq g \tag{3}$$
$$x = \textstyle\sum_{s \in \mathcal{S}} \lambda^s x^s, \quad \sum_{s \in \mathcal{S}} \lambda^s = 1, \quad \lambda \geq 0 \tag{7}$$
$$y = \textstyle\sum_{t \in \mathcal{T}} \gamma^t y^t, \quad \sum_{t \in \mathcal{T}} \gamma^t = 1, \quad \gamma \geq 0 \tag{8}$$

The second strategy, that of relaxing the flow constraints (2), does not yield a separable Lagrangian relaxation. That is, for the corresponding Lagrangian dual (LD_F) one has

$$v(LD_F) = \min\{\, cx + fy \,|\, (x,y) \in \mathcal{P}_F \cap conv(\mathcal{Q}_L \cap \mathcal{Q}_D)\,\}$$

and the relevant set is $\{(x^r, y^r)\}_{r \in \mathcal{R}}$, containing the extreme points of $conv(\mathcal{Q}_L \cap \mathcal{Q}_D)$, which yields the DW reformulation

$$v(LD_F) = \min \, cx + fy \tag{1}$$
$$Ax = b \tag{2}$$
$$(x,y) = \sum_{r \in \mathcal{R}} \theta^r (x^r, y^r), \quad \sum_{r \in \mathcal{R}} \theta^r = 1, \quad \theta \geq 0 \tag{9}$$

In Sect. 2, we show that, under mild assumptions that are satisfied for many network design models, (LD_F) can be reformulated in an "almost" separable form, which we call *quasi-separable*. Then, in Sect. 3 we compare the quasi-separable DW reformulation to a disaggregated DW reformulation, which we define. Finally, in Sect. 4 we illustrate our results on a special case of (ND), the *budget-constrained multicommodity capacitated unsplittable network design problem (BMCUND)*.

2   Quasi-Separable Lagrangian Dual

To derive the quasi-separable DW reformulation of (LD_F), we recall recently published results [1] concerning the Lagrangian subproblem associated with the relaxation of the flow constraints. If we denote by π the (unrestricted) Lagrange multipliers associated with the flow constraints (2), and with $\bar{c} = c - \pi$, the Lagrangian relaxation can be written as

$$v(LR_F^{\bar{c}}) = \min \left\{ \bar{c}x + fy \,|\, (x,y) \in \mathcal{X} \times \mathcal{Y}, \; Dx + Ey \geq g, \; Hy \geq p \right\}.$$

Using a Benders' decomposition strategy, we consider y as "complicating" variables and define the Benders subproblem

$$v(BS^{\bar{c}}(y)) = \min\{\, \bar{c}x \,|\, x \in \mathcal{Q}_L(y) \,\}$$

where $\mathcal{Q}_L(y) = \{\, x \in \mathcal{X} \,|\, Dx \geq g - Ey \,\}$; hence, the Lagrangian relaxation can be rewritten as

$$v(LR_F^{\bar{c}}) = \min \left\{ fy + v(BS^{\bar{c}}(y)) \,|\, y \in \mathcal{Q}_D^y \right\}.$$

The following assumption holds for many network design models where y are binary variables.

Assumption QS. *[1] $v(BS^{\bar{c}}(y))$ can be written as a linear function of $y \in \mathcal{Y}$: for any cost vector $\bar{c}$, there exists a cost vector $w^{\bar{c}}$ such that $v(BS^{\bar{c}}(y)) = w^{\bar{c}}y$.*

The following proposition, due to [1], shows that Assumption QS allows to decompose the Lagrangian relaxation by optimizing first over $x \in \mathcal{Q}_L(y)$, then over $y \in \mathcal{Q}_D^y$, giving rise to a quasi-separable Lagrangian dual.

Proposition 1 [1]. *Under Assumption QS, it holds*

$$conv(\mathcal{Q}_L \cap \mathcal{Q}_D) = conv(\mathcal{Q}_L) \cap conv(\mathcal{Q}_D).$$

Proof. We prove that, however chosen cost vectors $\bar{c}$ and $\bar{f}$, minimizing them over the two sets yields the same result. In fact

$$
\begin{aligned}
\min_{(x,y)\in conv(\mathcal{Q}_L\cap\mathcal{Q}_D)} \bar{c}x + \bar{f}y &= \min_{(x,y)\in \mathcal{Q}_L\cap\mathcal{Q}_D} \bar{c}x + \bar{f}y = \min_{y\in \mathcal{Q}_D^y} \bar{f}y + v(BS^{\bar{c}}(y)) \\
&= \min_{y\in \mathcal{Q}_D^y} (\bar{f} + w^{\bar{c}})y \qquad = \min_{y\in conv(\mathcal{Q}_D^y)} (\bar{f} + w^{\bar{c}})y \\
&= \min_{y\in conv(\mathcal{Q}_D^y)} \left(\bar{f}y + \min_{x\in \mathcal{Q}_L(y)} \bar{c}x\right) \\
&= \min_{y\in conv(\mathcal{Q}_D^y)} \left(\bar{f}y + \min_{x\in conv(\mathcal{Q}_L(y))} \bar{c}x\right) \\
&= \min_{y\in conv(\mathcal{Q}_D^y)} \left(\min_{(x,y)\in conv(\mathcal{Q}_L)} \bar{c}x + \bar{f}y\right) \\
&= \min_{(x,y)\in conv(\mathcal{Q}_L)\cap conv(\mathcal{Q}_D)} \bar{c}x + \bar{f}y.
\end{aligned}
$$

Corollary 1. *Under Assumption QS, it holds*

$$v(LD_F) = \min_{y\in conv(\mathcal{Q}_D^y)} \left(fy + \min_{x\in P_F^x \cap conv(\mathcal{Q}_L(y))} cx\right).$$

The next proposition gives sufficient conditions for a model of the form (ND) to satisfy Assumption QS.

Proposition 2. *Consider any model (ND) such that, for some set J, both $\mathcal{Y}$ and $\mathcal{X}$ decompose over J, i.e., $\mathcal{Y} = \times_{j\in J} \mathcal{Y}^j$ and $\mathcal{X} = \times_{j\in J} \mathcal{X}^j$. Thus, $x = [x^j]_{j\in J}$ and $y = [y^j]_{j\in J}$. Let $I(j)$ be the set of indices of the variables in y^j, i.e., $y^j = [y_i]_{i\in I(j)}$, and $I = \cup_{j\in J} I(i)$. If*

1. $\mathcal{Y}^j = \{y^j \in \{0,1\}^{|I(j)|} \mid \sum_{i\in I(j)} y_i \leq 1\}$;
2. $\mathcal{Q}_L(y)$ *decomposes over J: $\mathcal{Q}_L(y) = \times_{j\in J}\{x^j \in \mathcal{X}^j \mid D^j x^j \leq E^j y^j\}$ for rational matrices $D^j \geq 0$ and $E^j \geq 0$ of appropriate dimensions;*

then model (ND) satisfies Assumption QS.

Proof. Under the assumptions, we can write the Benders subproblem as

$$v(BS^{\bar{c}}(y)) = \sum_{j\in J} \min\{\bar{c}_j x^j \mid x^j \in \mathcal{X}^j, \ D^j x^j \leq E^j y^j\}.$$

For each $j \in J$, if $y^j = 0$ then the unique solution is $x^j = 0$. Otherwise, let $i \in I(j)$ be unique index of the nonzero variable in y^j: then, x^j can be obtained by solving

$$w_i^{\bar{c}} = \min\{\bar{c}_j x^j \mid x^j \in \mathcal{X}^j, \ D^j x^j \leq e_i\}$$

where e_i is the column of E^j corresponding to y_i. Note that this problem is feasible, as $x^j = 0$ is a solution, and bounded, since $\mathcal{X}^j$ is bounded. Thus, $v(BS^{\bar{c}}(y)) = \sum_{i\in I} w_i^{\bar{c}} y_i$, and Assumption QS is satisfied.   □

Using the same notation as in the proof of Proposition 2, we define for each $i \in I$

$$\mathcal{Q}_L^x(i) = \left\{ x^j \in \mathcal{X}^j \mid D^j x^j \le e_i \right\}$$

where j is the unique index such that $i \in I(j)$ (this should be denoted by "$j(i)$", but we will avoid it whenever i is clear from the context, as we will use y_i^j for y_i only if necessary). We then denote as $\{x^{j,s}\}_{s \in \mathcal{S}(i)}$ the set of extreme points of $conv(\mathcal{Q}_L^x(i))$.

Proposition 3. *Under the assumptions of Proposition 2, we have*

$$conv(\mathcal{Q}_L(y)) = \underset{j \in J}{\times} \left\{ x^j \in \mathcal{X}^j \; \middle| \; \begin{array}{ll} x^j = \sum_{i \in I(j)} \sum_{s \in \mathcal{S}(i)} \omega^s x^{j,s} & \\ y_i = \sum_{s \in \mathcal{S}(i)} \omega^s & i \in I(j) \\ \omega^s \ge 0 & i \in I(j) \,, \; s \in \mathcal{S}(i) \end{array} \right\}$$

Proof. Due to the assumptions we have

$$conv(\mathcal{Q}_L(y)) = \underset{j \in J}{\times} \left\{ \{0\} \cup \bigcup_{i \in I(j)} conv(\mathcal{Q}_L^x(i)) \right\} \,.$$

Clearly, we only need to discuss each $j \in J$ (with the corresponding y^j and x^j) separately. Consider any

$$x^j = \sum_{i \in I(j)} \sum_{s \in \mathcal{S}(i)} \omega^s x^{j,s} \,.$$

If $y^j = 0$, then $x^j = 0$. Otherwise, let i be the unique index in $I(j)$ such that $y_i = 1$. Clearly, $y_h = 0$ for $h \in I(j) \backslash \{i\}$; therefore, $\omega^s = 0$ for all $s \in \mathcal{S}(h)$. Consequently

$$\sum_{s \in \mathcal{S}(i)} \omega^s x^{j,s} = x^j \in conv(\mathcal{Q}_L^x(i))$$

which implies the result. $\qquad\qquad\qquad\qquad\qquad\qquad\qquad\qquad\qquad\qquad\square$

With the same notation as in Sect. 1, we can now write the quasi-separable DW reformulation of the flow relaxation dual:

$$v(LD_F) = \min cx + fy \tag{1}$$
$$Ax = b \tag{2}$$
$$y = \sum_{t \in T} \gamma^t y^t, \quad \sum_{t \in T} \gamma^t = 1, \quad \gamma \ge 0 \tag{8}$$
$$x^j = \sum_{i \in I(j)} \sum_{s \in \mathcal{S}(i)} \omega^s x^{j,s} \qquad\qquad j \in J \tag{10}$$
$$y_i = \sum_{s \in \mathcal{S}(i)} \omega^s \qquad\qquad i \in I \tag{11}$$
$$\omega^s \ge 0 \qquad\qquad i \in I \,, \; s \in \mathcal{S}(i) \tag{12}$$

This DW reformulation corresponds to the expression of (LD_F) given by Corollary 1. Indeed, constraints (8) correspond to $y \in conv(\mathcal{Q}_D^y)$, constraints (2) correspond to $x \in \mathcal{P}_F^x$, and, by Proposition 3, constraints (10)–(12) correspond to $x \in conv(\mathcal{Q}_L^x(y))$. The quasi-separable DW reformulation relies on the fact that the Benders subproblem derived from the Lagrangian subproblem decomposes by $j \in J$. As such, it bears close resemblance to a disaggregated DW reformulation that could be derived from the Lagrangian relaxation of constraints (2) and (4). Next, we compare these two reformulations, showing that they are essentially the same when $\mathcal{P}_D$ is an integral polyhedron.

3   Comparison with Disaggregated DW Reformulation

The disaggregated DW reformulation is obtained by relaxing in a Lagrangian way both the flow constraints (2) and the design constraints (4). The resulting Lagrangian subproblem decomposes by $j \in J$, i.e., its feasible domain is $\times_{j \in J} \mathcal{Q}_L(j)$, where

$$\mathcal{Q}_L(j) = \{ (x^j, y^j) \in \mathcal{X}^j \times \mathcal{Y}^j \mid D^j x^j \leq E^j y^j \}.$$

The disaggregated DW reformulation of the corresponding Lagrangian dual, called the *flow-design relaxation dual* and denoted (LD_{FD}), can then be written as follows, where $\{ (x^{j,r}, y^{j,r}) \}_{r \in \mathcal{R}^j}$ are the extreme points of $conv(\mathcal{Q}_L(j))$ excluding $(0, 0)$:

$$v(LD_{FD}) = \min cx + fy \tag{1}$$
$$Ax = b \tag{2}$$
$$Hy \geq p \tag{4}$$
$$x^j = \textstyle\sum_{r \in \mathcal{R}^j} \theta^{j,r} x^{j,r} \qquad j \in J \tag{13}$$
$$y^j = \textstyle\sum_{r \in \mathcal{R}^j} \theta^{j,r} y^{j,r} \qquad j \in J \tag{14}$$
$$\textstyle\sum_{r \in \mathcal{R}^j} \theta^{j,r} \leq 1, \;\; \theta^j \geq 0 \quad j \in J \tag{15}$$

Note that, for any $j \in J$, there is a one-to-one correspondence between the extreme points $\{ (x^{j,r}, y^{j,r}) \}_{r \in \mathcal{R}^j}$ of $conv(\mathcal{Q}_L(j)) \setminus \{ (0, 0) \}$ and the extreme points $\{ x^{j,s} \}_{s \in \mathcal{S}(i)}$ of $conv(\mathcal{Q}_L^x(i))$ for some $i \in I(j)$. Indeed, for each $s \in \cup_{i \in I(j)} \mathcal{S}(i)$, there exists a unique $r \in \mathcal{R}^j$ such that $x^{j,r} = x^{j,s}$ and $y_i = 1$. We denote as $r = r(s)$ this unique index.

In general, we have $v(LD_F) \geq v(LD_{FD})$ and the inequality can be strict if $conv(\mathcal{Q}_D) \subset \mathcal{P}_D$. However, if $\mathcal{P}_D$ is an integral polyhedron, then $v(LD_F) = v(LD_{FD})$. In fact, the next proposition show that when $conv(\mathcal{Q}_D) = \mathcal{P}_D$ the quasi-separable and disaggregated DW reformulations are essentially identical.

Proposition 4. *If $\mathcal{P}_D$ is an integral polyhedron, then there is a one-to-one correspondence between the solutions of the quasi-separable and disaggregated DW reformulations, given by*

$$\theta^{j,r} y_i^{j,r} = \omega^s \qquad for \qquad j \in J, \, i \in I(j), \, s \in \mathcal{S}(i), \, r = r(s). \tag{16}$$

Proof. The objective functions and the flow constraints are the same in the two models. In addition, since $\mathcal{P}_D$ is an integral polyhedron, (8) are equivalent to (4). Because of (16), nonnegativity constraints (12) and (15) are equivalent. For any $j \in J$, the identity $\sum_{i \in I(j)} y_i^{j,r} = 1$ is true for all $r \in \mathcal{R}^j$, hence:

1. (10) and (13) are identical, since

$$x^j = \textstyle\sum_{r \in \mathcal{R}^j} \theta^{j,r} x^{j,r} = \sum_{i \in I(j)} \sum_{r \in \mathcal{R}^j} \theta^{j,r} y_i^{j,r} x^{j,r} = \sum_{i \in I(j)} \sum_{s \in \mathcal{S}(i)} \omega^s x^{j,s}.$$

2. (11) and (14) are identical, since $y_i = \sum_{r \in \mathcal{R}^j} \theta^{j,r} y_i^{j,r} = \sum_{s \in \mathcal{S}(i)} \omega^s$.

3. (15) is implied by (11) and the definition of $\mathcal{Y}$:

$$\sum_{r \in \mathcal{R}^j} \theta^{j,r} = \sum_{i \in I(j)} \sum_{r \in \mathcal{R}^j} \theta^{j,r} y_i^{j,r} = \sum_{i \in I(j)} \sum_{s \in \mathcal{S}(i)} \omega^s = \sum_{i \in I(j)} y_i \leq 1 .$$

This concludes the proof. $\square$

This proposition implies that the quasi-separable DW reformulation really brings something more than the disaggregated DW reformulation only for problems where $\mathcal{P}_D$ is not an integral polyhedron. In the next section we present such a problem.

4   Illustration with the BMCUND

The Budget-Constrained Multicommodity Capacitated Unsplittable Network Design problem (BMCUND) is defined on a directed graph $G = (N, J)$, where N is the set of nodes and J is the set of arcs. For each node $n \in N$ we define the sets of outgoing and incoming arcs, J_n^+ and J_n^-, respectively. Each commodity $k \in K$ corresponds to an origin–destination pair such that d_k units of flow must travel between the origin $O(k)$ and the destination $D(k)$ using a single path; this is why the problem is termed *unsplittable*, to distinguish it from the splittable variant where the flow of each commodity can be split among several paths. There is a limited budget M on the global investment costs to select the arcs to be used, where using arc $j \in J$ incurs a fixed cost $f^j \geq 0$ and provides a capacity $u^j > 0$. The objective function to be minimized are the routing costs $c_k^j \geq 0$ for each unit of commodity $k \in K$ through arc $j \in J$. We introduce two sets of variables to model the problem: x_k^j is 1 if the demand d_k of commodity k flows on arc j, and 0, otherwise; y^j is 1, if arc j is used, and 0, otherwise. The model is then written as follows:

$$v(BND) = \min \sum_{j \in J} \sum_{k \in K} d_k c_k^j x_k^j \tag{17}$$

$$\sum_{j \in J_n^+} x_k^j - \sum_{j \in J_n^-} x_k^j = b_k^n \qquad n \in N, \ k \in K \tag{18}$$

$$\sum_{k \in K} d_k x_k^j \leq u^j y^j \qquad j \in J \tag{19}$$

$$x_k^j \leq y^j \qquad j \in J, \ k \in K \tag{20}$$

$$\sum_{j \in J} f^j y^j \leq M \tag{21}$$

$$x_k^j \in \{0, 1\} \qquad j \in J, \ k \in K \tag{22}$$

$$y^j \in \{0, 1\} \qquad j \in J \tag{23}$$

where b_k^n is the supply of node n for commodity k, i.e., 1 for $n = O(k)$, -1 for $n = D(k)$, and 0 otherwise. This model is a special case of (ND) for which the sets are defined as $\mathcal{X} = \{x = [x_k^j]_{j \in J, k \in K} \mid (22)\}$, $\mathcal{Y} = \{y = [y^j]_{j \in J} \mid (23)\}$, $\mathcal{Q}_F^x = \{x \in \mathcal{X} \mid (18)\}$, $\mathcal{Q}_L = \{(x,y) \in \mathcal{X} \times \mathcal{Y} \mid (19), (20)\}$, $\mathcal{Q}_D^y = \{y \in \mathcal{Y} \mid (21)\}$, and $I = J$. Relaxing the flow constraints in a Lagrangian way yields

$$\min \left\{ \sum_{j \in J} \sum_{k \in K} \bar{c}_k^j x_k^j \mid (x, y) \in \mathcal{Q}_L \cap \mathcal{Q}_D \right\},$$

where $\bar{c}_k^j = d_k c_k^j + \pi_k^{t(j)} - \pi_k^{h(j)}$, π_k^n are the Lagrange multipliers, and $h(j)$ and $t(j)$ are the head and the tail of arc j, respectively. The Benders subproblem $(BS^{\bar{c}}(y))$ decomposes by arcs: for each $j \in J$, if $y^j = 0$, we obtain the trivial solution where all variables take value 0. If $y^j = 1$ instead, a 0–1 knapsack subproblem must be solved. Let $\tilde{x}^j$ be the solution, with optimal value $w_j^{\bar{c}}$: if $w_j^{\bar{c}} < 0$, then $(\tilde{x}^j, 1)$ is the optimal solution, otherwise the all-0 solution is optimal. This shows that

$$v(BS^{\bar{c}}(y)) = \min\{\,\bar{c}x \mid x \in \mathcal{Q}_L^x(y)\,\} = \sum_{j \in J} w_j^{\bar{c}} y^j = w^{\bar{c}} y \ ,$$

i.e., Assumption QS is satisfied. Note that the assumption is also verified for the splittable version of the problem, which is analogous save that a continuous (rather than a 0–1) knapsack problem must be solved to compute $w_j^{\bar{c}}$.

Before presenting the DW reformulations of (LD_F), we note that the flow relaxation dual dominates both the linking relaxation dual (LD_L) and the flow-design relaxation dual (LD_{FD}). Indeed, $\mathcal{P}_F$ is an integral polyhedron, which implies that

$$
\begin{aligned}
v(LD_L) &= \min\{\,cx \mid (x,y) \in conv(\mathcal{Q}_F) \cap \mathcal{P}_L \cap conv(\mathcal{Q}_D)\,\} \\
&= \min\{\,cx \mid (x,y) \in \mathcal{P}_F \cap \mathcal{P}_L \cap conv(\mathcal{Q}_D)\,\} \\
&\leq \min\{\,cx \mid (x,y) \in \mathcal{P}_F \cap conv(\mathcal{Q}_L) \cap conv(\mathcal{Q}_D)\,\} \\
&= \min\{\,cx \mid (x,y) \in \mathcal{P}_F \cap conv(\mathcal{Q}_L \cap \mathcal{Q}_D)\,\} = v(LD_F) \ .
\end{aligned}
$$

The inequality can be strict if $conv(\mathcal{Q}_L) \subset \mathcal{P}_L$, which is possible since $\mathcal{P}_L$ is not an integral polyhedron, as the Benders subproblem reduces to a 0–1 knapsack problem. Also, since $\mathcal{P}_D$ is not an integral polyhedron, we have

$$
\begin{aligned}
v(LD_{FD}) &= \min\{\,cx \mid (x,y) \in \mathcal{P}_F \cap conv(\mathcal{Q}_L) \cap \mathcal{P}_D\,\} \\
&\leq \min\{\,cx \mid (x,y) \in \mathcal{P}_F \cap conv(\mathcal{Q}_L) \cap conv(\mathcal{Q}_D)\,\} \\
&= \min\{\,cx \mid (x,y) \in \mathcal{P}_F \cap conv(\mathcal{Q}_L \cap \mathcal{Q}_D)\,\} = v(LD_F) \ .
\end{aligned}
$$

The inequality can be strict if $conv(\mathcal{Q}_D) \subset \mathcal{P}_D$, which is possible since $\mathcal{Q}_D$ is a 0–1 knapsack set.

We now present and contrast the two DW reformulations of LD_F. With the notation set forth in Sect. 1, the standard DW reformulation is:

$$v(LD_F) = \min \sum_{j \in J} \sum_{k \in K} d_k c_k^j x_k^j \tag{17}$$

$$\sum_{j \in J_n^+} x_k^j - \sum_{j \in J_n^-} x_k^j = b_k^n \quad n \in N, \ k \in K \tag{18}$$

$$x_k^j = \sum_{r \in R} \theta^r x_k^{j,r} \qquad j \in J, \ k \in K \tag{24}$$

$$y^j = \sum_{r \in R} \theta^r y^{j,r} \qquad\qquad j \in J \tag{25}$$

$$\sum_{r \in R} \theta^r = 1, \ \theta \geq 0 \tag{26}$$

It is interesting to note that constraints (25) are redundant and can be removed. Indeed, any link between the flow and design variables is captured in the

Lagrangian subproblem, since the design variables do not appear in the objective function of the BMCUND.

To derive the quasi-separable DW reformulation, we use the same notation set forth in Sect. 2; note that, in this case, $\mathcal{Q}_L^x(i) = \{\, x^j \in \mathcal{X}^j \mid \sum_{k \in K} d_k x_k^j \le u^j \,\}$ and $i = j$, since each of the simplices in the general treatment is actually a single variable. Then,

$$v(LD_F) = \min \ \sum_{j \in J} \sum_{k \in K} d_k c_k^j x_k^j \tag{17}$$

$$\sum_{j \in J_n^+} x_k^j - \sum_{j \in J_n^-} x_k^j = b_k^n \quad n \in N, \ k \in K \tag{18}$$

$$y^j = \sum_{t \in T} \gamma^t y^{j,t} \qquad j \in J \tag{27}$$

$$\sum_{t \in T} \gamma^t = 1, \ \gamma \ge 0 \tag{8}$$

$$x_k^j = \sum_{s \in \mathcal{S}(j)} \omega^s x_k^{j,s} \qquad j \in J, \ k \in K \tag{28}$$

$$y^j = \sum_{s \in \mathcal{S}(j)} \omega^s \qquad j \in J \tag{29}$$

$$\omega^s \ge 0 \qquad j \in J, \ s \in \mathcal{S}(j) \tag{30}$$

Note that (28) is somehow simpler than the general (10), again due to the fact that $I(j) = \{\, j \,\}$. Compared to the standard DW reformulation, the quasi-separable DW reformulation is larger, but has an obvious advantage when applying column generation: the same extreme point y^t of $conv(\mathcal{Q}_D^y)$ can be recombined with different corresponding extreme points of $conv(\mathcal{Q}_L^x(y^t))$, while many more columns with the same y^t, but with different x values, might be needed to solve the standard DW reformulation. This should result in much less column generation iterations when solving the quasi-separable DW reformulation, as already shown for disaggregated DW reformulations (e.g., [9]). Computational results on large-scale instances of the BMCUND will soon be reported to verify this assertion.

References

1. Gendron, B.: Revisiting Lagrangian relaxation for network design. Discrete Appl. Math. **261**, 203–218 (2019)
2. Crainic, T.G., Frangioni, A., Gendron, B.: Bundle-based relaxation methods for multicommodity capacitated fixed charge network design. Discrete Appl. Math. **112**, 73–99 (2001)
3. Frangioni, A., Gorgone, E.: Bundle methods for sum-functions with "easy" components: applications to multicommodity network design. Math. Program. A **145**, 133–161 (2014)
4. Holmberg, K., Hellstrand, J.: Solving the uncapacitated network design problem by a Lagrangian heuristic and branch-and-bound. Oper. Res. **46**, 247–259 (1998)
5. Holmberg, K., Yuan, D.: A Lagrangian heuristic based branch-and-bound approach for the capacitated network design problem. Oper. Res. **48**, 461–481 (2000)
6. Kliewer, G., Timajev, L.: Relax-and-cut for capacitated network design. In: Brodal, G.S., Leonardi, S. (eds.) ESA 2005. LNCS, vol. 3669, pp. 47–58. Springer, Heidelberg (2005). https://doi.org/10.1007/11561071_7

7. Sellmann, M., Kliewe, G., Koberstein, A.: Lagrangian cardinality cuts and variable fixing for capacitated network design. In: Möhring, R., Raman, R. (eds.) ESA 2002. LNCS, vol. 2461, pp. 845–858. Springer, Heidelberg (2002). https://doi.org/10.1007/3-540-45749-6_73
8. Geoffrion, A.M.: Lagrangean relaxation for integer programming. Math. Program. Study **2**, 82–114 (1974)
9. Frangioni, A., Gendron, B.: A stabilized structured Dantzig-Wolfe decomposition method. Math. Program. B **140**, 45–76 (2013)

Dynamic Programming Approach to the Generalized Minimum Manhattan Network Problem

Yuya Masumura[1], Taihei Oki[2], and Yutaro Yamaguchi[1(✉)]

[1] Osaka University, Osaka 565-0871, Japan
`yutaro_yamaguchi@inf.kyushu-u.ac.jp`
[2] The University of Tokyo, Tokyo 113-8656, Japan

Abstract. We study the generalized minimum Manhattan network (GMMN) problem: given a set P of pairs of points in the Euclidean plane $\mathbb{R}^2$, we are required to find a minimum-length geometric network which consists of axis-aligned segments and contains a shortest path in the L_1 metric (a so-called Manhattan path) for each pair in P. This problem commonly generalizes several NP-hard network design problems that admit constant-factor approximation algorithms, such as the rectilinear Steiner arborescence (RSA) problem, and it is open whether so does the GMMN problem.

As a bottom-up exploration, Schnizler (2015) focused on the intersection graphs of the rectangles defined by the pairs in P, and gave a polynomial-time dynamic programming algorithm for the GMMN problem whose input is restricted so that both the treewidth and the maximum degree of its intersection graph are bounded by constants. In this paper, as the first attempt to remove the degree bound, we provide a polynomial-time algorithm for the star case, and extend it to the general tree case based on an improved dynamic programming approach.

1 Introduction

In this paper, we study a geometric network design problem in the Euclidean plane $\mathbb{R}^2$. For a pair of points s and t in the plane, a path between s and t is called a *Manhattan path* (or an *M-path* for short) if it consists of axis-aligned segments whose total length is equal to the Manhattan distance of s and t (in other words, it is a shortest s–t path in the L_1 metric). The *minimum Manhattan network (MMN) problem* is to find a minimum-length geometric network that contains an M-path for every pair of points in a given terminal set. In the *generalized minimum Manhattan network (GMMN) problem*, given a set P of pairs of terminals, we are required to find a minimum-length network that contains

---

The full version [9] of this paper is available at arXiv.

Y. Masumura has moved to Fast Retailing Co., Ltd., Tokyo 135-0063, Japan, and Y. Yamaguchi has moved to Kyushu University, Fukuoka 819-0395, Japan.

© Springer Nature Switzerland AG 2020
M. Baïou et al. (Eds.): ISCO 2020, LNCS 12176, pp. 237–248, 2020.
https://doi.org/10.1007/978-3-030-53262-8_20

an M-path for every pair in P. Throughout this paper, let $n = |P|$ denote the number of terminal pairs.

The GMMN problem was introduced by Chepoi, Nouioua, and Vaxès [2], and is known to be NP-hard as so is the MMN problem [3]. The MMN problem and another NP-hard special case named the *rectilinear Steiner arborescence (RSA) problem* admit polynomial-time constant-factor approximation algorithms, and in [2] they posed a question whether so does the GMMN problem or not, which is still open.

Das, Fleszar, Kobourov, Spoerhase, Veeramoni, and Wolff [4] gave an $O(\log^{d+1} n)$-approximation algorithm for the d-dimensional GMMN problem based on a divide-and-conquer approach. They also improved the approximation ratio for $d = 2$ to $O(\log n)$. Funke and Seybold [5] (see also [14]) introduced the *scale-diversity* measure $\mathcal{D}$ for (2-dimensional) GMMN instances, and gave an $O(\mathcal{D})$-approximation algorithm. Because $\mathcal{D} = O(\log n)$ is guaranteed, this also implies $O(\log n)$-approximation as with Das et al. [4], which is the current best approximation ratio for the GMMN problem in general.

As another approach to the GMMN problem, Schnizler [13] explored tractable cases by focusing on the intersection graphs of GMMN instances. The intersection graph represents for which terminal pairs M-paths can intersect. He showed that, when both the treewidth and the maximum degree of intersection graphs are bounded by constants, the GMMN problem can be solved in polynomial time by dynamic programming (see Table 1). His algorithm heavily depends on the degree bound, and it is natural to ask whether we can remove it, e.g., whether the GMMN problem is difficult even if the intersection graph is restricted to a tree without any degree bound.

In this paper, we give an answer to this question. Specifically, as the first tractable case without any degree bound in the intersection graphs, we provide a polynomial-time algorithm for the star case, and extend it to the general tree case based on a dynamic programming approach inspired by and improving Schnizler's algorithm [13].

Theorem 1. *There exists an $O(n^2)$-time algorithm for the GMMN problem when the intersection graph is restricted to a star.*

Theorem 2. *There exists an $O(n^5)$-time algorithm for the GMMN problem when the intersection graph is restricted to a tree.*

Due to the page limitation, we sometimes give only intuitions or high-level ideas with figures, and leave the details of them for the full version [9], which also contains several improvements (cf. Table 1).

Related Work

The MMN problem was first introduced by Gudmundsson, Levcopoulos, and Narashimhan [6]. They gave 4- and 8-approximation algorithms running in $O(n^3)$ and $O(n \log n)$ time, respectively. The current best approximation ratio is 2, which was obtained independently by Chepoi et al. [2] using an LP-rounding technique, by Nouioua [11] using a primal-dual scheme, and by Guo, Sun, and Zhu [7] using a greedy method.

Table 1. Exactly solvable cases classified by the class of intersection graphs, whose treewidth and maximum degree are denoted by tw and Δ, respectively.

Class	Time complexity
tw $= O(1)$, $\Delta = O(1)$	$O(n^{4\Delta(\Delta+1)(\mathrm{tw}+1)+2})$ [13]
Trees (tw $= 1$, $\Delta = O(1)$)	$O(n^{4\Delta^2+1})$ [13]
Cycles (tw $= \Delta = 2$)	$O(n^{25})$ [13]
Stars (tw $= 1$, $\Delta = n - 1$)	$O(n^2)$ (Theorem 1)
Trees (tw $= 1$)	$O(n^3)$ (Theorem 2 + speeding-up in [9])
Cycles (tw $= \Delta = 2$)	$O(n^4)$ (cf. the full version [9])

The RSA problem is another important special case of the GMMN problem. In this problem, given a set of terminals in $\mathbb{R}^2$, we are required to find a minimum-length network that contains an M-path between the origin and every terminal. The RSA problem was first studied by Nastansky, Selkow, and Stewart [10] in 1974. The complexity of the RSA problem had been open for a long time, and Shi and Su [15] showed that the decision version is strongly NP-complete after three decades. Rao, Sadayappan, Hwang, and Shor [12] proposed a 2-approximation algorithm that runs in $O(n \log n)$ time. Lu and Ruan [8] and Zachariasen [16] independently obtained PTASes, which are both based on Arora's technique [1] of building a PTAS for the metric Steiner tree problem.

Organization
The rest of this paper is organized as follows. In Sect. 2, we describe necessary definitions and notations. In Sect. 3, we present an algorithm for the star case and sketch a proof of Theorem 1. In Sect. 4, based on a dynamic programming approach, we extend our algorithm to the tree case and prove Theorem 2.

2   Preliminaries

2.1   Problem Formulation

For a point $p \in \mathbb{R}^2$, we denote by p_x and p_y its x- and y-coordinates, respectively, i.e., $p = (p_x, p_y)$. Let $p, q \in \mathbb{R}^2$ be two points. We write $p \le q$ if both $p_x \le q_x$ and $p_y \le q_y$ hold. We define two points $p \wedge q = (\min \{p_x, q_x\}, \min \{p_y, q_y\})$ and $p \vee q = (\max \{p_x, q_x\}, \max \{p_y, q_y\})$. We denote by pq the segment whose endpoints are p and q, and by $\|pq\|$ its length, i.e., $pq = \{\alpha p + (1-\alpha)q \mid \alpha \in [0, 1]\}$ and $\|pq\| = \sqrt{(p_x - q_x)^2 + (p_y - q_y)^2}$. We also define $d_x(p, q) = |p_x - q_x|$ and $d_y(p, q) = |p_y - q_y|$, and denote by $d(p, q)$ the *Manhattan distance* between p and q, i.e., $d(p, q) = d_x(p, q) + d_y(p, q)$. Note that $\|pq\| = d(p, q)$ if and only if $p_x = q_x$ or $p_y = q_y$, and then the segment pq is said to be *vertical* or *horizontal*, respectively, and *axis-aligned* in either case.

A *(geometric) network* N in $\mathbb{R}^2$ is a finite simple graph with a vertex set $V(N) \subseteq \mathbb{R}^2$ and an edge set $E(N) \subseteq \binom{V(N)}{2} = \{\{p, q\} \mid p, q \in V(N), \ p \ne q\}$, where we often identify each edge $\{p, q\}$ with the corresponding segment pq. The *length* of N is defined as $\|N\| = \sum_{\{p,q\} \in E(N)} \|pq\|$. For two points $s, t \in \mathbb{R}^2$, a path π between s and t (or an *s–t path*) is a network of the following form:

$$V(\pi) = \{s = p_0, p_1, p_2, \ldots, p_k = t\},$$
$$E(\pi) = \{\{p_{i-1}, p_i\} \mid i \in [k]\},$$

where $[k] = \{1, 2, \ldots, k\}$ for a nonnegative integer k. An *s–t* path π is called a *Manhattan path* (or an *M-path*) for a pair (s, t) if every edge $\{p_{i-1}, p_i\} \in E(\pi)$ is axis-aligned and $\|\pi\| = d(s, t)$ holds.

We are now ready to state our problem formally.

Problem (Generalized Minimum Manhattan Network (GMMN))

Input: A set P of n pairs of points in $\mathbb{R}^2$.
Goal: Find a minimum-length network N in $\mathbb{R}^2$ that consists of axis-aligned edges and contains a Manhattan path for every pair $(s, t) \in P$.

Throughout this paper, when we write a pair $(p, q) \in \mathbb{R}^2 \times \mathbb{R}^2$, we assume $p_x \leq q_x$ (by swapping if necessary). A pair (p, q) is said to be *regular* if $p_y \leq q_y$, and *flipped* if $p_y \geq q_y$. In addition, if $p_x = q_x$ or $p_y = q_y$, then there exists a unique M-path for (p, q) and we call such a pair *degenerate*.

2.2   Restricting a Feasible Region to the Hanan Grid

For a GMMN instance P, we denote by $\mathcal{H}(P)$ the *Hanan grid*, which is a grid network in $\mathbb{R}^2$ consisting of vertical and horizontal lines through each point appearing in P. More formally, it is defined as follows (see Fig. 1):

$$V(\mathcal{H}(P)) = \left(\bigcup_{(s,t) \in P} \{s_x, t_x\} \right) \times \left(\bigcup_{(s,t) \in P} \{s_y, t_y\} \right) \subseteq \mathbb{R}^2,$$
$$E(\mathcal{H}(P)) = \{\{p, q\} \in \binom{V(\mathcal{H}(P))}{2} \mid \|pq\| = d(p, q), \ pq \cap V(\mathcal{H}(P)) = \{p, q\}\}.$$

Note that $\mathcal{H}(P)$ is an at most $2n \times 2n$ grid network. It is not difficult to see that, for any GMMN instance P, at least one optimal solution is contained in the Hanan grid $\mathcal{H}(P)$ as its subgraph (cf. [5]).

For each pair $v = (p, q) \in V(\mathcal{H}(P)) \times V(\mathcal{H}(P))$, we denote by $\Pi_P(v)$ or $\Pi_P(p, q)$ the set of all M-paths for v that are subgraphs of the Hanan grid $\mathcal{H}(P)$. By the problem definition, we associate each n-tuple of M-paths, consisting of an M-path $\pi_v \in \Pi_P(v)$ for each $v \in P$, with a feasible solution $N = \bigcup_{v \in P} \pi_v$ on $\mathcal{H}(P)$, where the union of networks is defined by the set unions of the vertex sets and of the edge sets. Moreover, each minimal feasible (as well as optimal) solution on $\mathcal{H}(P)$ must be represented in this way. Based on this correspondence, we abuse the notation as $N = (\pi_v)_{v \in P} \in \prod_{v \in P} \Pi_P(v)$, and define Feas($P$) and

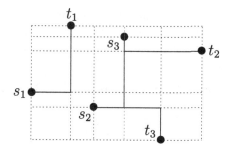

Fig. 1. An optimal solution (solid) to a GMMN instance $\{(s_1, t_1), (s_2, t_2), (s_3, t_3)\}$ lies on the Hanan grid (dashed), where (s_1, t_1) and (s_2, t_2) are regular pairs and (s_3, t_3) is a flipped pair.

Opt(P) as the sets of feasible solutions covering all minimal ones and of all optimal solutions, respectively, on $\mathcal{H}(P)$, i.e.,

$$\text{Feas}(P) = \prod_{v \in P} \Pi_P(v),$$

$$\text{Opt}(P) = \arg\min\{\|N\| \mid N \in \text{Feas}(P)\}.$$

Thus, we have restricted a feasible region of a GMMN instance P to the Hanan grid $\mathcal{H}(P)$. In other words, the GMMN problem reduces to finding a network $N = (\pi_v)_{v \in P} \in \text{Opt}(P)$ as an n-tuple of M-paths in Feas(P).

2.3   Specialization Based on Intersection Graphs

The *bounding box* of a pair $v = (p, q) \in \mathbb{R}^2 \times \mathbb{R}^2$ indicates the rectangle region

$$\{z \in \mathbb{R}^2 \mid p \wedge q \leq z \leq p \vee q\},$$

and we denote it by $B(v)$ or $B(p, q)$. Note that $B(p, q)$ is the region where an M-path for (p, q) can exist. For a GMMN instance P and a pair $v \in P$, we denote by $\mathcal{H}(P, v)$ the subgraph of the Hanan grid $\mathcal{H}(P)$ induced by $V(\mathcal{H}(P)) \cap B(v)$. We define the *intersection graph* IG[P] of P by

$$V(\text{IG}[P]) = P,$$

$$E(\text{IG}[P]) = \{\{u, v\} \in \binom{P}{2} \mid E(\mathcal{H}(P, u)) \cap E(\mathcal{H}(P, v)) \neq \emptyset\}.$$

The intersection graph IG[P] intuitively represents how a GMMN instance P is complicated in the sense that, for each $u, v \in P$, an edge $\{u, v\} \in E(\text{IG}[P])$ exists if and only if two M-paths $\pi_u \in \Pi_P(u)$ and $\pi_v \in \Pi_P(v)$ can share some segments, which saves the total length of a network in Feas(P). In particular, if IG[P] contains no triangle, then no segment can be shared by M-paths for three different pairs in P, and hence $N \in \text{Feas}(P)$ is optimal (i.e., $\|N\|$ is minimized) if and only if the total length of segments shared by two M-paths in N is maximized.

We denote by GMMN[$\cdots$] the GMMN problem with restriction on the intersection graph of the input; e.g., IG[P] is restricted to a tree in GMMN[Tree]. Each restricted problem is formally stated in the relevant section.

3   An $O(n^2)$-Time Algorithm for GMMN[Star]

In this section, as a step to GMMN[Tree], we present an $O(n^2)$-time algorithm for GMMN[Star], which is formally stated as follows.

Problem (GMMN[Star])

Input: A set $P \subseteq \mathbb{R}^2 \times \mathbb{R}^2$ of n pairs whose intersection graph IG[P] is a star, whose center is denoted by $r = (s, t) \in P$.
Goal: Find an optimal network $N = (\pi_v)_{v \in P} \in \mathrm{Opt}(P)$.

A crucial observation for GMMN[Star] is that an M-path $\pi_l \in \Pi_P(l)$ for each leaf pair $l \in P - r$ can share some segments only with an M-path $\pi_r \in \Pi_P(r)$ for the center pair r. Hence, minimizing the length of $N = (\pi_v)_{v \in P} \in \mathrm{Feas}(P)$ is equivalent to maximizing the total length of segments shared by two M-paths π_r and π_l for $l \in P - r$.

In Sect. 3.1, we observe that, for each leaf pair $l \in P - r$, once we fix where an M-path $\pi_r \in \Pi_P(r)$ for r enters and leaves the bounding box $B(l)$, the maximum length of segments that can be shared by π_r and $\pi_l \in \Pi_P(l)$ is easily determined. Thus, GMMN[Star] reduces to finding an optimal M-path $\pi_r \in \Pi_P(r)$ for the center pair $r = (s, t)$, and in Sect. 3.2, we formulate this task as the computation of a longest s–t path in an auxiliary directed acyclic graph (DAG), which is constructed from the subgrid $\mathcal{H}(P, r)$. As a result, we obtain an exact algorithm that runs in linear time in the size of auxiliary graphs, which are simplified so that it is always $O(n^2)$ in Sect. 3.3.

3.1   Observation on Sharable Segments

Without loss of generality, we assume that the center pair $r = (s, t)$ is regular, i.e., $s \leq t$. Fix an M-path $\pi_r \in \Pi_P(r)$ and a leaf pair $l = (s_l, t_l) \in P - r$. Obviously, if π_r is disjoint from the bounding box $B(l)$, then any M-path $\pi_l \in \Pi_P(l)$ cannot share any segment with π_r. Suppose that π_r intersects $B(l)$, and let $\pi_r[l]$ denote the intersection $\pi_r \cap \mathcal{H}(P, l)$. Let $v = (p, q)$ be the pair of two vertices on π_r such that $\pi_r[l]$ is a p–q path, and we call v the *in-out pair* of π_r for l. As $\pi_r \in \Pi_P(r)$, we have $\pi_r[l] \in \Pi_P(v)$, and v is also regular (recall the assumption $p_x \leq q_x$). Moreover, for any M-path $\pi_v \in \Pi_P(v)$, the network π_r' obtained from π_r by replacing its subpath $\pi_r[l]$ with π_v is also an M-path for r in $\Pi_P(r)$. Since $B(v) \subseteq B(l)$ does not intersect $B(l')$ for any other leaf pair $l' \in P \setminus \{r, l\}$, once $v = (p, q)$ is fixed, we can freely choose an M-path $\pi_v \in \Pi_P(v)$ instead of $\pi_r[l]$ for maximizing the length of segments shared with some $\pi_l \in \Pi_P(l)$. For each possible in-out pair $v = (p, q)$ of M-paths in $\Pi_P(r)$ (the sets of those vertices p and q are formally defined in Sect. 3.2 as $V_{\llcorner}(r, l)$ and $V_{\lnot}(r, l)$, respectively), we

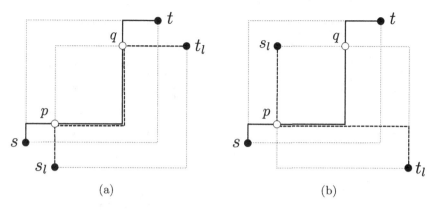

Fig. 2. (a) If $l = (s_l, t_l)$ is a regular pair, for any $\pi_v \in \Pi_P(p, q)$, some $\pi_l \in \Pi_P(l)$ completely includes π_v. (b) If $l = (s_l, t_l)$ is a flipped pair, while any $\pi_l \in \Pi_P(l)$ cannot contain both horizontal and vertical segments of any $\pi_v \in \Pi_P(p, q)$, one can choose $\pi_v \in \Pi_P(p, q)$ so that the whole of either horizontal or vertical segments of π_v can be included in some $\pi_l \in \Pi_P(l)$.

denote by $\gamma(l, p, q)$ the maximum length of segments shared by two M-paths for l and $v = (p, q)$, i.e.,

$$\gamma(l, p, q) = \max \left\{ \|\pi_l \cap \pi_v\| \mid \pi_l \in \Pi_P(l),\ \pi_v \in \Pi_P(p, q) \right\}.$$

Then, the following lemma is easily observed (see Fig. 2).

Lemma 1. *For every leaf pair $l \in P - r$, the following properties hold.*

(1) If l is a regular pair, $\gamma(l, p, q) = d(p, q)\ (= d_x(p, q) + d_y(p, q))$.
(2) If l is a flipped pair, $\gamma(l, p, q) = \max \{d_x(p, q),\ d_y(p, q)\}$.

3.2 Reduction to the Longest Path Problem in DAGs

In this section, we reduce GMMN[Star] to the longest path problem in DAGs. Let P be a GMMN[Star] instance and $r = (s, t) \in P$ $(s \le t)$ be the center of IG[P], and we construct an auxiliary DAG G from the subgrid $\mathcal{H}(P, r)$ as follows (see Fig. 3).

First, for each edge $e = \{p, q\} \in E(\mathcal{H}(P, r))$ with $p \le q$ (and $p \ne q$), we replace e with an arc (p, q) of length 0. For each leaf pair $l \in P - r$, let s'_l and t'_l denote the lower-left and upper-right corners of $B(r) \cap B(l)$, respectively, so that (s'_l, t'_l) is a regular pair with $B(s'_l, t'_l) = B(r) \cap B(l)$. If (s'_l, t'_l) is degenerate, then we change the length of each arc (p, q) with $p, q \in V(\mathcal{H}(P, r) \cap B(l))$ from 0 to $\|pq\|$, which clearly reflects the (maximum) sharable length in $B(l)$. Otherwise,

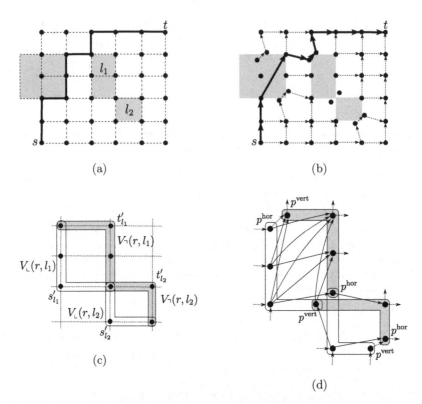

Fig. 3. (a) An M-path for r in the subgrid $\mathcal{H}(P, r)$. (b) The corresponding directed s–t path in the auxiliary DAG G, where the dashed arcs are of length 0. (c) The boundary vertex sets for leaf pairs. (d) The corresponding parts in G, where the length of each interior arc (p, q) is $\gamma(l, p, q)$ for $p \in V_{\llcorner}(r, l)$ and $q \in V_{\urcorner}(r, l)$.

the bounding box $B(s'_l, t'_l) \subseteq B(l)$ has a nonempty interior, and we define four subsets of $V(\mathcal{H}(P, r) \cap B(l))$ as follows:

$$V_{\llcorner}(r, l) = \{p \in V(\mathcal{H}(P, r) \cap B(l)) \mid p_x = (s'_l)_x \text{ or } p_y = (s'_l)_y\},$$
$$V_{\urcorner}(r, l) = \{q \in V(\mathcal{H}(P, r) \cap B(l)) \mid q_x = (t'_l)_x \text{ or } q_y = (t'_l)_y\},$$
$$V_{\bullet}(r, l) = V_{\llcorner}(r, l) \cap V_{\urcorner}(r, l),$$
$$\begin{aligned}V_{\blacksquare}(r, l) &= V(\mathcal{H}(P, r) \cap B(l)) \setminus (V_{\llcorner}(r, l) \cup V_{\urcorner}(r, l)) \\&= \{z \in V(\mathcal{H}(P, r) \cap B(l)) \mid (s'_l)_x < z_x < (t'_l)_x \text{ and } (s'_l)_y < z_y < (t'_l)_y\}.\end{aligned}$$

As r is regular, any M-path $\pi_r \in \Pi_P(r)$ intersecting $B(l)$ enters it at some $p \in V_{\llcorner}(r, l)$ and leaves it at some $q \in V_{\urcorner}(r, l)$, and the maximum sharable length $\gamma(l, p, q)$ in $B(l)$ is determined by Lemma 1. We remove all the interior vertices in $V_{\blacksquare}(r, l)$ (with all the incident arcs) and all the boundary arcs (p, q) with $p, q \in V_{\llcorner}(r, l) \cup V_{\urcorner}(r, l)$. Instead, for each pair (p, q) of $p \in V_{\llcorner}(r, l)$ and $q \in V_{\urcorner}(r, l)$ with $p \le q$ and $p \ne q$, we add an interior arc (p, q) of length $\gamma(l, p, q)$. Let $E_{\text{int}}(l)$ denote the set of such interior arcs for each nondegenerate pair $l \in P - r$.

Finally, we care about the corner vertices in $V_\bullet(r, l)$, which can be used for cheating if l is flipped as follows. Suppose that $p \in V_\bullet(r, l)$ is the upper-left corner of $B(l)$, and consider the situation when the in-out pair (p', q') of $\pi_r \in \Pi_P(r)$ for l satisfies $p'_x = p_x < q'_x$ and $p'_y < p_y = q'_y$. Then, (p', q') is not degenerate, and by Lemma 1, the maximum sharable length in $B(l)$ is $\gamma(l, p', q') = \max\{d_x(p', q'), d_y(p', q')\}$ as it is represented by an interior arc (p', q'), but one can take another directed p'-q' path that consists of two arcs (p', p) and (p, q') in the current graph, whose length is $d_y(p', p) + d_x(p, q') = d_y(p', q') + d_x(p', q') > \gamma(l, p', q')$. To avoid such cheating, for each $p \in V_\bullet(r, l)$, we divide it into two distinct copies p^{hor} and p^{vert} (which are often identified with its original p unless we need to distinguish them), and replace the endpoint p of each incident arc e with p^{hor} if e is horizontal and with p^{vert} if vertical (see Fig. 3 (d)). In addition, when p is not shared by any other leaf pair,[1] we add an arc $(p^{\text{hor}}, p^{\text{vert}})$ of length 0 if p is the upper-left corner of $B(s'_l, t'_l)$ and an arc $(p^{\text{vert}}, p^{\text{hor}})$ of length 0 if the lower-right, which represents the situation when $\pi_r \in \Pi_P(r)$ intersects $B(l)$ only at p.

Let G be the constructed directed graph, and denote by $\ell(e)$ the length of each arc $e \in E(G)$. The following two lemmas complete our reduction, whose proofs are left for the full version [9] (see Fig. 3 again).

Lemma 2. *The directed graph G is acyclic.*

Lemma 3. *Any longest s-t path π_G^* in G with respect to ℓ satisfies*

$$\sum_{e \in E(\pi_G^*)} \ell(e) = \max_{\pi_r \in \Pi_P(r)} \left(\sum_{l \in P-r} \max_{\pi_l \in \Pi_P(l)} \|\pi_l \cap \pi_r\| \right).$$

3.3 Computational Time Analysis with Simplified DAGs

A longest path in a DAG G is computed in $O(|V(G)| + |E(G)|)$ time by dynamic programming. Although the subgrid $\mathcal{H}(P, r)$ has $O(n^2)$ vertices and edges, the auxiliary DAG G constructed in Sect. 3.2 may have much more arcs due to $E_{\text{int}}(l)$, whose size is $\Theta(|V_\llcorner(r, l)| \cdot |V_\neg(r, l)|)$ and can be $\Omega(n^2)$. This, however, can be always reduced to linear by modifying the boundary vertices and the incident arcs appropriately in order to avoid creating diagonal arcs in $B(l)$. We just illustrate a high-level idea in Fig. 4 and leave the details for the full version [9]. Thus, the total computational time is $O(n^2)$, which completes the proof of Theorem 1.

4   An $O(n^5)$-Time Algorithm for GMMN[Tree]

In this section, we present an $O(n^5)$-time algorithm for GMMN[Tree], which is the main target in this paper and stated as follows.

---

[1] Note that p can be shared as corners of two different leaf pairs due to our definition of the intersection graph, and then leaving one bounding box means entering the other straightforwardly.

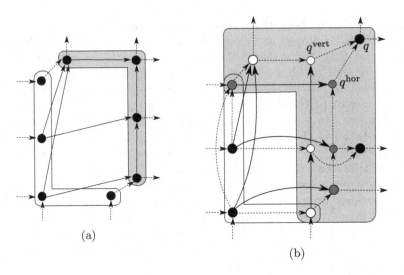

(a)

(b)

Fig. 4. (a) Simplification for a regular pair. (b) Simplification for a flipped pair, where the gray and white vertices distinguish sharing horizontal and vertical segments in $\mathcal{H}(P, r) \cap B(l)$, respectively, and the dashed arcs are of length 0.

Problem (GMMN[Tree])

Input: A set $P \subseteq \mathbb{R}^2 \times \mathbb{R}^2$ of n pairs whose intersection graph IG$[P]$ is a tree.
Goal: Find an optimal network $N = (\pi_v)_{v \in P} \in \mathrm{Opt}(P)$.

For a GMMN[Tree] instance P, we choose an arbitrary pair $r \in P$ as the root of the tree IG$[P]$; in particular, when IG$[P]$ is a star, we regard the center as the root. The basic idea of our algorithm is dynamic programming on the tree IG$[P]$ from the leaves toward r. Each subproblem reduces to the longest path problem in DAGs like the star case, which is summarized as follows.

Fix a pair $v = (s_v, t_v) \in P$. If $v \neq r$, then there exists a unique parent $u = \mathrm{Par}(v)$ in the tree IG$[P]$ rooted at r, and there are $O(n^2)$ possible in-out pairs (p_v, q_v) of $\pi_u \in \Pi_P(u)$ for v. We virtually define $p_v = q_v = \epsilon$ for the case when we do not care the shared length in $B(u)$, e.g., $v = r$ or π_u is disjoint from $B(v)$. Let P_v denote the vertex set of the subtree of IG$[P]$ rooted at v (including v itself). For every possible in-out pair (p_v, q_v), as a subproblem, we compute the maximum total length $\mathrm{dp}(v, p_v, q_v)$ of sharable segments in $B(P_v) = \bigcup_{w \in P_v} B(w)$, i.e.,

$$\mathrm{dp}(v, \epsilon, \epsilon) = \max \left\{ \sum_{w \in P_v - v} \|\pi_w \cap \pi_{\mathrm{Par}(w)}\| \ \middle| \ (\pi_w)_{w \in P} \in \mathrm{Feas}(P) \right\},$$

$$\mathrm{dp}(v, p_v, q_v) = \max \left\{ \sum_{w \in P_v} \|\pi_w \cap \pi_{\mathrm{Par}(w)}\| \ \middle| \ \begin{array}{l} (\pi_w)_{w \in P} \in \mathrm{Feas}(P), \\ \pi_u[v] \in \Pi_P(p_v, q_v) \end{array} \right\}.$$

By definition, the goal is to compute $dp(r, \epsilon, \epsilon)$. If v is a leaf in $IG[P]$, then $P_v = \{v\}$. In this case, $dp(v, p_v, q_v)$ is the maximum length of segments shared by two M-paths $\pi_v \in \Pi_P(v)$ and $\pi_u \in \Pi_P(u)$ with $\pi_u[v] \in \Pi_P(p_v, q_v)$, which is easily determined (cf. Lemma 1). Otherwise, using the computed values $dp(w, p_w, q_w)$ for all children w of v and all possible in-out pairs (p_w, q_w), we reduce the task to the computation of a longest s_v–t_v path in an auxiliary DAG, as with finding an optimal M-path for the center pair in the star case.

4.1 Constructing Auxiliary DAGs for Subproblems

Let $v = (s_v, t_v) \in P$, which is assumed to be regular without loss of generality. If $v = r$, then let $p_v = q_v = \epsilon$; otherwise, let $u = \mathrm{Par}(v)$ be its parent, and fix a possible in-out pair $u' = (p_v, q_v)$ of $\pi_u \in \Pi_P(u)$ for v (including the case $p_v = q_v = \epsilon$). Let $C_v \subseteq P_v$ be the set of all children of v. By replacing r and $P - r$ in Sect. 3.2 with v and $C_v + u'$ (or C_v if $p_v = q_v = \epsilon$), respectively, we construct the same auxiliary directed graph, denoted by $G[v, p_v, q_v]$. We then change the length of each interior arc $(p_w, q_w) \in E_{int}(w)$ for each child $w \in C_v$ from $\gamma(w, p_w, q_w)$ to $dp(w, p_w, q_w) - dp(w, \epsilon, \epsilon)$, so that it represents the difference of the total sharable length in $B(P_w) = \bigcup_{w \in P_w} B(w')$ between the cases when an M-path for v intersects $B(w)$ (enters at p_w and leaves at q_w) and when an M-path for v is ignored. As with Lemma 2, the graph $G[v, p_v, q_v]$ is acyclic. The following lemma completes the reduction of computing $dp(v, p_v, q_v)$ to finding a longest s_v–t_v path in $G[v, p_v, q_v]$, whose proof is left for the full version [9].

Lemma 4. *Let π_G^* be a longest s_v–t_v path in $G[v, p_v, q_v]$ with respect to ℓ. We then have*

$$dp(v, p_v, q_v) = \sum_{e \in E(\pi_G^*)} \ell(e) + \sum_{w \in C_v} dp(w, \epsilon, \epsilon).$$

4.2 Computational Time Analysis

This section completes the proof of Theorem 2. For a pair $v \in P$, suppose that $\mathcal{H}(P, v)$ is an $a_v \times b_v$ grid graph. For each possible in-out pair (p_v, q_v), to compute $dp(v, p_v, q_v)$, we find a longest path in the DAG $G[v, p_v, q_v]$ constructed in Sect. 4.1, which has $O(a_v b_v) = O(n^2)$ vertices and $O(\delta_v(a_v + b_v)^2) = O(\delta_v n^2)$ edges, where δ_v is the degree of v in $IG[P]$. Hence, for solving the longest path problem once for each $v \in P$, it takes $\sum_{v \in P} O(\delta_v n^2) = O(n^3)$ time in total (recall that $IG[P]$ is a tree). For each $v \in P - r$, there are respectively at most $a_v + b_v = O(n)$ candidates for p_v and for q_v, and hence $O(n^2)$ possible in-out pairs. Thus, the total computational time is bounded by $O(n^5)$, and we are done.

We remark that this can be improved to $O(n^3)$ by computing $dp(v, p_v, q_v)$ for many possible in-out pairs (p_v, q_v) at once using extra DPs. See the full version [9] for the details.

Acknowledgment. We are grateful to the anonymous reviewers for their careful reading and giving helpful comments.

References

1. Arora, S.: Approximation schemes for NP-hard geometric optimization problems: a survey. Math. Programm. **97**(1–2), 43–69 (2003)
2. Chepoi, V., Nouioua, K., Vaxès, Y.: A rounding algorithm for approximating minimum Manhattan networks. Theor. Comput. Sci. **390**(1), 56–69 (2008)
3. Chin, F.Y., Guo, Z., Sun, H.: Minimum Manhattan network is NP-complete. Discrete Comput. Geom. **45**(4), 701–722 (2011)
4. Das, A., Fleszar, K., Kobourov, S., Spoerhase, J., Veeramoni, S., Wolff, A.: Approximating the generalized minimum Manhattan network problem. Algorithmica **80**(4), 1170–1190 (2018)
5. Funke, S., Seybold, M.P.: The generalized minimum Manhattan network problem (GMMN) - scale-diversity aware approximation and a primal-dual algorithm. In: Proceedings of Canadian Conference on Computational Geometry (CCCG), vol. 26 (2014)
6. Gudmundsson, J., Levcopoulos, C., Narasimhan, G.: Approximating a minimum Manhattan network. Nordic J. Comput. **8**(2), 219–232 (2001)
7. Guo, Z., Sun, H., Zhu, H.: Greedy construction of 2-approximation minimum Manhattan network. In: Hong, S.-H., Nagamochi, H., Fukunaga, T. (eds.) ISAAC 2008. LNCS, vol. 5369, pp. 4–15. Springer, Heidelberg (2008). https://doi.org/10.1007/978-3-540-92182-0_4
8. Lu, B., Ruan, L.: Polynomial time approximation scheme for the rectilinear Steiner arborescence problem. J. Combinat. Optim. **4**(3), 357–363 (2000)
9. Masumura, Y., Oki, T., Yamaguchi, Y.: Dynamic programming approach to the generalized minimum Manhattan network problem (2020). arXiv:2004.11166
10. Nastansky, L., Selkow, S.M., Stewart, N.F.: Cost-minimal trees in directed acyclic graphs. Zeitschrift für Oper. Res. **18**(1), 59–67 (1974)
11. Nouioua, K.: Enveloppes de Pareto et Réseaux de Manhattan: Caractérisations et Algorithmes. Ph.D. thesis, Université de la Méditerranée (2005)
12. Rao, S.K., Sadayappan, P., Hwang, F.K., Shor, P.W.: The rectilinear Steiner arborescence problem. Algorithmica **7**(1–6), 277–288 (1992)
13. Schnizler, M.: The Generalized Minimum Manhattan Network Problem. Master's thesis, University of Stuttgart (2015)
14. Seybold, M.P.: Algorithm Engineering in Geometric Network Planning and Data Mining. Ph.D. thesis, University of Stuttgart (2018)
15. Shi, W., Su, C.: The rectilinear Steiner arborescence problem is NP-complete. SIAM J. Comput. **35**(3), 729–740 (2005)
16. Zachariasen, M.: On the approximation of the rectilinear Steiner arborescence problem in the plane (2000)

On Finding Shortest Paths
in Arc-Dependent Networks

P. Wojciechowski[1], Matthew Williamson[2], and K. Subramani[1(✉)]

[1] LDCSEE, West Virginia University, Morgantown, WV, USA
{pwojciec,k.subramani}@mail.wvu.edu
[2] MCS, Marietta College, Marietta, OH, USA
williamm@marietta.edu

Abstract. This paper is concerned with the design and analysis of algorithms for the arc-dependent shortest path (ADSP) problem. A network is said to be arc-dependent if the cost of an arc a depends upon the arc taken to enter a. These networks are fundamentally different from traditional networks in which the cost associated with an arc is a fixed constant and part of the input. The ADSP problem is also known as the suffix-1 path-dependent shortest path problem in the literature. This problem has a polynomial time solution if the shortest paths are not required to be simple. The ADSP problem finds applications in a number of domains including highway engineering, turn penalties and prohibitions, and fare rebates. In this paper, we are interested in the ADSP problem when restricted to simple paths. We call this restricted version the simple arc-dependent shortest path (SADSP) problem. We show that the SADSP problem is **NP-complete**. We present inapproximability results and an exact exponential algorithm for this problem. Additionally, we explore the problem of detecting negative cost cycles in arc-dependent networks and discuss its algorithmic complexity.

1 Introduction

This paper studies the problem of computing shortest paths in arc-dependent networks. In an arc-dependent network, the cost of an arc a is not a fixed value. Instead, the cost of the arc depends on the arc taken to enter it. This differs from traditional networks, where the cost of an arc is a fixed constant. The problem of finding shortest paths in such a network is known as the arc-dependent shortest path (ADSP) problem. This problem is also known as the suffix-1 path-dependent shortest path problem and has been previously studied in the literature [1,4,7,10,11]. The problem has a polynomial-time solution as long as the shortest paths do not need to be simple.

An extension of this problem is known as the path-dependent shortest path (PDSP) problem, where the cost of arc a depends on the path taken to the

K. Subramani—This research was supported in part by the Air-Force Office of Scientific Research through Grant FA9550-19-1-017. This research was also supported in part by the Air-Force Research Laboratory Rome through Contract FA8750-17-S-7007.

M. Baïou et al. (Eds.): ISCO 2020, LNCS 12176, pp. 249–260, 2020.
https://doi.org/10.1007/978-3-030-53262-8_21

arc. This problem is claimed to be **NP-complete** [9,10]. A variant of the path-dependent shortest path problem is the suffix-k PDSP problem, where the cost of arc a depends on the last k arcs of the preceding path instead of the entire preceding path [10].

In this paper, we are interested in the ADSP problem where the shortest paths must be simple paths. Recall that a simple path is a path without repeated vertices or repeated arcs. We refer to this as the simple arc-dependent shortest path problem (SADSP). In the general suffix-k PDSP problem, the shortest paths are not required to be simple. In fact, in some cases, non-simple paths are actually shorter than simple paths depending on the path taken [10]. We show that when the ADSP problem is restricted to simple paths, the problem is actually **NP-complete**. We also provide special case algorithms for the SADSP problem.

Arc-dependent networks are motivated by a number of applications. In highway engineering, it is desirable to find routes between points in a city's street and/or freeway network that are minimized with respect to time and distance. While efficient algorithms for this problem exist [3,5], it is possible that a street network may include "turn penalties" [4]. These penalties increase either the time or distance (or both) of a route based on the turns taken at intersections. Turn prohibitions [7] can also complicate transportation planning. These prohibitions eliminate some possible solutions to the shortest path problem. For example, some intersections do not allow a driver to make a left turn even though traffic flows in both directions along that road. Turn prohibitions can be modeled by having an infinite turn penalty. Fare rebates [10] provide an additional layer of complexity in public transportation since taking specific routes in a transportation network may result in a discount while other routes may not provide such a discount.

The principal contributions of this paper are as follows:

1. Showing that the decision version of the SADSP problem is **NP-complete**.
2. Showing that the problem of detecting simple negative cycles in an arc-dependent network is **NP-complete**.
3. Showing that the optimization version of the SADSP problem is **NPO-complete**.
4. The design and analysis of an exact exponential algorithm for the SADSP problem.
5. Providing an integer programming representation of the SADSP problem.

The rest of this paper is organized as follows: Sect. 2 details a formal description of the SADSP problem. Section 3 establishes the complexity of the SADSP problem. We also discuss a polynomial time special case. Section 4 examines the problem of finding negative cycles in arc-dependent networks. Section 5 shows that the problem of approximating the SADSP is **NPO-complete**. In Sect. 6, we give an exact algorithm for the SADSP problem that runs in $O(n^3 \cdot 2^n)$ time. In Sect. 7, we discuss an integer programming representation of the path and cycle problems discussed in this paper. We conclude in Sect. 8 by summarizing our contributions in this paper and outlining avenues for future research.

2   Statement of Problems

Let $\mathbf{G} = \langle \mathbf{V}, \mathbf{E}, \mathbf{C} \rangle$ denote a directed network. $\mathbf{V}$ is the vertex set with n vertices, and $\mathbf{E} = \{e_1, e_2, \ldots, e_m\}$ is the set of arcs. Let $s \in \mathbf{V}$ be the source vertex.

The cost structure is represented by the matrix $\mathbf{C}$, where entry (i, j) corresponds to the cost of arc e_j assuming that e_j was entered through arc e_i. The matrix $\mathbf{C}$ has $(m + 1)$ rows and m columns. The $(m + 1)^{\text{th}}$ row of $\mathbf{C}$ contains the cost of arcs that do not have any incoming arcs. We use the phantom arc e_0 entering s to account for these costs. We refer to $\mathbf{G}$ as an arc-dependent network.

Let P_{st} denote the path $(e_1 - e_2 - e_3 - \cdots - e_k)$, where e_1 is an arc leaving vertex s and e_k is an arc entering node t. Note that $\mathbf{C}[e_j, e_k]$ only matters when the head of arc e_j is the tail of arc e_k. Otherwise, e_j cannot be the predecessor of e_k. In cases where the head of e_j is not the tail of e_k, we define $\mathbf{C}[e_j, e_k] = 0$. As a result, we note that $\mathbf{C}$ does not represent the connectivity of $\mathbf{G}$.

The cost of a path P_{st} between vertices s and t is given by:

$$cost(P_{st}) = \mathbf{C}[e_0, e_1] + \mathbf{C}[e_1, e_2] + \cdots + \mathbf{C}[e_{k-1}, e_k].$$

A shortest path in $\mathbf{G}$ is known as an *arc-dependent shortest path* (ADSP).

Example 1. Consider a network with four vertices and the following arcs:

1. Arc e_1 : (s, v_1) with cost 0.
2. Arc e_2 : (s, v_2) with cost -1.
3. Arc e_3 : (v_1, v_2) with cost 1.
4. Arc e_4 : (v_1, t) with cost 1.
5. Arc e_5 : (v_2, t) with costs $\mathbf{C}[e_2, e_5] = 3$ and $\mathbf{C}[e_3, e_5] = 0$.

The resulting network $\mathbf{G}$ and cost matrix $\mathbf{C}$ is shown in Fig. 1.

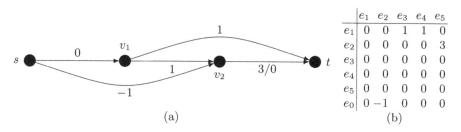

Fig. 1. Example of an Arc-Dependent Network. (a) is the network $\mathbf{G}$. (b) is the cost matrix $\mathbf{C}$.

In this paper, we explore two variants of the SADSP problem. These are:

1. SADSP$_D$: Given an arc-dependent network G, source vertex s, target vertex t, and cost k, does G have an arc-dependent simple path from s to t with cost less than k?.
2. SADSP$_{Opt}$: Given an arc-dependent network G, source vertex s, and target vertex t, what is the shortest arc-dependent simple path from s to t in G?

3   On the Computational Complexity of the SADSP Problem

In this section, we explore the complexity of the $SADSP_D$ problem. The general ADSP problem can be solved in polynomial time by transforming the network into its respective dual network and then applying a polynomial time shortest path algorithm [1,10]. This holds because the ADSP is not necessarily simple.

We now show that the $SADSP_D$ problem is **NP-complete**. We accomplish this using a reduction from Hamiltonian Path.

Let **G** be an unweighted directed network with n vertices. We construct an arc-dependent weighted network **G'** as follows:

1. Create the vertex s_0.
2. For each vertex v_i in **G**, create the vertices w_i and s_i.
3. For each vertex w_i in **G'**, create the arc (s_0, w_i), which always costs 0.
4. For each pair of vertices w_i and s_k, where $1 \leq i, k \leq n$, create the following arcs:
   a. (w_i, s_k) with cost 0 if the preceding arc is (s_{k-1}, w_i) and 1 otherwise.
   b. (s_k, w_i) with cost 0 if the preceding arc is (w_j, s_k), where arc (v_j, v_i) is in **G**, and 1 otherwise.

With this reduction, we can now prove that the $SADSP_D$ problem is **NP-complete**.

Note that once a path is found, the cost of the path can be computed in polynomial time. Thus, the $SADSP_D$ problem is in **NP**. All that remains is to show that the problem is **NP-hard**.

Theorem 1. **G'** *has a 0-cost simple arc-dependent path from* s_0 *to* s_n *if and only if* **G** *has a Hamiltonian Path.*

Proof. We first assume that **G** has a Hamiltonian Path.

Without loss of generality, assume that the Hamiltonian Path is $(v_1, v_2) \rightarrow (v_2, v_3) \rightarrow \cdots \rightarrow (v_{n-1}, v_n)$. By construction, the path

$$(s_0, w_1) \rightarrow (w_1, s_1) \rightarrow (s_1, w_2) \rightarrow (w_2, s_2) \rightarrow (s_2, w_3) \rightarrow \cdots \rightarrow (w_n, s_n)$$

has cost 0. Thus **G'** has a SADSP of cost 0 from s_0 to s_n.

We now assume that G' has a 0-cost SADSP P from s_0 to s_n.

By construction, the only arcs in **G'** are of the form (s_k, w_i) or (w_i, s_k). In other words, there are no arcs of the form (s_i, s_j) or (w_i, w_j). We define $P(k)$, where $0 \leq k \leq n - 1$, such that vertex $w_{P(k)}$ immediately follows vertex s_k in path P. In other words, let $(s_k, w_{P(k)}) \in P$. By construction, the only 0-cost outgoing arc from $w_{P(k)}$ is $(w_{P(k)}, s_{k+1})$. Thus, P must be of the form:

$$(s_0, w_{P(0)}) \rightarrow (w_{P(0)}, s_1) \rightarrow (s_1, w_{P(1)}) \rightarrow (w_{p(1)}, s_2) \rightarrow \cdots \rightarrow (w_{P(n-1)}, s_n).$$

Since P is a simple path, vertices $w_{P(0)}, w_{P(1)}, \ldots, w_{P(n-1)}$ must be distinct. Since the total cost of path P is 0, each arc in P must also cost 0. By construction, each arc $(s_k, w_{P(k)})$, for $1 \leq k \leq n - 1$, has cost 0 only if arc $(v_{P(k-1)}, v_{P(k)})$ is in **G**. This implies that the path

$$(v_{P(0)}, v_{P(1)}) \rightarrow (v_{P(1)}, v_{P(2)}) \rightarrow \cdots \rightarrow (v_{P(n-2)}, v_{P(n-1)})$$

is in **G**. Since vertices $w_{P(0)}, w_{P(1)}, \ldots, w_{P(n-1)}$ are distinct in **G'**, it follows that $v_{P(0)}, v_{P(1)}, \ldots, v_{P(n-1)}$ are distinct in **G**. Therefore, by definition, the above path must be a Hamiltonian Path in **G**. □

Corollary 1. *The SADSP$_D$ problem is* **NP-complete** *for the class of bipartite graphs in which the arc costs belong to the set* $\{0, 1\}$.

Proof. Note that the graph **G'** is bipartite. Thus, the SADSP$_D$ problem is **NP-complete** even when restricted to bipartite graphs. Additionally, the arc costs in **G'** are restricted to the set $\{0, 1\}$. Thus, the SADSP$_D$ problem is **NP-compete** when **G** is bipartite, there are only two possible arc costs, and all arc costs are non-negative. □

We now examine a case where the SADSP$_{Opt}$ problem can be solved in polynomial time.

Theorem 2. *The SADSP$_{Opt}$ problem can be solved in polynomial time for directed acyclic graphs.*

Proof. Let **G** be a directed acyclic graph with an arc-dependent cost function. Let s and t be vertices in **G**. Since **G** is acyclic, any path from s to t is necessarily simple. Thus, the problem of finding a SADSP from s to t in **G** is equivalent to the problem of finding any ADSP from s to t. This problem is known to be in **P** [1,10]. □

4 Negative Cycles in Arc-Dependent Networks

In this section, we discuss the problem of finding simple negative cycles in arc-dependent networks. We call this problem the simple arc-dependent negative cycle problem and it is defined as follows:

Definition 1. *Simple arc-dependent negative cycle (SADNC) problem: Given an arc-dependent network* **G**, *does* **G** *contain a simple cycle NC consisting of arcs e_1 through e_k, such that:*

$$\mathbf{C}[e_k, e_1] + \sum_{i=2}^{k} \mathbf{C}[e_{i-1}, e_i] < 0?$$

Note that the cost of e_1, the first arc in NC, depends on e_k, the last arc in NC. Thus, the negative cost of the cycle depends only on the cycle itself and not how the cycle was reached initially.

We now show that the SADNC problem is **NP-complete**. We do this by modifying the reduction from Hamiltonian Path to SADSP from Sect. 3.

Let $\mathbf{G}$ be an unweighted directed network, and let $\mathbf{G}'$ be the arc-dependent network constructed using the reduction described in Sect. 3. If we add the arc (s_n, s_0) with cost -1 to $\mathbf{G}'$, then $\mathbf{G}'$ has a negative cost arc-dependent simple cycle if and only if $\mathbf{G}$ has a Hamiltonian Path. This gives us the following result which follows from Theorem 1.

Theorem 3. *The SADNC problem is* **NP-complete**.

Proof. Note that once a cycle is found, the cost of the cycle can be computed in polynomial time. Thus, the SADNC problem is in **NP**. All that remains is to show that the problem is **NP-hard**.

Let $\mathbf{G}$ be an unweighted directed network. Let $\mathbf{G}'$ be the corresponding arc-dependent network with the additional arc (s_n, s_0) with cost -1. We now show that $\mathbf{G}$ has a Hamiltonian Path if and only if $\mathbf{G}'$ has a negative-cost arc-dependent simple cycle.

First assume that $\mathbf{G}$ has a Hamiltonian Path. From Theorem 1, we know that $\mathbf{G}'$ has a 0-cost arc-dependent simple path from s_0 to s_n. If we include arc (s_n, s_0) with this path, we obtain a cycle with a total cost of -1. Thus, we have the desired negative cost cycle.

Now assume that $\mathbf{G}'$ has a simple arc-dependent negative cost cycle. The only negative cost arc in $\mathbf{G}'$ is (s_n, s_0). Thus, this arc must be in the cycle. Since this arc has cost -1, the remainder of the cycle must be a 0-cost SADSP from s_0 to s_n. Therefore, from Theorem 1, $\mathbf{G}$ has a Hamiltonian Path. □

Note that $\mathbf{G}'$ has only one arc which costs -1 and that the cost of this arc is *not* arc-dependent. Thus, the problem of finding a negative cost arc-dependent simple cycle is still **NP-complete** even in this restricted case.

5   Inapproximability

In this section, we show that the SADSP$_{Opt}$ problem is **NPO-complete** [2].

Note that the length of a simple path is limited by the number of vertices in $\mathbf{G}$. Additionally, once a path is found, the cost of the path can be computed in polynomial time. Thus, the SADSP$_{Opt}$ problem is in **NPO**. All that remains is to show that the problem is **NPO-hard**. This will be done via reduction from the Traveling Salesman problem.

Let $\mathbf{G}$ be a complete weighted directed network with n vertices and cost function $\mathbf{c}$ such that $c(v_i, v_j)$ is the cost of arc (v_i, v_j). Let c_{max} be the highest (largest) cost of any arc in $\mathbf{G}$. We construct an arc-dependent network $\mathbf{G}'$ as follows:

1. Create the vertex s_0.
2. For each vertex v_i in $\mathbf{G}$, create the vertices w_i and s_i.
3. For each vertex w_i in $\mathbf{G}'$, create the arc (s_0, w_i), which always costs 0.
4. For each pair of vertices w_i and s_k, where $1 \le i, k \le n$, create the following arc:

a. (w_i, s_k) with cost 0 if the preceding arc is (s_{k-1}, w_i) and cost $(n \cdot c_{max} + 1)$ otherwise.

b. (s_k, w_i) with cost $c(v_j, v_i)$ if the preceding arc is (w_j, s_k).

Theorem 4. *For any $h \leq n \cdot c_{max}$, $\mathbf{G'}$ has a SADSP from s_0 to s_n of cost h if and only if $\mathbf{G}$ has a Traveling Salesman Path of cost h.*

Proof. We first assume that $\mathbf{G}$ has a Traveling Salesman Path of cost h.

Without loss of generality, assume that the path is $(v_1, v_2) \rightarrow (v_2, v_3) \rightarrow \cdots \rightarrow (v_{n-1}, v_n)$. By construction, the path

$$(s_0, w_1) \rightarrow (w_1, s_1) \rightarrow (s_1, w_2) \rightarrow (w_2, s_2) \rightarrow (s_2, w_3) \rightarrow \cdots \rightarrow (w_n, s_n)$$

has cost $\sum_{i=1}^{n-1} c(v_i, v_{i+1}) = h$. Thus, $\mathbf{G'}$ has a SADSP of cost h from s_0 to s_n.

We now assume that G' has a SADSP P of cost h from s_0 to s_n.

By construction, the only arcs in $\mathbf{G'}$ are of the form (s_k, w_i) or (w_i, s_k). In other words, there are no arcs of the form (s_i, s_j) or (w_i, w_j). We define $P(k)$, where $0 \leq k \leq n - 1$, such that vertex $w_{P(k)}$ immediately follows vertex s_k in path P. In other words, let $(s_k, w_{P(k)}) \in P$. By construction, the cost of going from $w_{P(k)}$ to any vertex $s_j \neq s_{k+1}$ is $n \cdot c_{max} + 1 > h$. Thus, P must be of the form:

$$(s_0, w_{P(0)}) \rightarrow (w_{P(0)}, s_1) \rightarrow (s_1, w_{P(1)}) \rightarrow (w_{p(1)}, s_2) \rightarrow \cdots \rightarrow (w_{P(n-1)}, s_n).$$

Since P is a simple path, vertices $w_{P(0)}, w_{P(1)}, \ldots, w_{P(n-1)}$ must be distinct. By construction, each arc $(s_k, w_{P(k)})$, for $1 \leq k \leq n - 1$, has cost $c(v_{P(k-1)}, v_{P(k)})$. Also, by construction, each arc $(w_{P(k)}, s_{k+1})$ has cost 0. Thus,

$$c(v_{P(0)}, v_{P(1)}) + c(v_{P(1)}, v_{P(2)}) + \cdots + c(v_{P(n-2)}, v_{P(n-1)}) = h.$$

Since vertices $w_{P(0)}, w_{P(1)}, \ldots, w_{P(n-1)}$ are distinct in $\mathbf{G'}$, it must be the case that $v_{P(0)}, v_{P(1)}, \ldots, v_{P(n-1)}$ are distinct in $\mathbf{G}$. Therefore, by definition, the above path must be a Traveling Salesman Path in $\mathbf{G}$ of cost h. □

Therefore, this reduction from Traveling Salesman is a strict reduction. Since TSP is **NPO-complete** [8], so is the SADSP$_{Opt}$ problem. In other words, the presence of a polynomial approximation for the SADSP$_{Opt}$ problem would imply that $\mathbf{P} = \mathbf{NP}$.

6   An Exact Exponential Algorithm

In this section, we provide an exact exponential algorithm for the SADSP$_{Opt}$ problem.

6.1   A Naive Approach

We first describe a naive approach to solving the problem. This approach proceeds as follows:

1. For each ordering of the vertices in **G** such that $v_1 = s$:
   a. Let k be the index of t in the ordering. This means that $v_k = t$.
   b. The cost of the path from s to t under this ordering is

$$\mathbf{C}[v_2, s] + \sum_{i=2}^{k} \mathbf{C}[(v_{i-1}, v_i), (v_i, v_{i+1})].$$

2. Return the st-path with the least cost.

It is easy to see that every possible st-path is considered using this procedure. However, there are $n!$ possible orderings of the vertices in **G**. Once an ordering is chosen, the cost of the st-path can be computed in polynomial time. Thus, this procedure runs in time $O^*(n!)$. However, a more efficient procedure is possible.

6.2   An Improved Exponential Algorithm

We now provide a more efficient method for solving the SADSP$_{Opt}$ problem.

Let us consider a method that constructs the simple path backwards, starting from the destination t. Finding the shortest one arc path to t can be done by simply looking at the costs of the arcs going into t. We can then look further back to consider all two-arc paths to t. As we continue backtracking, we need to keep track of the set of intermediate vertices so that we do not reuse these vertices.

Suppose we know, for vertices v_i and v_j and set of vertices R, the shortest path p from v_i to t with predecessor v_j such that the set of intermediate vertices is exactly R. Then, from the perspective of further backtracking, it does not matter what order the vertices of R are visited by p. We only need to know that these vertices cannot be used for further backtracking. Thus, we only need to know the shortest path for each start vertex, predecessor vertex, and set of intermediate vertices. This leads us to a dynamic programming based exact exponential algorithm.

Let $P(v_i, v_j, R)$ be the least cost of any path from v_i to t with predecessor v_j and set of intermediate vertices R. Note that the cost of any path from v_i to t depends only on the vertex v_j that precedes v_i. Thus, $P(v_i, v_j, R)$ is well defined. If there are no intermediate vertices, then the only possible path from v_i to t is the arc (v_i, t). Thus, $P(v_i, v_j, \emptyset) = \mathbf{C}[(v_j, v_i), (v_i, t)]$. We will now show that

$$P(v_i, v_j, R) = \min_{v_k \in R} \left\{ \mathbf{C}[(v_j, v_i), (v_i, v_k)] + P(v_k, v_i, R \setminus \{v_k\}) \right\}.$$

Theorem 5. *Let* $\mathbf{G} = \langle \mathbf{V}, \mathbf{E}, \mathbf{c} \rangle$ *be an arc-dependent network. For each* $v_i, v_j \in \mathbf{V}$ *and each* $R \subseteq \mathbf{V} \setminus \{s, t, v_i, v_j\}$,

$$P(v_i, v_j, R) = \min_{v_k \in R} \left\{ \mathbf{C}[(v_j, v_i), (v_i, v_k)] + P(v_k, v_i, R \setminus \{v_k\}) \right\}.$$

Proof. If $|R| = 1$, then there is only one vertex $v_k \in R$. By definition of P, this must be the only intermediate vertex between v_i and t. Thus, the only possible path from v_i to v_t is $(v_i, v_k) \rightarrow (v_k, t)$. The cost of this path is

$$P(v_i, v_j, R) = \mathbf{C}[(v_i, v_k), (v_k, t)] + \mathbf{C}[(v_j, v_i), (v_i, v_k)]$$
$$= \mathbf{C}[(v_j, v_i), (v_i, v_k)] + P(v_k, v_i, \emptyset).$$

Now assume that this holds true for all sets R of size h. Let R' be a set of size $(h + 1)$.

Let p be the shortest path from v_i to t with predecessor v_j and set of intermediate vertices R'. We know that some $v_k \in R'$ immediately follows v_i on p. Thus, p can be broken up into the arc (v_i, v_k) and a path p' from v_k to t. Note that p' has $R' \setminus \{v_k\}$ as its set of intermediate vertices and has v_i as its predecessor. This means that the cost of p' is at least $P(v_k, v_i, R \setminus \{v_k\})$.

If there is a path p^* from v_k to t with set of intermediate vertices $R' \setminus \{v_k\}$ and predecessor v_i that is shorter than p', then the path consisting of the arc (v_i, v_k) followed by p^* is a simple path from v_i to t with set of intermediate vertices R' that is shorter than p. Since this violates the optimality of p, p' must be the shortest path from v_k to t with set of intermediate vertices $R' \setminus \{v_k\}$ and predecessor v_i. Thus, the cost of p' is $P(v_k, v_i, R \setminus \{v_k\})$.

This means that

$$P(v_i, v_j, R') = \min_{v_k \in R'} \left\{ \mathbf{C}[(v_j, v_i), (v_i, v_k)] + P(v_k, v_i, R' \setminus \{v_k\}) \right\}.$$

$\square$

From Theorem 5, $P(v_i, v_j, R)$ can be found in $O(n)$ time once P is known for every pair of vertices and every subset of R. Note that we only need to find $P(v_i, v_j, R)$ when $(v_j, v_i) \in \mathbf{E}$. Thus, there are $O(m \cdot 2^n)$ possible inputs to P. This means that P can be computed in $O(m \cdot n \cdot 2^n)$ time using a dynamic program. Once P is computed, it is easy to see that the SADSP from s to t in $\mathbf{G}$ has cost

$$\min_{R \subseteq \mathbf{V} \setminus \{s,t\}} P(s, \_, R),$$

where $\_$ implies that s does not have a preceding arc in the shortest path from s to t. The same dynamic program can be easily modified to return the shortest path. Thus, the SADSP$_{Opt}$ problem can be solved in $O(m \cdot n \cdot 2^n)$ time.

Note that this algorithm can be extended to solve the SADNC problem by checking if for each arc (v_i, v_j) in $\mathbf{G}$, the cost of the shortest path p from v_j to v_i is less than $-\mathbf{C}[e_k, (v_i, v_j)]$ where e_k is the last arc of p.

6.3   A Lower Bound

We now utilize the reduction from the Traveling Salesman Problem used in Sect. 5 to establish a likely lower bound on the running time of any algorithm that would solve the SADSP$_{Opt}$ problem.

From Theorem 4, we can solve the Traveling Salesman Problem on a graph with n vertices by solving the SADSP_{Opt} problem on a graph with $(2 \cdot n + 1)$ vertices. This means that it is unlikely for there to be a $o^*(\sqrt{2}^{n'})$ algorithm for solving the SADSP_{Opt} problem on a network with n' vertices. If such an algorithm existed, then it would solve the Traveling Salesman Problem on a graph with n vertices in time $o^*(\sqrt{2}^{2 \cdot n + 1}) = o^*(2^n)$. This would constitute an improvement on the bound of $O(n^2 \cdot 2^n)$ obtained by the Heldman-Karp algorithm [6].

7   Integer Programming Formulation

In this section, we provide a **compact** integer programming formulation for the SADSP_{Opt} problem.

Let **G** be an arc-dependent network. We construct the corresponding integer program **I** as follows:

1. For each node v_i, create the variable $v_i \in \{0, \ldots, n\}$.
2. For each arc e_j in **G**, create the variable $x_j \in \{0, 1\}$.
3. If arc e_j goes from node v_i to $v_{i'}$, create the constraint $v_{i'} \geq v_i + (n+1) \cdot x_j - n$.
4. For each pair of arcs e_j and e_k such that the head of e_j is the tail of e_k:
   (a) Create the variable $w_{jk} \in \{0, 1\}$.
   (b) Create the constraint $x_j + x_k - 1 \leq w_{jk}$.
5. For each arc e_k leaving s, create the variable w_{0k} and the constraint $w_{0k} = x_k$.
6. For each arc e_k, create the constraint $\sum w_{jk} \leq x_k$.
7. For each node v_i, let H_i be the set of arcs with head v_i, and let T_i be the set of arcs with tail v_i.
8. For each node $v_i \notin \{s, t\}$, create the constraints

$$\sum_{e_j \in H_i} x_j = \sum_{e_j \in T_i} x_j \leq 1.$$

9. Create the constraints $\sum_{e_j \in H_t} x_j = 1$ and $\sum_{e_j \in T_s} x_j = 1$.
10. Create the objective function $\min \sum \mathbf{C}[e_j, e_k] \cdot w_{jk}$.

We now show that **I** solves the SADSP_{Opt} problem.

Theorem 6. **G** *has an arc-dependent path p from s to t of cost c if and only if* **I** *has a solution such that the objective function has value c.*

Proof. First assume that **G** has an arc-dependent path p of cost c. For each arc e_j, set $x_j = 1$ if $e_j \in p$ and $x_j = 0$, otherwise.

Let e_k be an arbitrary arc in p. If e_k is the first arc in p, then by construction e_k leaves s. Thus, the variable $w_{0,k} = 1$ and arc e_k contributes $\mathbf{C}[e_0, e_k]$ to the value of the optimization function. Recall that e_0 is a phantom arc.

If e_j is the arc that precedes arc e_k in p, then we have the following:

1. The head of e_j is the tail of e_k. Thus, the variable w_{jk} exists.
2. $x_j = x_k = 1$. This implies that the constraints $x_j + x_k - 1 \le w_{jk}$ and $w_{jk} \le 1$ force $w_{jk} = 1$.
3. Let $e_{j'} \ne e_j$ be an arc such that the head of $e_{j'}$ is the tail of e_k.
4. Since p is simple, $e_{j'}$ cannot be in p, so $x_{j'} = 0$. Thus, $w_{j'k} = 0$ satisfies the constraint $x_{j'} + x_k - 1 \le w_{j'k}$.
5. Since $x_k = w_{jk}$, the constraint $\sum w_{jk} \le x_k$ forces $w_{j'k} = 0$. Thus, arc e_k contributes exactly $\mathbf{C}[e_j, e_k]$ to the value optimization function. This is precisely the cost of arc e_k in the path p.

Since each arc in p contributes its cost to the optimization function, we have that the value of the optimization function is the total cost of p as desired.

Now assume that $\mathbf{I}$ has a solution for which the value of the optimization function is c. For each arc e_j in $\mathbf{G}$, add e_j to p if and only if $x_j = 1$. We now show that p is a simple path with arc-dependent cost c.

Observe the following:

1. The constraint $\sum_{e_j \in T_s} x_j = 1$ ensures that exactly one arc leaves s. Thus, p starts at s.
2. The constraint $\sum_{e_j \in T_s} x_j = 1$ ensures that exactly one arc enters t. Thus, p ends at t.
3. For each v_i, the constraints $\sum_{e_j \in H_i} x_j = \sum_{e_j \in T_i} x_j \le 1$ ensure that at most one arc leaves v_i, at most one arc enters v_i, and that the an arc leaves v_i if and only if an arc enters x_i.
4. For each arc $e_j \in p$, the constraint $v_{i'} \ge v_i + (n+1) \cdot x_j - n$ becomes $v_{i'} \ge v_i + 1$. This prevents p from containing a cycle. Thus, p is a simple path.

As argued previously, each arc in p contributes its cost to the optimization function. Therefore, p has arc-dependent cost c as desired. $\square$

If we remove the constraints created in steps 5 and 9 from the construction of integer program $\mathbf{I}$, then this becomes an integer programming formulation of the SADNC problem.

8   Conclusion

In this paper, we discussed the arc-dependent shortest path problem in arbitrarily weighted networks. In particular, we focused on the SADSP_D, SADSP_{Opt}, and SADNC problems in such networks. We established several complexity results. In particular, we showed that the SADSP_D problem is **NP-complete** and that the SADSP_{Opt} problem is **NPO-complete**. We designed an exact exponential algorithm for the SADSP_{Opt} problem. Finally, we provided integer programming representations of the SADSP_{Opt} and SADNC problems.

We plan on investigating graph classes for which the SADSP_{Opt} problem can be solved in polynomial time. We have not yet studied the complexity of this problem for planar graphs or graphs with bounded degree. We may be able to obtain better results for the SADSP_{Opt} problem for these restricted graph types.

References

1. Añez, J., De La Barra, T., Pérez, B.: Dual graph representation of transport networks. Transp. Res. Part B: Methodol. **30**(3), 209–216 (1996)
2. Ausiello, G., Crescenzi, P., Gambosi, G., Kann, V., Marchetti-Spaccamela, A., Protasi, M.: Complexity and Approximation: Combinatorial Optimization and Their Approximability Properties, 1st edn. Springer, Heidelberg (1999). https://doi.org/10.1007/978-3-642-58412-1
3. Bellman, R.E.: Dynamic Programming. Princeton University Press, Princeton (1957)
4. Caldwell, T.: On finding minimum routes in a network with turn penalties. Commun. ACM **4**(2), 107–108 (1961)
5. Dijkstra, E.W.: A note on two problems in connexion with graphs. Numer. Math. **1**, 269–271 (1959)
6. Held, M., Karp, R.M.: A dynamic programming approach to sequencing problems. J. Soc. Ind. Appl. Math. **10**(1), 196–210 (1962)
7. Kirby, R.F., Potts, R.B.: The minimum route problem for networks with turn penalties and prohibitions. Transp. Res. **3**(3), 397–408 (1969)
8. Orponen, P., Mannila, H.: On approximation preserving reductions: complete problems and robustmeasures. Technical report, Department of Computer Science, University of Helsinki (1987)
9. Tan, J.: On path-dependent shortest path and its application in finding least cost path in transportation networks. Master's thesis, School of Computing, National University of Singapore (2003)
10. Tan, J., Leong, HW.: Least-cost path in public transportation systems with fare rebates that are path- and time-dependent. In: Proceedings of the 7th International IEEE Conference on Intelligent Transportation Systems (IEEE Cat. No. 04TH8749), pp. 1000–1005, October 2004
11. Ziliaskopoulos, A.K., Mahmassani, H.S.: A note on least time path computation considering delays and prohibitions for intersection movements. Transp. Res. Part B: Methodol. **30**(5), 359–367 (1996)

Heuristics

The Knapsack Problem with Forfeits

Raffaele Cerulli, Ciriaco D'Ambrosio, Andrea Raiconi$^{(\boxtimes)}$, and Gaetano Vitale

University of Salerno, 84084 Fisciano, Italy
{raffaele,cdambrosio,araiconi,gvitale}@unisa.it

Abstract. In this paper we introduce and study the Knapsack Problem with Forfeits. With respect to the classical definition of the problem, we are given a collection of pairs of items, such that the inclusion of both in the solution involves a reduction of the profit. We propose a mathematical formulation and two heuristic algorithms for the problem. Computational results validate the effectiveness of our approaches.

Keywords: Knapsack Problem · Conflicts · Forfeits · Carousel Greedy

1 Introduction

In many optimization problems, contrasting or mutually exclusive choices often arise. In combinatorial optimization, this issue has been often faced by the introduction of disjunctive constraints or conflicts, meaning that there exists a collection of pairs of items, such that for each pair at most one of them can be included in the solution. Variants of classical problems with the addition of conflicts include the Minimum Spanning Tree Problem [4,7,13,20,21], the Maximum Flow Problem [17], the Knapsack Problem [1,10,12,16,18] and the Bin Packing Problem [8,9,15,19]. In [2,3] disjunctive constraints are used to model interference constraints in the Maximum Lifetime Problem on Wireless Sensor Networks.

In this paper, we study a variant of the 0/1 Knapsack Problem that considers *soft conflict* constraints, or *forfeits*. In more detail, we introduce a *forfeit cost* to be paid each time that both objects in a so-called *forfeit pair* are chosen to be part of the solution. This variant can be of use in scenarios in which strict conflicts may lead to infeasible solutions, or the drawback caused to avoid all conflicts may impact the result more than allowing some of them. We can think of several applications of the problem, including:

- Each object is a machine which needs a worker to be operated. Forfeit pairs represent machines that can only be operated by a worker that we are currently paying, and hence the activation of two such machines requires hiring a new worker, i.e. another salary;

The original version of this chapter was revised: the second author's family name was corrected. The correction to this chapter is available at
https://doi.org/10.1007/978-3-030-53262-8_25

© Springer Nature Switzerland AG 2020, corrected publication 2020
M. Baïou et al. (Eds.): ISCO 2020, LNCS 12176, pp. 263–272, 2020.
https://doi.org/10.1007/978-3-030-53262-8_22

- The chosen items represent the work shift assigned to an employee, and forfeit pairs represent tasks that would involve extras on the salary if assigned together;
- In deciding a series of investments, a cost could derive from making two investment decisions at the same time.

We introduce a mathematical formulation and two heuristic algorithms to solve the problem. The first algorithm is a constructive greedy, while the second is based on the recently introduced Carousel Greedy paradigm. In [5], Carousel Greedy was proposed and applied with success to the Minimum Label Spanning Tree Problem, the Minimum Vertex Cover Problem, the Maximum Independent Set Problem and the Minimum Weight Vertex Cover Problem. It has been subsequently used in several other contexts, including a coverage problem in Wireless Sensor Networks [2], a generalization of the Minimum Label Spanning Tree Problem [6], sentiment analysis in the context of Big Data [11] and community detection [14].

The paper is organized as follows. In Sect. 2, we formally define the problem by presenting notation and the mathematical formulation. The two proposed heuristic algorithms are introduced and discussed in Sect. 3. Computational results are reported and commented in Sect. 4. Conclusions and final remarks are included in Sect. 5.

2   Notation and Mathematical Formulation

Let n be the number of objects, composing the set X. Each object $i \in X$ has an associated profit $p_i > 0$ in the set P and positive weight $w_i > 0$ in the set W, $i = 1, \ldots, n$. Let F be a set of l distinct forfeit pairs

$$F = \{F_k\}_{k=1,\ldots,l}, \quad F_k \subseteq X, |F_k| = 2 \,\forall F_k \in F$$

and let $d_k > 0$, in the set D, be the forfeit cost associated with F_k, $k = 1, \ldots, l$.

Finally, let $b > 0$ be the available budget, that is, the upper bound on the maximum weight of the items chosen to be part of the solution. The problem can be formulated as follows:

$$\max \sum_{i=1}^{n} p_i x_i - \sum_{k=1}^{l} d_k v_k \tag{1}$$

$$\text{s.t.} \tag{2}$$

$$\sum_{i=1}^{n} w_i x_i \le b \tag{3}$$

$$x_i + x_j - v_k \le 1 \qquad \forall F_k = \{i, j\}, k = 1, \ldots, l \tag{4}$$

$$x_j \in \{0, 1\} \qquad \forall i = 1, \ldots, n \tag{5}$$

$$0 \le v_k \le 1 \qquad \forall k = 1, \ldots, l \tag{6}$$

where:

- variable x_i is equal to 1 if object i is selected, and 0 otherwise;
- variable v_k assumes value 1 if the forfeit cost d_k is to be paid according to the chosen objects, and 0 otherwise.

This formulation is a natural generalization of the one reported in [16]. In fact, we relax the strict conflict constraints of the type $x_i + x_j \leq 1 \ \forall F_k = \{i, j\}, k = 1, \ldots, l$ by introducing variables v_j (see constraints (4)). These variables allow us to choose both the elements of the forfeit pair. In this case, v_k is forced to value 1, and therefore, according to objective function (1), the related cost d_k is paid. Constraint (3) makes sure that the total budget is not violated.

The problem includes the classical 0–1 Knapsack Problem as special case and is therefore NP-Hard.

In the next section two heuristic algorithms for the problem are proposed.

3   Algorithms

In this section, two heuristic algorithms for the problem are proposed. We first propose a constructive greedy algorithm in Sect. 3.1. Furthermore, we present an enhancement of this algorithm based on the Carousel Greedy paradigm in Sect. 3.2.

3.1   GreedyForfeits Algorithm

The pseudocode of our constructive greedy algorithm is reported in Algorithm 1.

The algorithm takes as input the items set X, the profit and weight sets P and W, the budget value b, the forfeits set F and forfeit costs set D. The set $S \subseteq X$ initialized in line 1 will contain the items chosen to be included in the solution, while the b_{res} value, introduced in line 2, corresponds in any phase of the algorithm to the residual budget, that is, $b_{res} = b - \sum_{i \in S} w_i$. The main loop of the algorithm is contained in lines 3–28. In each iteration, we first build the set X_{iter} (lines 4–9), containing the items that can still be added to S. That is, X_{iter} contains any item $i \in X$ which does not currently belong to S, and such that its weight w_i is not greater than b_{res}. If X_{iter} is empty, clearly no more items can be added, and the algorithm stops returning S (lines 10–12). Otherwise, we evaluate the most promising element of X_{iter} to be added to S. The main idea is to evaluate each item $i \in X_{iter}$ according to the ratio between profit and weight, $\frac{p_i}{w_i}$. However, for any forfeit pair $F_k = \{i, j\}$ containing i and such that the other item j is already in S, we subtract from p_i the related cost d_k. The updated profit value, indicated as p'_i, reflects the forfeit costs that would have to be paid if i is added to S. For each $i \in X_{iter}$, the computation of p'_i is described in lines 14–19, while the computation of the ratio value $ratio_i$ is reported in line 20. Then, the element $i^* \in X_{iter}$ corresponding to the maximum ratio value is identified (line 22). We note that it is possible for p'_{i^*} (and therefore for $ratio_{i^*}$) to be a negative value. If this is true, it means that it is not convenient to add any other item to S, and the set is returned (lines 23–25). Otherwise, both S and b_{res} are updated to reflect the addition of i^* to the solution (lines 26–27), and

---

Algorithm 1. GreedyForfeits

    Input: (X, W, P, b, F, D)

  1: $S \leftarrow \emptyset$
  2: $b_{res} \leftarrow b$
  3: **while** $X \setminus S \neq \emptyset$ **do**
  4:      $X_{iter} \leftarrow \emptyset$
  5:      **for** $i \in X$ **do**
  6:          **if** $w_i \leq b_{res}$ & $i \notin S$ **then**
  7:              $X_{iter} \leftarrow X_{iter} \cup \{i\}$
  8:          **end if**
  9:      **end for**
10:      **if** $X_{iter} = \emptyset$ **then**
11:          **return** S
12:      **end if**
13:      **for** $i \in X_{iter}$ **do**
14:          $p'_i \leftarrow p_i$
15:          **for** $F_k = \{i, j\} \in F$ **do**
16:              **if** $j \in S$ **then**
17:                  $p'_i \leftarrow p'_i - d_k$
18:              **end if**
19:          **end for**
20:          $ratio_i \leftarrow \frac{p'_i}{w_i}$
21:      **end for**
22:      $i^* \leftarrow argmax[ratio]$
23:      **if** $ratio_{i^*} < 0$ **then**
24:          **return** S
25:      **end if**
26:      $S \leftarrow S \cup \{i^*\}$
27:      $b_{res} \leftarrow b_{res} - w_{i^*}$
28: **end while**
29: **return** S

---

current iteration ends. Finally, if the main loop ends without encountering any of the two mentioned stopping conditions (meaning that trivially all elements of X could be added to S), the set S is returned.

As in many constructive greedy algorithms, a limit of the proposed Greedy-Forfeits is that the contribution of each item composing the solution is evaluated in the moment in which it is added to it. An item appearing attractive in the first iterations could actually lead to many forfeit costs to be added later on. A way to overcome this issue is represented by the carousel greedy paradigm, which we used to enhance the GreedyForfeits heuristic, as described in Sect. 3.2.

3.2 CarouselForfeits Algorithm

The Carousel Greedy (CG) paradigm, originally proposed in [5], provides a generalized framework to improve constructive greedy algorithms, posing itself as a trade-off (in terms of computational time and solution quality) among such greedy procedures and meta-heuristics. The main intuition is that, generally, the choices taken according to the greedy criteria in the first steps of the algorithm could be not very effective due to the lack of knowledge about the subsequent structure of the solution. Therefore, such early choices could end up compromising the quality of the final solution. In order to overcome this phenomenon, earlier choices are iteratively reconsidered and eventually replaced with new ones. Given a basic constructive heuristic, a CG is composed of three main steps:

1. Using the greedy algorithm, a solution is first built, and then some of its latest choices are discarded, obtaining a partial solution.
2. For a predefined number of iterations, the oldest choice is discarded, and a new one is taken according to the greedy criteria of the basic algorithm.
3. Finally, the solution is completed by applying the greedy algorithm, starting from the partial solution obtained at point 2.

The pseudocode of the CG algorithm that we developed, CarouselForfeits, is reported in Algorithm 2.

Algorithm 2. CarouselForfeits

Input: $(X, W, P, b, F, D, \alpha, \beta)$

1: $S \leftarrow GreedyForfeits(X, W, P, b, F, D)$
2: $S' \leftarrow RemoveLastChoices(S, \beta)$
3: $size \leftarrow |S'|$
4: **for** $i = 1 \rightarrow \alpha \times size$ **do**
5:     $S' \leftarrow RemoveOldestChoice(S')$
6:     $i^* \leftarrow GreedyForfeitsSingle(X, W, P, b, F, D, S')$
7:     $S' \leftarrow S' \cup \{i^*\}$
8: **end for**
9: $S'' \leftarrow GreedyForfeitsInit(X, W, P, b, F, D, S')$
10: **return** S''

The algorithm takes the same input of GreedyForfeits, plus two parameters, α and β, such that $0 \leq \beta \leq 1$ and $\alpha \geq 1$.

Lines 1–2 correspond to the first CG step. We first use our greedy to obtain a feasible solution $S \subseteq X$. We then obtain a partial solution S' by dropping some of the last choices; more precisely, the last $\beta|S|$ added items are dropped. Let $size$ be $|S'|$ at this point; the second CG step (lines 4–8) is iterated $\alpha \times size$ times. In each iteration, we first drop from S' the oldest choice. We then execute a variant of GreedyForfeits called GreedyForfeitsSingle. It initializes the solution with S' instead of the empty set, executes a single iteration of the main loop, identifying the best element to be added i^* according to our greedy criterion, and returns it. S' is then updated to include i^*. Finally, in the third and last CG step (line 9), we complete S' by executing a second variant of our greedy,

GreedyForfeitsInit, which again initializes the solution with S', and completes the solution iterating the main loop until no more items can be added. The resulting solution S'' is returned (line 10).

4 Computational Results

In this section we compare the results of our two heuristics with values provided by solving the mathematical formulation reported in Sect. 2, on a set of benchmark instances. We have generated instances according to the following parameters: number of items n in the set $\{500, 700, 800, 1000\}$, number of randomly chosen forfeit pairs $l = n \times 6$, budget $b = n \times 3$. Item weights are integer values assigned randomly, chosen in the interval $[3, \ldots, 20]$. Item profits are random integers in the interval $[5, \ldots, 25]$, while forfeit values are random integers in the interval $[2, \ldots, 15]$. We generated randomly 10 instances for each value of n, for a total of 40 test cases. For the CG parameters, after a preliminary tuning phase, the values $\alpha = 2$, $\beta = 0.05$ were chosen.

All tests were executed on a workstation with an Intel Xeon CPU E5-2650 v3 running at 2.30 GHz with 128 GB of RAM. GreedyForfeits and CarouselForfeits were coded in C++. CPLEX 12.10 was used to solve the mathematical formulation, considering a time limit of one hour, and collecting the best solution found whenever it is violated.

The results on "small" instances ($n = 500$ and $n = 700$) are reported in Table 1, while the results for the "large" ones ($n = 800$ and $n = 1000$) are contained in Table 2. In the tables, each row refers to the results collected for a given instance. The first two columns contain the value of n and an identifier (between 1 and 10) for the instance. The following three columns contain the results obtained by CPLEX, namely the solution value, the related number of paid forfeits and the computational time in seconds. Whenever the time limit (3600 s) is reached, this is indicated with "TL". The following four columns refer to the GreedyForfeits algorithm, and contain the solution value, the gap between such solution and the one found by CPLEX, the number of paid forfeits and the computational time, respectively. Finally, the last four columns contain the same values for CarouselForfeits. Each value reported under the "%gap" columns is computed using the formula $100 - 100 \frac{sol_{heu}}{sol_{opt}}$, where sol_{opt} is the solution value found by CPLEX on the related instance, and sol_{heu} the solution value found by GreedyForfeits or CarouselForfeits.

Let us consider the results on the small instances (Table 1). We note that 18 out of 20 instances were solved to optimality within the time limit. Fon $n = 500$, all instances were solved, requiring up to around 1000 s in the worst case (id = 3 and id = 7). A single instance required less than 100 s (id = 8). For $n = 700$, two instances could not be solved (id = 2 and id = 9), while two other instances (id 8 and 10) required around 3000 s. A single instance (id = 6) was solved in less than 1000 s. According to expectations, the computational time required to solve the instances grows with their size.

Table 1. Results on small instances

Instance type		Model			Greedy				CG			
n	id	sol	#forf.	time(s)	sol	%gap	#forf.	time(s)	sol	%gap	#forf.	time(s)
500	1	2626	66	402.84	2309	12.07	121	0.61	2510	4.42	86	1.74
	2	2660	57	262.91	2428	8.72	109	0.50	2556	3.91	79	1.31
	3	2516	53	1002.02	2335	7.19	106	0.49	2400	4.61	84	1.26
	4	2556	46	222.16	2327	8.96	91	0.50	2441	4.50	73	1.26
	5	2625	58	375.77	2328	11.31	121	0.51	2502	4.69	91	1.29
	6	2615	36	337.87	2309	11.70	95	0.53	2500	4.40	62	1.33
	7	2627	63	958.39	2316	11.84	131	0.51	2470	5.98	90	1.36
	8	2556	50	75.97	2282	10.72	111	0.49	2471	3.33	82	1.28
	9	2613	53	221.67	2337	10.56	113	0.51	2524	3.41	91	1.34
	10	2558	53	662.10	2346	8.29	94	0.49	2439	4.65	82	1.30
700	1	3589	57	1613.96	3218	10.34	144	1.40	3448	3.93	95	3.51
	2	3422	68	TL	2865	16.28	166	1.40	3253	4.94	98	3.45
	3	3679	64	1031.14	3374	8.29	127	1.36	3449	6.25	110	3.47
	4	3664	69	2302.12	3380	7.75	135	1.36	3512	4.15	116	3.51
	5	3647	67	1748.07	3332	8.64	138	1.39	3457	5.21	106	3.51
	6	3596	68	698.76	3330	7.40	138	1.39	3447	4.14	125	3.65
	7	3542	72	1067.65	3086	12.87	154	1.38	3319	6.30	118	3.65
	8	3619	60	3339.49	3183	12.05	142	1.38	3389	6.36	94	3.61
	9	3553	80	TL	3040	14.44	175	1.43	3363	5.35	119	3.64
	10	3652	64	2971.29	3167	13.28	152	1.40	3462	5.20	96	3.60

Looking at the results for GreedyForfeits, we observe that the algorithm is remarkably fast, requiring around 0.5 s for $n = 500$ and 1.4 s for $n = 700$. The gap among the solution values found by the algorithm and the ones returned by CPLEX is within 15% in 19 out 20 cases, and within 10% in 8 cases. In the worst case ($n = 700$, id = 2) the gap is 16.28%. Looking at the number of forfeits, we note that the greedy solutions contain in all cases around twice of such penalties to be paid. In the worst cases ($n = 500$, id=6 and $n = 700, id = 1$) such ratio is 2.64 and 2.53, respectively. These results justify the introduction of the CarouselForfeits as a mean to improve these solutions.

Indeed, we note that CarouselForfeits allows us to obtain significantly reduced solution gaps. For $n = 500$, such gap is between 3.33% (id = 8) and 5.98% (id = 7), and it is under 5% in 9 out of 10 cases. For $n = 700$, the gap varies from 3.93% to 6.36%. We may note that this performance enhancement is reflected in a reduction in the number of forfeits. With respect to GreedyForfeits, CarouselForfeits brings a reduction in the number of paid forfeit costs that is between 9.42% ($n = 700$, id = 6) and 40.96% ($n = 700$, id = 2). The reduction is above 20% in 14 out of 20 cases.

Table 2. Results on large instances

Instance type		Model			Greedy				CG			
n	id	sol	#forf.	time(s)	sol	%gap	#forf.	time(s)	sol	%gap	#forf.	time(s)
800	1	4184	84	TL	3877	7.34	152	2.12	4024	3.82	126	5.31
	2	4065	64	TL	3649	10.23	147	2.09	3827	5.85	106	5.25
	3	4101	89	TL	3684	10.17	189	2.15	3886	5.24	148	5.47
	4	4051	82	TL	3633	10.32	164	2.17	3850	4.96	132	5.48
	5	4085	95	TL	3622	11.33	197	2.14	3900	4.53	129	5.48
	6	4249	90	TL	3790	10.80	175	2.15	4084	3.88	123	5.68
	7	4117	79	TL	3682	10.57	195	2.21	3897	5.34	145	5.70
	8	4063	94	TL	3636	10.51	180	2.15	3859	5.02	135	5.64
	9	4080	75	TL	3703	9.24	149	2.11	3853	5.56	123	5.58
	10	4124	91	TL	3794	8.00	169	2.19	4050	1.79	125	5.71
1000	1	4925	89	TL	4455	9.54	181	4.72	4655	5.48	151	12.04
	2	4966	123	TL	4562	8.14	212	4.77	4756	4.23	176	12.14
	3	5170	108	TL	4724	8.63	199	4.87	4897	5.28	158	12.35
	4	5139	115	TL	4602	10.45	202	4.76	4916	4.34	159	12.12
	5	5134	97	TL	4693	8.59	194	4.82	4935	3.88	152	12.16
	6	5063	99	TL	4606	9.03	169	4.74	4858	4.05	130	12.35
	7	5100	108	TL	4482	12.12	206	4.75	4876	4.39	156	12.51
	8	5178	100	TL	4780	7.69	189	4.83	4916	5.06	165	12.61
	9	5108	97	TL	4536	11.20	193	4.76	4890	4.27	153	12.50
	10	5174	118	TL	4724	8.70	217	4.89	4998	3.40	160	12.66

Finally we note that, while more complex than the greedy algorithm, CarouselForfeits remains a fast algorithm, running in 1.74 s in the worst case for $n = 500$ (id $= 1$) and in 3.65 s in the worst case for $n = 700$ (id $= 6$ and id $= 7$).

We now turn to large instances (Table 2). We note that no instances of this size could be solved to certified optimality within the time limit, emphasizing the need for good heuristic algorithms. Looking at GreedyForfeits, we note that in terms of percentage gap from the CPLEX solutions there is not a noticeable difference with respect to the case of small instances. Indeed, the gap is below 10% for 10 out of 20 instances, and equal to 12.12% in the worst case ($n = 1000$, id $= 7$). Again, the number of paid forfeit costs for greedy solutions are around twice those produced by CPLEX with a peak of this ratio equal to 2.47 ($n = 800$, id $= 7$). In terms of computational times, all instances required around 2 s for $n = 800$ and less than 5 s for $n = 1000$.

Looking at CarouselForfeits, we observe that for $n = 800$ gaps are between 1.79% (id $= 10$) and 5.85% (id $= 2$), while for $n = 1000$ they are between 3.40% (id $= 10$) and 5.48% (id $= 1$). The reduction in the number of paid forfeits with respect to GreedyForfeits is between 12.70% ($n = 1000$, id $= 8$) and 34.52% ($n = 800$, id $= 5$). As in the case of small instances, this gap is above 20% in 14 out of 20 cases. Finally, we observe that the computational times of CarouselForfeits remain very reasonable, since all instances with $n = 800$ are solved in less than 6 s, and all instances with $n = 1000$ in less than 13 s.

5 Conclusions

In this work we introduced and studied the Knapsack Problem with soft conflict constraints, or forfeits. In this variant, a cost must be paid each time that both elements of a so-called forfeit pair are chosen together to be included in the solution. A mathematical formulation and two heuristic approaches have been proposed. In particular, we designed a Carousel Greedy algorithm which is able to extend and improve our constructive greedy, and produces solutions of good quality in fast computational times.

Future research efforts will be spent on developing new exact and metaheuristic approaches for the problem, as well as applying the concept of forfeits to other problems.

References

1. Bettinelli, A., Cacchiani, V., Malaguti, E.: A branch-and-bound algorithm for the knapsack problem with conflict graph. INFORMS J. Comput. **29**(3), 457–473 (2017)
2. Carrabs, F., Cerrone, C., D'Ambrosio, C., Raiconi, A.: Column generation embedding carousel greedy for the maximum network lifetime problem with interference constraints. In: Sforza, A., Sterle, C. (eds.) ODS 2017. SPMS, vol. 217, pp. 151–159. Springer, Cham (2017). https://doi.org/10.1007/978-3-319-67308-0_16
3. Carrabs, F., Cerulli, R., D'Ambrosio, C., Raiconi, A.: Prolonging lifetime in wireless sensor networks with interference constraints. In: Au, M.H.A., Castiglione, A., Choo, K.-K.R., Palmieri, F., Li, K.-C. (eds.) GPC 2017. LNCS, vol. 10232, pp. 285–297. Springer, Cham (2017). https://doi.org/10.1007/978-3-319-57186-7_22
4. Carrabs, F., Cerulli, R., Pentangelo, R., Raiconi, A.: Minimum spanning tree with conflicting edge pairs: a branch-and-cut approach. Ann. Oper. Res. 1–14 (2018). https://doi.org/10.1007/s10479-018-2895-y
5. Cerrone, C., Cerulli, R., Golden, B.: Carousel greedy: a generalized greedy algorithm with applications in optimization. Comput. Oper. Res. **85**, 97–112 (2017)
6. Cerrone, C., D'Ambrosio, C., Raiconi, A.: Heuristics for the strong generalized minimum label spanning tree problem. Networks **74**(2), 148–160 (2019)
7. Darmann, A., Pferschy, U., Schauer, J.: Minimal spanning trees with conflict graphs. Optimization online (2009)
8. Epstein, L., Levin, A.: On bin packing with conflicts. SIAM J. Optim. **19**(3), 1270–1298 (2008)
9. Gendreau, M., Laporte, G., Semet, F.: Heuristics and lower bounds for the bin packing problem with conflicts. Comput. Oper. Res. **31**(3), 347–358 (2004)
10. Gurski, F., Rehs, C.: The knapsack problem with conflict graphs and forcing graphs of bounded clique-width. In: Fortz, B., Labbé, M. (eds.) Operations Research Proceedings 2018. ORP, pp. 259–265. Springer, Cham (2019). https://doi.org/10.1007/978-3-030-18500-8_33
11. Hadi, K., Lasri, R., El Abderrahmani, A.: An efficient approach for sentiment analysis in a big data environment. Int. J. Eng. Adv. Technol. **8**(4), 263–266 (2019)

12. Hifi, M., Otmani, N.: An algorithm for the disjunctively constrained knapsack problem. Int. J. Oper. Res. **13**(1), 22–43 (2012)
13. Kanté, M.M., Laforest, C., Momège, B.: Trees in graphs with conflict edges or forbidden transitions. In: Chan, T.-H.H., Lau, L.C., Trevisan, L. (eds.) TAMC 2013. LNCS, vol. 7876, pp. 343–354. Springer, Heidelberg (2013). https://doi.org/10.1007/978-3-642-38236-9_31
14. Kong, H., Kang, Q., Li, W., Liu, C., Kang, Y., He, H.: A hybrid iterated carousel greedy algorithm for community detection in complex networks. Physica A: Stat. Mech. Appl. **536** (2019). Article Number 122124
15. Muritiba, A.E.F., Iori, M., Malaguti, E., Toth, P.: Algorithms for the bin packing problem with conflicts. Informs J. Comput. **22**(3), 401–415 (2010)
16. Pferschy, U., Schauer, J.: The knapsack problem with conflict graphs. J. Graph Algorithms Appl. **13**(2), 233–249 (2009)
17. Pferschy, U., Schauer, J.: The maximum flow problem with conflict and forcing conditions. In: Pahl, J., Reiners, T., Voß, S. (eds.) INOC 2011. LNCS, vol. 6701, pp. 289–294. Springer, Heidelberg (2011). https://doi.org/10.1007/978-3-642-21527-8_34
18. Pferschy, U., Schauer, J.: Approximation of knapsack problems with conflict and forcing graphs. J. Comb. Optim. **33**(4), 1300–1323 (2016). https://doi.org/10.1007/s10878-016-0035-7
19. Sadykov, R., Vanderbeck, F.: Bin packing with conflicts: a generic branch-and-price algorithm. INFORMS J. Comput. **25**(2), 244–255 (2013)
20. Samer, P., Urrutia, S.: A branch and cut algorithm for minimum spanning trees under conflict constraints. Optim. Lett. **9**(1), 41–55 (2014). https://doi.org/10.1007/s11590-014-0750-x
21. Zhang, R., Kabadi, S.N., Punnen, A.P.: The minimum spanning tree problem with conflict constraints and its variations. Discret. Optim. **8**(2), 191–205 (2011)

An Efficient Matheuristic
for the Inventory Routing Problem

Pedro Diniz[1], Rafael Martinelli[2(✉)], and Marcus Poggi[1]

[1] Departamento de Informática, PUC-Rio, Rio de Janeiro, Brazil
{pfonseca,poggi}@inf.puc-rio.br
[2] Departamento de Engenharia Industrial, PUC-Rio, Rio de Janeiro, Brazil
martinelli@puc-rio.br

Abstract. We consider the general multi-vehicle and multi-period Inventory Routing Problem (IRP). A challenging aspect of solving IRPs is how to capture the relationship among the periods where the routing takes place. Once the routes are defined, computing the optimal inventory at each customer on each period amounts to solving a network flow problem. We investigate the impact of efficiently solving this recurring network problem on the solutions found by the devised algorithm. A very significant impact is observed when solving 638 instances in a classical benchmark set, improving 113 upper bounds through assembling the network optimization into an ILS-RVND algorithm. In particular, the results suggested this approach performs better for larger instances with more periods, obtaining speed-ups of about ten times. A detailed comparison against nine of the most prominent exact and heuristic methods favors the proposed approach.

Keywords: Inventory routing Problem · Vendor-Managed Inventory · Matheuristics · Network Simplex · Iterated Local Search

1 Introduction

Logistics decisions are recognized for having a significant impact on every organization's strategic planning. With the advent of the internet and the ever-increasing globalization, this impact is growing and fostering companies in a continuous search to reduce costs and increase logistics efficiency. Global optimization of the supply chain is one of the efforts that have been employed in the late years and is gaining in popularity due to improved results when compared to traditional models. Vendor-Managed Inventory (VMI) systems are a big step towards this objective. It centralizes the decisions on the suppliers allowing them to reduce both production and distribution costs at the same time by combining and coordinating transportation and demand for multiple customers. The Inventory Routing Problem [10] is then an application of VMI to define routes to one or more vehicles to service a set of customers during a planning horizon. In every period, all customers' demands should be met with products from the

© Springer Nature Switzerland AG 2020
M. Baïou et al. (Eds.): ISCO 2020, LNCS 12176, pp. 273–285, 2020.
https://doi.org/10.1007/978-3-030-53262-8_23

limited customers' inventories and/or from the production of the single depot, where all vehicles are located. Therefore, besides the customers' visits to each customer, the amount of product delivered must also be decided for each vehicle and period. The problem then asks for the minimum overall cost, considering the vehicles' routing costs, and inventory costs from the customers and depot.

This problem models three simultaneous supply chain decisions into one single (global) optimization problem: (i) decide the periods in which a customer should be visited, (ii) decide how much should be delivered for each customer, and (iii) decide the best delivery routes. Considering also variations of the problem, usually, one out of these two inventory policies are applied regarding customer visits: (i) order-up-to level policy (OU), where customer inventory is filled to its maximum capacity at every visit; and (ii) maximum level policy (ML), where customer inventory may be below but never above its limit at every visit.

The IRP is gaining much attention recently due to its importance, complexity, many variations, and the lack of an exact algorithm capable of solving instances of reasonable size in a short time. Still, considering it belongs to the class of the NP-hard problems, which can be proved by a reduction to the Traveling Salesman Problem, advances in the literature over the last ten years are very promising. Over the years, different methods were proposed for the single and, more recently, multi-vehicle versions of the problem. For the single-vehicle version, heuristic methods include a two-step heuristic algorithm [9], an Adaptive Large Neighborhood Search (ALNS) [13], a hybrid tabu search method [4], Simulated Annealing [2,20] and Iterated Local Search (ILS) [2,20]. The first exact method for the single-vehicle IRP was a branch-and-cut algorithm [5]. Methods for the multi-vehicle version are more recent but are increasing in a fast-pace. Heuristic methods include an ALNS [12], a hybrid matheuristic [7], a kernel search matheuristic from [18], an ILS [22], a Simulated Annealing [2] and unified matheuristic [11]. Exact methods for this variant include branch-and-cut [1,14], a branch-cut-and-price [16] and, more recently, a single-period cutting planes [8].

This paper considers a class of the multi-vehicle and multi-period Inventory Routing (IRP). The main result is a matheuristic composed of an Iterated Local Search, with Random Variable Neighborhood Descent, that explores a modification on a Network Flow algorithm to efficiently find the optimal inventory flow and costs, given the routes to be performed in each period. The basic idea was already explored in [12], where the authors report the computation time to solve the Network Flow problem as an issue. It limited the approach performance significantly in terms of solution quality versus computation time. In the resolution of IRPs, capturing the distribution of goods relation over the periods is a challenge. This research shows that efficiently solving this Network Flow problem pays off. A very significant impact is observed. When solving 638 instances in the small benchmark set from [5], for the ML policy, we improve 113 upper bounds through plugging the efficient network optimization into an ILS-RVND algorithm. In particular, the results suggest this approach performs better on larger instances, with more periods, customers, and vehicles. The proposed algorithm achieves about ten times faster execution time by updating and reusing the underlying structure used by the Network Simplex algorithm. As the final algorithm intensively uses an enhanced Network Simplex method, we classify this heuristic algorithm as a *Matheuristic*.

This paper is organized as follows. The next section presents the metaheuristic used. Section 3 explains the changes in the Network Simplex algorithm to speed up the solution. In Sect. 4, we show the computational experiments and analyses. Section 5 concludes the work and lists future research.

2   Iterated Local Search

We start by describing the basic Iterated Local Search (ILS) algorithm. We address the fundamental constructs of the heuristic and the main components that build up the algorithm: the neighborhoods and the local search. The search-space used during the algorithm execution comprehends both feasible and infeasible solutions. A collection of neighborhoods is defined to explore this search-space. These neighborhoods are commonly used on several classes of vehicle routing problems, except the insert and remove neighborhoods, which are particular for the IRP since a customer may or may not be visited in a given period of the planning horizon. The neighborhoods are the following:

- $\texttt{insert}(c, v, p)$: Insert customer c into route v in period p.
- $\texttt{remove}(c, v, p)$: Remove customer c from route v in period p.
- $\texttt{relocate}(c_1, v_1, p_1, c_2, v_2, p_2)$: Remove c_1 from route v_1 and period p_1, and insert into route v_2 in period p_2 before customer c_2.
- $\texttt{swap}(v, c_1, c_2)$: Swap customer at position c_1 with customer at position c_2 in rout v.
- $\texttt{shift}(v, c, k)$: Move customer at position c to position $c + k$ in route v.
- $\texttt{reverse-subtour}(s)$: Reverse the subtour s. This movement is equivalent to the well-known 2-opt move.

The ILS uses a Randomized Variable Neighborhood Descent (RVND) [20], which chooses, at each iteration, among the above neighborhoods. Given the current solution, represented by a set of customers tours for each vehicle and period, it randomly selects a neighborhood and performs a local search using the best improvement strategy. The new solution cost is then obtained by the sum of the routing and inventory costs.

The routing cost can be calculated in constant time, as for most of VRPs, given the changes performed by a move on the routes. Regarding the inventory, the modified routes may impose different deliveries and may forbid fulfilling all customers' demands. Therefore, changes in the amount delivered for each customer and period may be required. Our algorithm finds the optimal inventory or detects no feasible inventory exists by solving a Network Flow problem. The inventory cost either corresponds to the optimal inventory cost or a penalty cost in case of infeasibility. The next section presents the Network Flow problem that determines the optimal delivery amounts.

An outline of the ILS is presented in Algorithm 1. It starts with an empty initial solution as the current solution, i.e., no customer is visited in any period by any vehicle. This solution is infeasible, therefore penalized inventory costs will be associated. Then, in each iteration, the current solution is perturbed by applying $max\_perturb$ random moves followed by a complete run of the local search. During the perturbation, after each random move, the new solution is accepted if it is an improving solution or if it passes an acceptance criterion.

Algorithm 1. Outline of the Iterated Local Search

```
 1: function ILS
 2:     sol ← localSearch(empty_sol)
 3:     best ← cur ← sol
 4:     no_imp ← 0
 5:     for i ← 1 to max_iter do
 6:         for j ← 1 to max_perturb do
 7:             new ← randMove(sol)
 8:             if new.cost < sol.cost or testAccept(new, sol) then
 9:                 sol ← new
10:             end if
11:         end for
12:         sol ← localSearch(sol)
13:         if sol.cost < cur.cost or testAccept(sol, cur) then
14:             cur ← sol
15:             if cur.cost < best.cost then
16:                 best ← cur
17:                 no_imp ← 0
18:             end if
19:         else
20:             sol ← cur
21:         end if
22:         no_imp ← no_imp + 1
23:         if no_imp > max_no_imp or sol.cost > (1 + max_perc)best.cost then
24:             sol ← cur ← best
25:         end if
26:     end for
27: end function
```

After the local search, this resulted solution is tested against the current best solution. It will be accepted if it is an improving solution or if it passes the acceptance criterion. If accepted, it will be tested for improvement against the global best solution and replace it in case of improvement. If no improvement was observed for a given number of iterations or the current solution value is higher than a given percentage of the global best, the current solution and current best solution are replaced with the global best solution.

The acceptance criterion is implemented based on the solution value obtained after the first call to the local search. In the first iteration, it will accept a solution

20% worse with a probability of 50%. This chance is then further decreased based on the number of iterations max_iter to accept a solution 10% worse with a probability of 10% on the last iteration [21].

3   Network Simplex

While the reduction to Minimum Cost Flow [19] is not something new to the literature, other studies that tried this same decomposition [12] reported significant running times where approximately 65% of the total time was spent solving only the subproblem. Our approach achieves, on average, ten times faster running times by reusing the underlying Network Simplex structure. Every optimal inventory is obtained starting the Network Simplex algorithm from the optimal solution of the previous iteration. The model and the update procedure are now described.

3.1   Formulation

The subproblem is defined on a directed acyclic graph $G = (V, A)$. In this graph, the vertex set is composed of $|T|$ copies of the supplier, customers, and vehicles, plus an artificial vertex representing the excess of product that may exist at the end of the time horizon for some customer or the supplier. Every supplier vertex in this graph is a source of flow equal to its production on the given period. Similarly, every customer demands a flow equal to its demand in the given period. The artificial vertex demands the difference between the total supply and the total demand.

The arc set has four different types of arcs. The first type contains arcs from the supplier vertex to vehicle vertices. The capacity of each arc is the vehicle's capacity, and the cost is zero. The second type contains arcs from the vehicle vertices to customers vertices. Each arc has unlimited capacity and cost zero. The third type contains arcs between consecutive periods for the supplier and the customers. They represent the inventory that may wait from one period to the next. Their cost and capacity are the inventory cost and capacity of each customer or supply. Finally, the last type contains the arcs from the customers or supplier in the last period to the excess vertex. For each arc, its capacity is the supplier or customer inventory capacity, but in this case, its cost is zero. An example of this graph, with three customers, two vehicles, and two periods, is presented in Fig. 1.

If a vehicle visits a customer in a given period, we keep the original cost of the corresponding vehicle-customer arc. On the other hand, if the vehicle does not visit the customer in the period, we set the arc cost to infinity. This approach is more efficient than to rebuild graph G or to remove and add arcs before every call to the Network Flow algorithm.

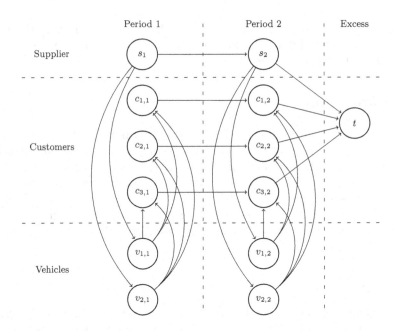

Fig. 1. Proposed network flow model for 2 customers, vehicles and periods.

3.2   Fast Flow Calculation

Once the graph is built, we use the Network Simplex algorithm to calculate the total inventory cost for a given matheuristic solution. During the local search, several move evaluations are performed in sequence by simple changes in the previous solution. By analyzing the neighborhoods presented in Sect. 2, it is possible to control the impact of each one in the Network Flow graph. For example, when an insert is evaluated, we have to add an arc from the corresponding vehicle to the customer (i.e., setting its cost to zero). For a remove evaluation, we have to remove the vehicle-customer arc (i.e., setting its cost to infinity). The modifications required by all vehicle moves can be represented by changes in costs on arcs.

The Network Simplex works using a spanning tree structure. It defines three sets, T, L and U. T contains the set of arcs that compose the spanning tree, i.e., the basic variables. L contains the set of arcs where the flow is zero, and U contains the set of arcs where flow equals the edge capacity. Both sets contain the non-basic variables. A Network Simplex iteration starts by calculating the reduced cost of every arc in L and U based on the dual information of the vertices, represented by the dual variables π_i. The reduced cost is then calculated as $\bar{c}_{ij} = c_{ij} + \pi_i - \pi_j$. If any non-basic arc has a negative reduced cost, the procedure moves this arc to set T, and removes another arc from T following the Simplex rules. At the end of execution, the optimal solution is found by joining the arc sets with positive flow, i.e., the arc sets T and U. We refer the reader to [15] for a complete description of the method.

Usually, any change in the solution would require executing the entire algorithm again because (i) some arcs may have been introduced (or removed), (ii) arcs' capacities may have changed, (iii) node supplies may have been modified, (iv) the maximum flow may have changed, and (v) arcs costs may have changed. Taking advantage of the IRP structure, the model we propose addresses most of these issues, and reduces changes to be the equivalent of deciding which arcs must be moved between T, L and U. In this model, change (i) does not happen because arcs are never introduced or removed, only their costs can change between zero or infinity. Changes (ii) and (iii) are also not present because the vehicle and inventory constraints are the same, and the production and demands values remain unchanged. All previous reasons implicate that change (iv) cannot occur too. The only change left is (v), and we show how to deal with it.

In Network Simplex, the arc cost is used to calculate the arc's dual cost and the dual variables π_i of the nodes below this arc if it is in T. These changes are sufficient to avoid having to execute the entire Network Simplex again and let the algorithm continue from where it stopped on the previous solution. The dual cost of each edge can be recalculated while verifying if the edge is a candidate to enter the tree. Otherwise, there is no need to check its reduced cost. Updating node potentials, on the other hand, requires more effort because they must be updated on a specific order. While other strategies may be used, we propose a procedure that iterates over all nodes starting from the root and continues node by node following the order used for tree construction (commonly referred to as the "thread order"). Algorithm 2 illustrates the procedure.

Algorithm 2. Update π values

1: **function** UPDATE_PI( )
2:     $node \leftarrow thread[root]$
3:     **while** $node \neq root$ **do**
4:         $arc \leftarrow pred[node]$
5:         $tgt \leftarrow arc.target$
6:         $src \leftarrow arc.source$
7:         **if** $tgt = node$ **then**
8:             $\pi[node] \leftarrow \pi[src] + cost[arc]$
9:         **else**
10:             $\pi[node] \leftarrow \pi[tgt] - cost[arc]$
11:         **end if**
12:     **end while**
13: **end function**

---

4 Computational Experiments

The NSIRP was implemented in C++ on Ubuntu Linux. For the inventory cost calculation (Minimum Cost Flow), we modified and used the Network Simplex algorithm from the LEMON C++ library [17], adding the proposed modifications. Computational experiments were performed with a single thread on an Intel Core i7-8700K 3.7 GHz with 64 GB RAM. The algorithm was tested over the benchmark instances proposed in [5]. They are composed of 160 files organized into two classes of instances (*low* and *high* inventory cost) and cover scenarios where the horizon H can be equal to 3 or 6 periods. The number of clients n is $n = 5t$, with $t = 1, ..., 10$ when $H = 3$ and $t = 1, ..., 6$ when $H = 6$. All instances were tested from two up to five identical vehicles, dividing the vehicle capacity by the number of vehicles (and rounding down), and resulting in a total of 640 instances (but two are known to be infeasible). For each instance, we run the matheuristic algorithm ten times.

We first present the results on our main contribution, the Fast Flow Calculation. We performed a complete run on all instances with the regular Network Simplex algorithm (NSA) and with the Fast Flow Calculation (FFC). Figure 2 shows the average fraction of the original time that our modification obtains, calculated for each instance as ub_{FFC}/ub_{NSA}, for each of the four types of instance while the number of vehicle grows: low inventory cost with three periods (L3), high inventory cost with three periods (H3), low inventory cost with six periods (L6) and high inventory cost with six periods (H6). The smallest improvement was found for instances with two vehicles, three periods, five customers, and high cost. Our method runs in 21.1% of the original time. From the figure, we can notice a clear tendency. As the number of vehicles or the number of customers, or even the number of periods goes up, the improvement is more significant. The best improvement is when the method obtains a running time of 6.0% from the original one. It is noteworthy the correlation between the same characteristics, but different cost types (high and low). The results show that they have no impact on the proposed method.

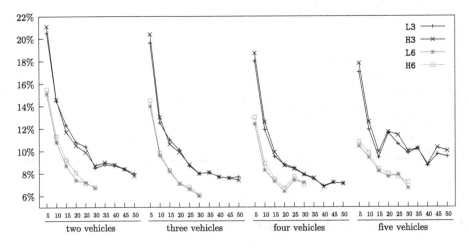

Fig. 2. Time fraction of fast flow, for each size, vehicles, periods and cost type.

Using fast flow, we now compare the results of NSIRP against the best known upper bounds and most prominent methods from the literature comprising nine methods: four exact, three matheuristics, and two metaheuristics. The exact methods are ABIS (Branch-and-Bound) from Archetti et al. (2007) [6], CL (Branch-and-Cut) from Coelho and Laporte (2014) [14], DRC (Branch-Cut-and-Price) from Desaulniers et al. (2016) [16], and ABW (Branch-and-Cut) from Avella et al. (2018) [8]. The matheuristic methods comprehend ABS Archetti et al. (2017) [7], the unified decomposition CCJ from Chitsaz et al. (2019) [11], and a kernel search AGMS from Archetti et al. (2019) [3,18]. The metaheuristics are an Iterated Local Search SOSG from Santos et al. (2016) [22], and a Simulated Annealing AMM from Alvarez et al. (2018) [2]. The parameters used in the experiments were $max\_iter = 500$, $maxperturb = 15$, $maxnoimp = 25$ and $maxperc = 1.2$.

Figure 3 shows the average gap for each instance type when the number of vehicles grows. Each gap is calculated against the best-known upper bound from the literature (LIT) as $(ub_{FFC} - ub_{LIT})/ub_{LIT}$. It becomes negative as the number of periods goes up with the worst, never exceeding 4.0%. The gap growth for each series was minimal reinforcing the correlation between same characteristics, except for L3. Table 1 shows the detailed results. The overall average gap was 0.77%, 0.01% on the best, and 1.84% on the worst. Of the 638 instances, NSIRP improved or at least matched 463 (72.0%) of the best-known upper bounds, with 113 improvements (51.0% of the 221 open instances) and 350 matches.

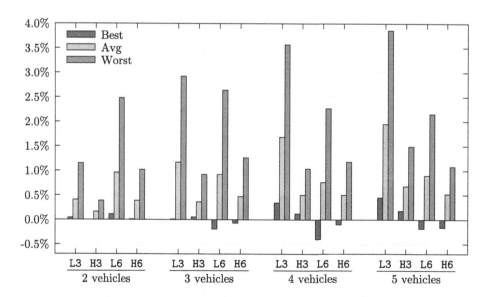

Fig. 3. Average UB gap for each group and number of vehicles.

Table 1. Summary results for classical IRP instances

Type	Veh	Best	Avg	Worst	Time	Better	Equal	Worse
L3	2	0.04%	0.41%	1.15%	47.88	0	45	5
H3	2	0.00%	0.16%	0.39%	58.85	2	41	7
L6	2	0.11%	0.96%	2.48%	75.17	3	18	9
H6	2	0.01%	0.39%	1.02%	89.78	7	14	9
Avg/Total		0.04%	0.48%	1.26%	67.92	12	118	30
L3	3	0.01%	1.17%	2.92%	74.74	2	44	4
H3	3	0.05%	0.36%	0.92%	92.79	3	35	12
L6	3	−0.19%	0.92%	2.64%	107.85	14	11	5
H6	3	−0.07%	0.47%	1.26%	133.40	14	9	7
Avg/Total		−0.05%	0.73%	1.94%	102.20	33	99	28
L3	4	0.34%	1.68%	3.57%	101.56	6	30	14
H3	4	0.12%	0.50%	1.04%	127.87	10	25	15
L6	4	−0.40%	0.76%	2.27%	140.04	15	5	10
H6	4	−0.10%	0.50%	1.18%	176.22	13	6	11
Avg/Total		-0.01%	0.86%	2.01%	136.42	44	66	50
L3	5	0.45%	1.95%	3.86%	129.59	0	31	19
H3	5	0.18%	0.68%	1.49%	165.29	3	26	21
L6	5	−0.18%	0.90%	2.15%	156.41	8	6	15
H6	5	−0.16%	0.52%	1.08%	221.56	13	4	12
Avg/Total		0.07%	1.01%	2.15%	168.21	24	67	67
Overall		0.01%	0.77%	1.84%	118.69	113	350	175

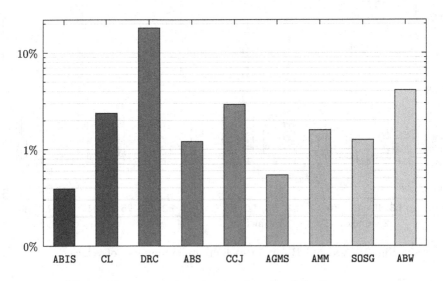

Fig. 4. Gap comparison from each work of the literature.

A best upper bound gap comparison against each method from the literature is presented in Fig. 4. For each method (MET), this gap is calculated against our best upper bound for each instance as $(ub_{MET} - ub_{FFC})/ub_{FFC}$. This comparison comprehends only the instances where the method provided an upper bound. From the figure, we can notice that, on average, NSIRP outperforms all nine methods we compare from the literature.

5 Conclusion

We have defined a new model to solve the multi-vehicle Inventory Routing. We have demonstrated how this model enables the development of inventory-exact solutions with more than ten times faster running times by extending current state-of-the-art Network Simplex implementation and proposing a fast flow procedure. We have implemented an ILS-RVND based matheuristic, entitled NSIRP, to assess the performance of the model on well-known instances from the literature and reported the results.

Computational experiments using well-known instances demonstrated that NSIRP could be widely applied between different instances, with an average upper bound gap of 0.77%. On the 638 tested instances, the method improved or at least matched 463 (72.0%) of the best-known upper bounds, improving 113 (51.0%) of the 221 open instances. Running time results indicates that there is no correlation between fast flow calculation times and the number of vehicles, customers, or periods. It is demonstrated to be a scalable method, well suited for the cases when one of these variables grows. Compared to other methods from the literature NSIRP was capable of decreasing the average upper bound gap for all of them, in the worst case by at least 0.4%.

Considering we are capable of solving the inventory subproblem using an exact algorithm that implements the simplex algorithm with significantly reduced running times, further work would be to use dual information to create more problem-specific neighborhoods. Use this information, such as the nodes π_i or the arcs' reduced costs, to reduce the search space, and navigate between solutions that otherwise would be composed of multiple moves.

Acknowledgements. This research was partially supported by the Conselho Nacional de Desenvolvimento Científico e Tecnológico (CNPq), grants 140084/2017-7, 313521/2017-4, 425962/2016-4 and 311954/2017-0, and by the Coordenação de Aperfeiçoamento de Pessoal de Nível Superior - Brasil (CAPES), Financing Code 001. All support is gratefully acknowledged.

References

1. Adulyasak, Y., Cordeau, J.F., Jans, R.: Formulations and branch-and-cut algorithms for multivehicle production and inventory routing problems. INFORMS J. Comput. **26**(1), 103–120 (2013)
2. Alvarez, A., Munari, P., Morabito, R.: Iterated local search and simulated annealing algorithms for the inventory routing problem. Int. Trans. Oper. Res. **25**(6), 1785–1809 (2018)
3. Archetti, C., Guastaroba, G., Huerta-Muñoz, D., Speranza, M.: A kernel search heuristic for the multi-vehicle inventory routing problem (2019). http://or-brescia.unibs.it/instances
4. Archetti, C., Bertazzi, L., Hertz, A., Speranza, M.G.: A hybrid heuristic for an inventory routing problem. INFORMS J. Comput. **24**(1), 101–116 (2012)
5. Archetti, C., Bertazzi, L., Laporte, G., Speranza, M.G.: A branch-and-cut algorithm for a vendor-managed inventory-routing problem. Transp. Sci. **41**(3), 382–391 (2007)
6. Archetti, C., Bianchessi, N., Irnich, S., Speranza, M.G.: Formulations for an inventory routing problem. Int. Trans. Oper. Res. **21**(3), 353–374 (2014)
7. Archetti, C., Boland, N., Speranza, M.G.: A matheuristic for the multivehicle inventory routing problem. INFORMS J. Comput. **29**(3), 377–387 (2017)
8. Avella, P., Boccia, M., Wolsey, L.A.: Single-period cutting planes for inventory routing problems. Transp. Sci. **52**(3), 497–508 (2018)
9. Bertazzi, L., Paletta, G., Speranza, M.G.: Deterministic order-up-to level policies in an inventory routing problem. Transp. Sci. **36**(1), 119–132 (2002)
10. Campbell, A., Clarke, L., Kleywegt, A., Savelsbergh, M.: The inventory routing problem. In: Crainic, T.G., Laporte, G. (eds.) Fleet Management and Logistics. CRT, pp. 95–113. Springer, Boston, MA (1998). https://doi.org/10.1007/978-1-4615-5755-5_4
11. Chitsaz, M., Cordeau, J.F., Jans, R.: A unified decomposition matheuristic for assembly, production, and inventory routing. INFORMS J. Comput. **31**(1), 134–152 (2019)
12. Coelho, L.C., Cordeau, J.F., Laporte, G.: Consistency in multi-vehicle inventory-routing. Transp. Res. Part C: Emerg. Technol. **24**, 270–287 (2012)
13. Coelho, L.C., Cordeau, J.F., Laporte, G.: The inventory-routing problem with transshipment. Comput. Oper. Res. **39**(11), 2537–2548 (2012)
14. Coelho, L.C., Laporte, G.: Improved solutions for inventory-routing problems through valid inequalities and input ordering. Int. J. Prod. Econ. **155**, 391–397 (2014)
15. Cunningham, W.H.: A network simplex method. Math. Program. **11**(1), 105–116 (1976)
16. Desaulniers, G., Rakke, J.G., Coelho, L.C.: A branch-price-and-cut algorithm for the inventory-routing problem. Transp. Sci. **50**(3), 1060–1076 (2015)
17. Dezső, B., Jüttner, A., Kovács, P.: LEMON - an open source C++ graph template library. Electron. Notes Theor. Comput. Sci. **264**(5), 23–45 (2011)
18. Huerta-Muñoz, D., Archetti, C., Guastaroba, G., Speranza, M.: A kernel search for the inventory routing problem (2019). http://redloca.ulpgc.es/images/workshop/2019/Slides_2019/Huerta_Munoz.pdf
19. Orlin, J.B.: A polynomial-time parametric simplex algorithm for the minimum cost network flow problem. Working papers 1484-83. Massachusetts Institute of Technology (MIT), Sloan School of Management (1983)

20. Peres, I.T., Repolho, H.M., Martinelli, R., Monteiro, N.J.: Optimization in inventory-routing problem with planned transshipment: a case study in the retail industry. Int. J. Prod. Econ. **193**, 748–756 (2017)
21. Ropke, S., Pisinger, D.: An adaptive large neighborhood search heuristic for the pickup and delivery problem with time windows. Transp. Sci. **40**(4), 455–472 (2006)
22. Santos, E., Ochi, L.S., Simonetti, L., González, P.H.: A hybrid heuristic based on iterated local search for multivehicle inventory routing problem. Electron. Notes Discret. Math. **52**, 197–204 (2016)

Solving a Real-World Multi-attribute VRP Using a Primal-Based Approach

Mayssoun Messaoudi[1,3]([✉]), Issmail El Hallaoui[1,3], Louis-Martin Rousseau[1,2], and Adil Tahir[1,3]

[1] Mathematics and Industrial Engineering Department,
Polytechnique Montréal, Montréal, QC H3C 3A7, Canada
`mayssoun.messaoudi@polymtl.ca`
[2] Interuniversity Research Centre on Enterprise Networks,
Logistics and Transportation (CIRRELT), Montréal, QC H3C 3A7, Canada
[3] Group for Research in Decision Analysis (GERAD),
Montréal, QC H3C 3A7, Canada

Abstract. Through this paper we focus on a real-life combinatorial problem arising in emergent logistics and transportation field. The main objective is to solve a realistic multi-attribute rich Vehicle Routing Problem using a primal-based algorithm embedded in column generation framework. The mathematical model is formulated as a Set Partitioning Problem (SPP) while the subproblem is the shortest path problem with resource constraints (SPPRC). The numerical study was carried out on real instances reaching 140 customers. The successful results show the effectiveness of the method, and highlight its interest.

Keywords: VRP · Column generation · Primal algorithm

1 Introduction

Supply chain encompasses several integrated activities and hand-offs allowing physical and information flows to be routed from the source (supplier) to the final destination (customer). Many believe that nearly two thirds of the supply chain total cost is related to transportation, more specifically, trucking is the dominant spend component.

Indeed, in a customer-centric era, transportation industry is becoming increasingly complex, and firms are facing a serious imperative: being able to deliver efficiently a whatever-whenever-wherever while respecting time, cost and quality. However, traditional networks, inefficient and fragile systems with limited computing performance are crippling firms to provide high-quality service, and maintain growth. In that respect, logistics providers are called upon to sharpen their practices through resilient tailor-made approaches.

Our main objective is to solve a real-world routing problem subject to a set of constraints and specific business rules commonly encountered in logistics industry markets. Our solution approach is based on a primal-based algorithm in

M. Baïou et al. (Eds.): ISCO 2020, LNCS 12176, pp. 286–296, 2020.
https://doi.org/10.1007/978-3-030-53262-8_24

column generation framework. The aim is not only to efficiently generate optimal dispatching plans, but also to shed light on the opportunity offered by such methods, especially when deployed on large-scale problems. In the remainder of this paper, note that all the mathematical formulations will be presented in their maximization form, but results also hold for a minimization scenario.

2   Related Literature

Vehicle Routing Problems (VRP) are so popular, and are the subject of an intensive literature due to their wide application in logistics and freight industry. Since its introduction by [1,2], several approaches regarding modelling and solution methods have been proposed for many VRP variants. Classical VRP aims to design minimal cost routes for a fleet of identical vehicles such that each customer is served exactly once, the capacity is respected and each vehicle starts and ends its route at the depot. In a-demand-driven-supply chain, many requirements and constraints arise, thus we switched from the classical VRP to new and more combinatorial and difficult models [3] which combine not only the usual restrictions such as time windows and fleet structure, but also specific business rules that vary according to the context. A detailed study of those variants can be found in [4]. Since VRP is NP-hard [5], heuristics and metaheuristics are more suitable than exact methods which are difficult to implement on large-scale problems. According to the classification of [6], solution methods are classified in two main classes: dual fractional and primal methods. Dual fractional methods [6] maintain iteratively optimality and feasibility until integrality is achieved. Whereas primal or augmentation methods maintain both feasibility and integrality and stop when optimality is reached.

One of the most known dual fractional methods is the branch-and-price (B&P) algorithm which combines branch-and-bound ($B\&B$) and column generation. The latter is an iterative process that solves the linear relaxation of the problem called a master problem (MP) for a restrictive subset of variables (columns), then called the restricted master problem (RMP). The duals related to the RMP's constraints are sent to a subproblem (SP) in order to generate positive-reduced cost variables, to be added into the RMP. The process stops when no such variables exist, and an optimal solution is obtained. If the latter is fractional, $B\&B$ is applied using a suitable branching strategy. However, despite its overall success, convergence problems can occur and affect the method's efficiency [7], they could be circumvented by using the stabilization strategies found in the literature and a fine-tuning strategy.

For VRPs, column generation is one of the notable exact methods that have successfully solved large and complex problems [8]. In such a context, the master problem is often modelled as a Set Partitioning Problem (SPP) and the subproblem is the Shortest Path Problem with Resource Constraints (SPPRC) defined on a directed graph [9], and usually solved with the dynamic programming algorithm introduced by [10].

Interestingly, exact primal methods have attracted very little interest in the literature, furthermore, concrete and adapted realizations of these approaches remain marginal. In a simple way, a primal procedure moves from one integer solution to an improving adjacent one until optimality, and such a sequence of moves leading to an improving solution is called a descent direction. With respect to the usual notation, a descent direction refers to augmentation direction in a maximization context as well. The first primal approach was introduced by [11], then [12] set the concept of the integral augmentation problem. An interesting combined approach was proposed by [13,14] that showed how integral simplex can be properly embedded in column generation context. They used an adequate branching strategy to obtain an optimal solution. Although, these models are restricted to small instances as they haven't been able to overcome the high degeneracy of SPPs. Integral Simplex Using Decomposition (ISUD) is one of the most promising primal methods, it was introduced by [15] to solve large-scale SPPs. Based on the improved primal simplex algorithm [16], ISUD takes advantage of degeneracy and finds strict descent directions at each iteration leading to an optimal solution. ISUD's performance has been enhanced by adding secant plans to penalize fractional descent directions [17], and also by exploring neighborhood search [18]. ISUD performed excellent computational results for SPPs with up to 500.000 variables. Recently, [19] introduced integral column generation (ICG). This three-stage sequential algorithm combines ISUD and column generation to solve SPPs. Experiments on large-scale instances of Vehicle and Crew Scheduling Problem (VCSP) and Crew Pairing Problem (CPP) showed that ICG exceeds two well-known column generation-based heuristics. The authors invoked the possibility to adapt ICG even on SPP with side constraints.

The remarkable performance of ISUD and ICG algorithms makes them worth pursuing. We believe that such methods have to be experimented on complex and well-known problems of literature such as rich routing problems.

2.1   Organization of the Paper and Contributions

To the best of our knowledge, it is the first time that a primal algorithm based on an exact method has been used to solve a rich vehicle routing problem. It is also an opportunity to discuss the procedure's performance on a real-world combinatorial problem. On the one hand, experimentation on real instances showed that the used primal method finds very good results in a short computing time. Indeed, it outperforms a well-known branch-and-price heuristic. On the other hand, the solution has led to a positive impact on the company's outcome indicators. This paper is organized as follows. In Sect. 3, we describe our problem and give the related notation. In Sect. 4, we give the mathematical formulation of master problem and subproblem. In Sect. 5, we introduce some theoretical notions related to the primal approach. The solution method (ICG) is described in the Sect. 6. The computational results are reported in the Sect. 7. Finally, concluded remarks are presented in the Sect. 8.

3   Problem Statement

In the remainder of the paper, we use the notation organized in the Table 1 bellow.

Table 1. Definition of the parameters and variables

Notation	Type	Description
Ω	–	Set of feasible routes
N	–	Set of customers to visit
K	–	Set of heterogeneous k-type vehicles
a_{ir}	Binary	Is equal to 1 if and only if customer i is served by route r
d_i	Real	Demand associated with customer $i \in N$
l_k	Real	Travelled distance between origin and destination by k-vehicle $\in K$
p_r	Real	Profit collected by the route $r \in \Omega$
q_k	Real	Capacity of vehicle $k \in K$
s_i	Real	Service time at customer $i \in N$
w^k	Real	Accumulated working time of k-vehicle $\in K$
θ_r	Binary	Is equal to 1 if and only if route r is selected

Given a set of customers $i \in N$ geographically scattered, a logistic hub O handles their transport operations. The daily task is to ensure next-day deliveries within specific time frames imposed by either customer convenience or/and urban traffic regulation, using heterogeneous vehicles with different capacities, types and operating costs. In addition, specific business rules such as the driving hours set up by work unions, loading rate, and urban accessibility regulation must be satisfied.

The objective is to maximize the profit collected by routes, resulting in the difference between freight rates and freight costs, while satisfying the following constraints:

1. Each customer $i \in N$ is visited by a single route $r \in \Omega$
2. Each customer $i \in N$ is visited within the time window $[a_i, b_i]$
3. Each customer $i \in N$ is visited by an allowed k-vehicle
4. The total load $\sum_{i \in N} d_i$ on the route travelled by k-vehicle does not exceed the vehicle's capacity q_k
5. The total travelling time l_k, including service times s_i, does not exceed the allowed working time w^k of k-vehicle

4   Mathematical Models

The master problem and the subproblem are formulated as SPP and SPPRC, respectively.

4.1 Master Problem

With respect to the below-mentioned notation, we formulate the problem as follows:

$$(SPP): \quad \max_{\theta} \quad \sum_{r \in \Omega} p_r \theta_r \tag{1}$$

$$\text{s.t.} \quad \sum_{r \in \Omega} a_{ir} \theta_r = 1 \qquad \forall i \in N, \tag{2}$$

$$\theta_r \in \{0, 1\} \qquad \forall r \in \Omega \tag{3}$$

Each variable θ_r is associated with a feasible route $r \in \Omega$ which specifies a sequence of customers $i \in N$ to be served. The objective function (1) aims to maximize the profit made by the feasible route r. The constraints (2) guarantee that each customer is delivered exactly once. The choice of binary variables is imposed by (3).

4.2 SPPRC on $G(V, A)$

The subproblem is modelized as a shortest path problem with resource constraints, and is solved using a labelling algorithm as shown in [20]. We have one supbroblem for each $k-$vehicle, put simply we omit the k-index. The reduced cost of feasible route r travelled by $k-$vehicle, starting and ending at the depot O and visiting a sequence of customers $i \in N$ is computed as:

$$\bar{p}_r = p_r - \sum_{i \in N} a_{ir} \pi_i > 0$$

The dual variable π is associated to the partitioning constraints (2). If all columns have negative reduced cost, the algorithm stops and an optimal solution is obtained for the linear relaxation of (SPP) (1–2). For each k-vehicle, SPPRC is defined on a cyclic graph $G = (V, A)$. V contains $|N| + 2$ vertices, one vertex for each customer $i \in N$, and (s, t) pair where s and t both refer to the depot O. A contains departure arcs (s, j), $\forall j \in N$, arrival arcs (i, t), $\forall i\ N$ and connecting arcs (i, j), $\forall (i, j) \in N$ so that the client j can be reached after the client i by a realistic route as indicated by the actual road map. Let $\Gamma = \{\mu_1, \mu_2, \ldots, \mu_{|\Gamma|}\}$ be the set of resource constraints, and L_i^{μ} be the label related to the resource $\mu \in \Gamma$ and associated to the vertex $i \in N$ such that the resource window $[a_i^{\mu}, b_i^{\mu}]$ is respected.

In our case, we consider the following resources:

- $L_i^T \in [a_i, b_i]$: The time resource indicates the arrival time at customer $i \in V$. L_i^T could be less than a_i, i.e., driver might arrive before the starting delivery period.
- $L_i^D \in [0, q_{max}]$: The demand resource specifies the accumulated load until customer $i \in V$.

- $L_i^W \in [0, 480 \, \text{min}]$: The working time resource specifies the total time travelled by the driver, from $\{O\}$ to customer $i \in V$.
- L_i^c: The cost resource is unconstrained, and used to compute the reduced cost of every travel arc $(i, j) \in A$.

The label represents a feasible partial path if: $L_i^\mu \in [a_i^\mu, b_i^\mu], \, \forall \, \mu \in \Gamma$. In such case, the label is feasible and extended along the $(i, j) \in A$ by calling for a resources-extension function, denoted $f_{ij}(L_i^\mu)$. While the labelling algorithm solves the subproblem, the SPP calls for ICG algorithm as explained in the Sect. 6.

5   Preliminaries

For the sake of clarity, we introduce some preliminaries concerning the primal approach principles, and the ISUD algorithm used in our solution procedure. We remind that detailed literature and examples could be found in [15, 16, 19].

As mentioned earlier, the primal methods have paid particular attention to SPPs. In addition to their popularity in routing and scheduling, these problems satisfy the Trubin theory. If we denote X the SPP polytope and X_S the set of its integer solutions, [21] shows that every edge of $Conv(X_S)$ is an edge of X, then SPP is said quasi-integral. This property makes it possible to use linear-programming pivots to reach integer extreme points.

Let consider the (SPP):

$$z^* = \max_{\theta} \left\{ \mathbf{p}^\top \boldsymbol{\theta} \, \middle| \, A\boldsymbol{\theta} = \mathbf{e}, \, \boldsymbol{\theta} \in \{0, 1\}^n \right\} \tag{4}$$

Let $A \in \{0, 1\}^{m \times n}$ be the binary constraint matrix, and let $A_1, A_2, \ldots A_n$ be the columns in A indexed in $J = \{1, 2, \ldots, n\}$ such that A_j denotes the jth column in A.

$$A = \begin{bmatrix} A_1 & A_2 & \cdots & A_n \end{bmatrix}$$

$\mathcal{F}_{SPP} \subseteq \{0, 1\}^n$ denotes the feasible region of SPP, while $\mathcal{F}_{SPP}^R$ denotes the feasible region of the linear relaxation of SPP. $\theta_r^* \in \mathcal{F}_{SPP}$ denotes the optimal solution and z_{SPP}^* is the optimal solution value.

Definition 1. *A column A_j is said to be **compatible** with S if $A_j \in Span(S)$, i.e., it can be written as a linear combination of the columns in S, otherwise, it is said to be incompatible.*

Definition 2. *The **incompatibility degree** δ_j of a column A_j towards a given integer solution is a metric measure that represents a distance of A_j from the solution. We note that δ_j of a compatible column is zero.*

Based on the definition of compatibility, ISUD decomposes the columns in A into three sets such that:

$$A = \begin{bmatrix} S & C & I \end{bmatrix}$$

Where S, C and I denote respectively the working basis (the support of the current integer solution), compatible columns subset and incompatible columns subset.

As described in Algorithm 1, the RMP is decomposed into two small subproblems: The complementary problem (CP) handles the incompatible columns and finds a descent direction $\mathbf{d}$ leading to an improved solution, while the reduced problem (RP) handles the compatible columns and seeks to improve the current solution.

Algorithm 1: $ISUD$ pseudocode

---

1 Find initial solution θ^0 and set $\bar{\theta} \leftarrow \theta^0$
2 $[S \quad C \quad I] \leftarrow$ Partition the binary matrix A into columns subsets
3 **do**
4   Solve $RMP(\bar{\theta}, C)$ to improve the current solution
5     $(Z^{CP}, d) \leftarrow$ Solve CP to find a descent direction
6 **while** $Z^{CP} > 0$ **and** $\mathbf{d}$ *is not integer*
7 $\bar{\theta} = \bar{\theta} + d$
8 **return** $\bar{\theta}$

Given a current solution $\bar{\theta} \in \mathcal{F}_{SPP}$, let $\mathbf{d}$ be the direction leading to the next solution such that $A\mathbf{d} = \mathbf{0}$. In fact, the set of the decent directions generates the null space of A which could be an infinite cone. Thus, normalization constraints $\mathbf{W}^\top \mathbf{d} = 1$ are added to bound the problem where $\mathbf{W}$ denotes the weight vector. The linear program CP finds, if possible, the combination of columns to obtain a feasible integer descent direction $\mathbf{d}$, i.e., that satisfies the following conditions:

1. $\mathbf{p}_r^\top \mathbf{d} > 0$ (improving)
2. $\boldsymbol{\theta} + \alpha\mathbf{d} \in \mathcal{F}_{SPP}, \quad \alpha > 0$ (integer)

6   Solution Method

ICG is a three-stage sequential algorithm, which merges primal approach in a column generation context.

The Algorithm 2 summarizes the major steps of the solution method:

1. The initialization step builds an artificial initial solution (θ^0, π^0). The initial primal solution θ^0 is built such that, in Step 1, each customer i is visited by a single vehicle k that bears a very large cost. In the initial dual solution π^0, each dual value is set to this large cost.
2. The first step starts by solving the subproblems $SP(\pi^t)$. Using the duals π^t corresponding to the current solution θ, positive-reduced cost routes are included in RMP. If no such routes are generated, we stop the algorithm and the best solution found θ^* is returned.
3. In the second step, ISUD solves the RMP to improve the solution. The criterion $minImp$ decides whether the improvement is sufficient or not. If so, neighborhood search is explored around θ^t by invoking a small MIP. This improvement step is iterated until the number of consecutive improvement failures $consFail$ reaches $maxConsFail$.

Algorithm 2: ICG pseudo-code

Parameters: $maxConsFail$, $minImp$.
Initialize   : $t \leftarrow 0$; $(\theta, \pi) \leftarrow (\theta^0, \pi^0)$; $consFail \leftarrow 0$
Output   : (z^*, θ^*)

1 **repeat**

   Step 1: CG
2 | $\Omega' \leftarrow$ Solve the $SP(\pi^t)$
3 | **if** $\Omega' = \varnothing$ **then**
4 | | **break**
5 | **end**
6 | $\Omega \leftarrow \Omega' \cup \Omega$
7 | $t \leftarrow t + 1$
   Step 2: RMP
8 | $(\theta^t, z^t, \pi^t) \leftarrow$ Solve the RMP using ISUD
9 | **if** $\frac{z^{t-1}-z^t}{z^{t-1}} \leq minImp$ **then**
10 | | $consFail \leftarrow consFail + 1$
11 | | $\theta_{NS}^t \leftarrow$ search an improved solution around θ^t by solving a restricted MIP
12 | | **if** $z_{NS}^t > z^t$ **then**
13 | | | $\theta^t \leftarrow \theta_{NS}^t$
14 | | | $(\theta^t, z^t, \pi^t) \leftarrow$ Resolve RMP using ISUD
15 | | **end**
16 | **else**
17 | | $consFail \leftarrow 0$
18 | **end**
19 | $(z^*, \theta^*) \leftarrow (z^t, \theta^t)$
20 **until** $consFail \geq maxConsFail$
21 **return** (z^*, θ^*)

Remark 1. The theoretical observations and empirical study made by [17], concluded that there is a strong correlation between the choice of the normalization weights vector $(W^\top \mathbf{d} = \mathbf{1})$ and the descent direction obtained. In practice, we use incompatibility degree δ_j as a weight vector to favor integrality.

7   Experimentation

Through this section, we discuss the results obtained by the ICG algorithm on real-life instances. ICG is compared to a well-know branch-and-price diving heuristic (DH) which is a dual-fractional heuristic based on column generation [22]. DH uses a depth-first search by exploring a single branch of the search tree. At each node, candidate columns for selection are evaluated, and the most fractional variable is set to 1. The process stops when the solution of the master problem is integer. The computing was performed on 7 real-life instances using a C++ implementation, under Linux on workstations with 3.3 GHz Intel Xeon

E3-1226 processors, and 8 GB RAM. The algorithms were implemented using *IBM CPLEX* commercial solver (version 12.4). The SPPRC was solved by dynamic programming using the *Boost* library version 1.54.

7.1 Instance Description

The real-life instances are provided by a major logistics provider. The instances correspond to home appliances' distribution of 7 weekdays and involve from $n = 34$ to $n = 140$ customers. We consider heterogeneous fleet of 6 types of vehicles. The fleet size is unlimited since the company can use external vehicles.

For each day, the order form indicates the customer index, his location, the quantity requested, the delivery deadline, and the allowed time frame. Following several tests, the ICG algorithm was implemented with the following parameters values: $minImp = 0.0025$, $maxConsFail = 5$. The Table 2 shows the computational performance of both ICG and DH. Column 1 indicates the name of the instance (*Name*). Column 2 indicates the number of customers (n) in each instance. For both ICG and DH runs, columns 3 and 7 display the number of iterations (*Iter*). Columns 4 and 8 display the total computing time in seconds (*Time*). Column 6 indicates the total number of integer solutions found (*nb.Sol*). Finally, columns 5 and 9 report the optimality gap in percentage (*Gap*) between the cost of the best integer solution found and the linear relaxation optimal value.

7.2 Numerical Results

Table 2. Computational results on 7 realistic instances

Instances		ICG				DH		
Name	n	Iter	Time	Gap%	nb.Sol	Iter	Time	Gap%
J2	34	4	0.95	0	5	7	0.23	0
J3	40	7	4.48	0.94	8	8	1.14	1.3
J7	66	9	22.6	1.3	28	33	14.4	1.77
J18	106	7	32	0.14	16	17	31.6	1.3
J25	136	12	515	1.62	51	94	1181.4	2.3
J19	140	9	168	0	20	18	321	0.05
J23	126	6	126	0.09	66	48	804	1.03

One can observe that ICG clearly outperforms DH. Indeed, the primal-based method has successfully obtained a feasible solution for all instances, within a competitive computing time $\in [0.95\,s, 515\,s]$. Optimality gap varies from 0.00% to 1.6%. For DH, the computing time $\in [0.23\,s, 1181.4\,s]$, and optimality gap varies from 0.00% to 1.77%. ICG reduces the DH computing time by a factor of

2.7 on average. ICG performs fewer iterations (at most 12) than DH (between 7 and 94). This can be explained by the ISUD algorithm which generates a large set of columns at each iteration. The Fig. 1 displays the time evolution according to the number of customers. ICG shows a remarkable performance on large instances. In fact, DH time was sped up by a factor of 3.5 on average. The authors [19] also noticed this finding while experimenting ICG on large VCSP and CPP instances.

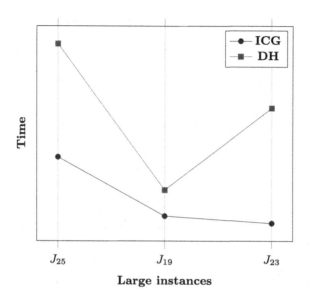

Fig. 1. Comparative computing times on large instances $n = 126, 136, 140$

We easily notice that ICG yields a large number of integer solutions (from 5 to 66), furthermore, ISUD generates a large number of columns. Thus, this important behaviour helps to improve the objective function since the first iterations. We notice that, for all instances solved with ICG algorithm, no branching method has been activated to obtain integer solutions. One recall that branch-and-bound methods could obtain a good solution with a good tuning and a proper branching strategy.

8   Conclusion and Further Work

In this paper, we have experimented for the first time a primal algorithm (ICG) on a complex routing problem. ICG combines the primal algorithm ISUD into column generation framework, and considers a set partitioning formulation. The computational indicators have shown the effectiveness of ICG algorithm while tested on real data reaching 140 customers and a realistic network.

Since the subproblem is a time-consuming, we would like to propose a new intelligent network modelling based on realistic data analysis. We are testing another ICG version that dynamically augments the search space to quickly find integer solutions without any branching recourse.

References

1. Dantzig, G.B., Ramser, J.H.: Manag. Sci. **6**(1), 80 (1959)
2. Clarke, G.U., Wright, J.W.: Oper. Res. **12**(4), 568 (1964)
3. Coelho, L.C., Renaud, J., Laporte, G.: Road-based goods transportation: a survey of real-world applications from 2000 to 2015. Technical report, FSA-2015-007, Québec, Canada (2015)
4. Toth, P., Vigo, D.: Vehicle Routing: Problems, Methods, and Applications, vol. 18. SIAM (2014)
5. Lenstra, J.K., Kan, A.H.G.: Networks **11**(2), 221 (1981)
6. Letchford, A.N., Lodi, A.: Math. Methods Oper. Res. **56**(1), 67 (2002)
7. Vanderbeck, F.: Implementing mixed integer column generation. In: Desaulniers, G., Desrosiers, J., Solomon, M.M. (eds.) Column Generation, pp. 331–358. Springer, Boston (2005). https://doi.org/10.1007/0-387-25486-2_12
8. Lübbecke, M.E., Desrosiers, J.: Oper. Res. **53**(6), 1007 (2005)
9. Irnich, S., Desaulniers, G.: Shortest path problems with resource constraints. In: Desaulniers, G., Desrosiers, J., Solomon, M.M. (eds.) Column Generation, pp. 33–65. Springer, Boston (2005). https://doi.org/10.1007/0-387-25486-2_2
10. Desrochers, M., Soumis, F.: INFOR: Inf. Syst. Oper. Res. **26**(3), 191 (1988)
11. Ben-Israel, A., Charnes, A.: J. Soc. Ind. Appl. Math. **11**(3), 667 (1963)
12. Young, R.D.: Oper. Res. **16**(4), 750 (1968)
13. Thompson, G.L.: Comput. Optim. Appl. **22**(3), 351 (2002)
14. Rönnberg, E., Larsson, T.: Eur. J. Oper. Res. **192**(1), 333 (2009)
15. Zaghrouti, A., Soumis, F., El Hallaoui, I.: Oper. Res. **62**(2), 435 (2014)
16. Elhallaoui, I., Metrane, A., Desaulniers, G., Soumis, F.: INFORMS J. Comput. **23**(4), 569 (2011)
17. Rosat, S., Elhallaoui, I., Soumis, F., Lodi, A.: Integral simplex using decomposition with primal cuts. In: Gudmundsson, J., Katajainen, J. (eds.) SEA 2014. LNCS, vol. 8504, pp. 22–33. Springer, Cham (2014). https://doi.org/10.1007/978-3-319-07959-2_3
18. Zaghrouti, A., El Hallaoui, I., Soumis, F.: Annals of Operations Research (2018). https://doi.org/10.1007/s10479-018-2868-1
19. Tahir, A., Desaulniers, G., El Hallaoui, I.: EURO J. Transp. Logist. 1–32 (2018)
20. Feillet, D., Dejax, P., Gendreau, M., Gueguen, C.: Networks **44**(3), 216 (2004)
21. Trubin, V.: Soviet Mathematics Doklady, vol. 10, pp. 1544–1546 (1969)
22. Joncour, C., Michel, S., Sadykov, R., Sverdlov, D., Vanderbeck, F.: Electron. Notes Discret. Math. **36**, 695 (2010)

Correction to: The Knapsack Problem with Forfeits

Raffaele Cerulli, Ciriaco D'Ambrosio, Andrea Raiconi,
and Gaetano Vitale

Correction to:
Chapter "The Knapsack Problem with Forfeits"
in: M. Baïou et al. (Eds.): *Combinatorial Optimization*,
LNCS 12176, https://doi.org/10.1007/978-3-030-53262-8_22

The original version of this chapter was revised. A typo in the second author's family name was inadvertently introduced during the publication process. The family name has been corrected to "D'Ambrosio."

The updated version of this chapter can be found at
https://doi.org/10.1007/978-3-030-53262-8_22

Author Index

Printed in the United States
By Bookmasters